Stadium and Arena Design

Second edition

Dedicated to Ron Taylor (1916–2002) and Kisho Kurokawa (1934–2007)

ice
Institution of Civil Engineers
publishing

Stadium and Arena Design

Second edition

Edited by
Peter Culley and John Pascoe

Published by ICE Publishing, One Great George Street, Westminster, London SW1P 3AA

Full details of ICE Publishing sales representatives and distributors can be found at: www.icevirtuallibrary.com/printbooksales

Other titles by ICE Publishing:

ICE Manual of Structural Design.
J Bull (ed.). ISBN 978-0-7277-4144-8
Structural Dynamics for Engineers, 2nd edition.
HA Buchholdt and E Moossavi-Nejad. ISBN 978-0-7277-6002-9
Intelligent Buildings, 2nd edition.
D Clements-Croome. ISBN 978-0-7277-5734-0

www.icevirtuallibrary.com

A catalogue record for this book is available from the British Library

ISBN 978-0-7277-5790-6

ICE Publishing is a division of Thomas Telford Ltd, a wholly-owned subsidiary of the Institution of Civil Engineers (ICE).

Commissioning Editor: Jo Squires
Production Editor: Colin Morgan
Market Development Executive: Elizabeth Hobson

FSC
www.fsc.org
MIX
Paper from responsible sources
FSC® C013056

Typeset by Academic + Technical, Bristol
Index created by Indexing Specialists (UK) Ltd, Hove, East Sussex
Printed and bound in Great Britain by TJ International Ltd, Padstow

Contents

Preface

Few engineers have the opportunity to become involved in the stadium development of an entire country. This is what happened to me in the 1990s. Sadly it was not the result of a country being selected to host a joyous event such as the Olympic Games or a World Championships. It was the result of the UK sports industry embracing the need for change following the Hillsborough Stadium disaster in which 96 football spectators lost their lives.

Integral to the national spirit of change was a transformation from a mixture of seated and standing spectator accommodation towards the all-seater ideal. I was asked to be the link between the steel industry and the owning, planning, designing, fabricating, constructing, commissioning, operating and maintaining functions making up the stadium business. I became integral to the massive growth in stadium developments, redevelopments and refurbishments that began taking shape in the UK. My involvements ranged from wide-span roof structures and cladding down to seat frames, gantries, safety fencing, gates and turnstiles.

By the end of the 1990s I had become well-known in the international stadium design community. I'd become involved in stadium designs for mainland Europe and beyond. And I'd become involved in advising on the types of closing-roof designs that had gained popularity in North America from the 1980s. This all led to me being encouraged – by the Institution of Civil Engineers among others – to write a book on the subject. This appealed to me because there had been no reference work for me when I started out.

The resulting book, *Stadium Engineering*, was published by the Institution of Civil Engineers in 2005. It included specialist stadium engineering contributions from world experts in their fields as well as a lot of content creation by me and my co-editor John Pascoe. This new edition of *Stadium and Arena Design* is by the *Stadium Engineering* team – a now expanded team – but is very different from its predecessor. A greater proportion of the content is by experts in the various facets of stadium planning, design and construction – with less input from John and me.

As editors, John and I have tried to ensure the book content delivers on the big stadium issues and incorporates a lot of valuable information that is unavailable elsewhere. We have also included a lot of references and web links so that *Stadium and Arena Design* is not constrained by its covers. Our wish is for the book to help bring people together in the pursuit of excellence and best practice. We like to think it will be used worldwide by professionals in the stadium business, by academics and by stadium enthusiasts.

We have dedicated this book to Professor Ron Taylor (1916–2002) and to Kisho Kurokawa (1934–2007). They both thought big – and bigger – while paying the utmost attention to the detailing in their designs. Ron Taylor inspired us and his influence is imbued in all our writing. Kisho Kurokawa wrote with us and gave *Stadium Engineering* its direction. They would have been enthusiastic and authoritative supporters of this new book.

Peter Culley

Foreword

The Maracana (front cover) is a magical place. It glows in the memory of all who know it. Here, at the original Maracana, 199 854 spectators shared the experience of the 1950 World Cup Final between Brazil and Uruguay. South American teams continued to win World Cups and the Maracana continued its reign as the world's biggest and best stadium. Throughout the 1950s it was unrivalled for luxury, function and safety – with its elegant elliptical frame, two-tiered seating bowl and fast emergency evacuation capability to surrounding parkland. These attributes also made it a venue of choice for concerts – for example, Frank Sinatra performed here in 1980 before an audience of 180 000. The Maracana demonstrates Brazil's ability to host international events – it has in recent years accommodated the 2007 Pan American Games and the 2013 Fédération Internationale de Football Association (FIFA) Confederations Cup. We remodelled the Maracana for the Brazil World Cup 2014 and the Rio 2016 Summer Olympics. We kept all the magic and added some of our own. The original two-tier seating bowl is replaced with a single-tier bowl and the original concrete roof is replaced with a beautifully translucent lightweight fibreglass tensioned membrane. The Maracana is today an all-seater stadium with a capacity of 78 838 spectators. It continues to be the biggest stadium in Brazil and the biggest in South America. It is a building that demonstrates well the changes to stadium and arena design which this book is all about – the availability of new construction materials, new computer-aided design capabilities, playing surfaces design, meeting broadcasting requirements, achieving spectator comfort and safety, managing efficient stadium ingress and egress, and bringing it all together holistically. Magic isn't something that just happens.

Among Brazil's other venues first used to host 2014 FIFA World Cup™ matches is the Itaipava Arena Pernambuco at Recife. Here we also did many things which further demonstrate topics of this book: integration of the stadium with the natural environment; rainwater drainage design for water re-use; design and specification of water-saving equipment throughout; use of trees and shrubs for surfacewater control; installation of rooftop solar panels; and natural light and ventilation. At Rio and Recife we considered Leadership in Energy and Environmental Design (LEED) certification at design concept stage. The United States Green Building Council (USGBC) granted both stadiums LEED certification for New Construction in the Silver category.

Our use of ethylene tetrafluoroethylene (ETFE) at Arena Pernambuco was the first use of this material in construction in Latin America. The lateral finish of the stadium facades is a system of pneumatic cushions made of ETFE film with edges attached to a metal frame by aluminium profiles. A combination of transparent and opaque films was feasible through the application of various types of screen printing on the surface. We used this potential to create a constant dialogue between internal and external areas and to facilitate diverse light control inside the stadium, depending on stadium usage. Membranes made with this polymer are 100% recyclable, with low coefficient of friction and non-stick properties which prevent dirt and dust accumulation.

For me it is impossible to see the beauty of the Maracana on the front cover and not want to be designing stadiums. The content of this book then provides plenty of ideas for practitioners in the industry to think about. I hope too that the book will help inspire a new generation of stadium designers who may first pick this book off their university or college library shelf.

Daniel Fernandes
Fernandes Arquitetos Associados

Award

Stadium and Arena Design is by the team that produced the award-winning book *Stadium Engineering* (Thomas Telford Publishing, London, 2005). In the photograph, Alabama architect Dudley Flotte (right), Chairman of the Competition, presents *Stadium Engineering* co-editor John Pascoe with a Construction Specifications Institute (CSI) Award at Knoxville, Tennessee, on 14 May 2005. In December 2005 *Stadium Engineering* was the sole UK winner in the Society for Technical Communication (STC) Trans European Competition for Technical Communication (the other five winners were from Belgium, Finland, France and Germany).

Introduction

Stadium and Arena Design looks at the technical and specialist expertise behind developing major sports and entertainment projects. Historically, stadium development focused on the sport, with an engineering-led design and, indeed, this was where I started – trained as a civil engineer. However, today, as reflected in this book and paralleled through my own career, sport facility development is a much more complex process demanding a fully integrated, all-encompassing delivery approach.

The quality of technical details in stadium design and construction had and always will have a huge impact on operation, maintenance and financial viability. However, the specialist skills required today are more comprehensive. What were purely sports venues are now comparable to mixed-use entertainment, retail and residential developments, and stadium designers are now also responsible for integrating hotel and conferencing facilities or retail and leisure outlets to make any sport development as sustainable as possible. There is also demand to integrate the latest smart technologies and recognise the need to think bigger to activate sport's potential to become an integral part of urban regeneration, a catalyst to renewal. It is as much about place-making as it is about fan and athlete experiences and therefore our project parameters are complex and the number of specialists involved has changed significantly: more than ever, behind every successful stadium or arena project is a coherent, fully integrated team.

A significant shift for sports is the focus on legacy. Legacy comes first and major sporting events work as initiators of larger scale development, an early milestone in the whole project process. This approach demands a greater and more in-depth understanding of the city and local community's long-term social, cultural and economic needs. Different objectives will operate for different locations but at the heart is sustainability, from delivery (the choice of contractors and local labour demands, materials, water and energy) to strategy (creating relevant sporting-led environments with a complementary mix of uses and coordinating often public-led funding).

All of this means working with a complex web of stakeholders from the outset, with the public as the most crucial – not only to gain buy in, but also to make stadiums and arenas viable in the long term and important parts of our urban fabric. Sports planning is about the bigger picture and we apply this principle to each of our projects as represented in our masterplans for the recently completed Singapore Sports Hub and for the London 2012 and Rio 2016 Olympic and Paralympic Games, as well as on our FIFA World Cup work in South Africa 2010 – where we delivered over five venues – and Qatar 2022 on the design of Al Wakrah stadium and precinct.

The result of this widening of our roles as professionals is fantastic. It creates sporting facilities in which we can all take pride – locally, regionally and internationally. And as the then manager of Moses Mabhida stadium in Durban, South Africa, for the 2010 FIFA World Cup, my pride was not only in the technical delivery of a legacy-led stadium but the memory of my son, Arthur, shaking hands with the German national team in the stadium's first World Cup match.

Sports development has social responsibility, international resonance and personal relevance for us all. I, for one, am glad that my professional career and technical skills have allowed me to play a part in the making of our sustainable cities across the world.

Alf Oschatz, AECOM
Director, Head of Sports, Europe, Middle East and Africa

Part I

Past, present and future

Stadium and Arena Design
ISBN 978-0-7277-5790-6

http://dx.doi.org/10.1680/saad.57906.003

Chapter 1
Evolution of stadium and arena engineering

John Pascoe

Roll up, roll up for the mystery tour! Our tour through time takes in 40 centuries and five continents. We have 32 stops at places hosting events that range from the foot race of the ancient world to Formula 1 motor racing in the 21st century. We hope you enjoy the ride.

Panathenaic Stadium, Athens, Greece

The Greek writer Homer probably lived in the 8th century BC but may have lived as far back as the 12th century BC. He wrote that competitors in the foot race ran to a mark in the distance, turned around it and ran back to the starting point – the ancient Greeks raced up and down a straight track, not around bends. The ancient Greek stadium gained its name from the fact that it had a sprinting track one stade (180 m, 600 ft) long, with all other races being multiples of that. However, the tracks at Olympia (192.27 m, 630.8 ft), Epidaurus (181.3 m, 594.8 ft), Delphi (177.5 m, 582.3 ft) and Pergamon (210 m, 689 ft) all differ in length. The ancient Greek stadium is, therefore, non-standard in size. It is, however, in the common form of a long parallelogram some 30 m (98.4 ft) wide. The enclosure containing the stadium would allow approximately 15 m (49.2 ft) at either end of the track, and might be square, as at Epidaurus, or curved, as at Delphi and at Panathinaea, Athens. Here then we have the derivation of the shape of the modern stadium.

The Panathenaic Stadium, Athens (Figure 1.1), was the venue for the first Olympic Games of the modern era, held in 1896. It was originally a natural hollow in the ground between the two hills of Agra and Ardettos. Lykourgos the Logographer (396–323 BC) created the stadium in 330–329 BC for the athletic competitions of the Great Panathenaea festivities. Between 140 and 144 AD, Herodes Atticus restored the stadium, giving it the horseshoe form as excavated in 1870, with a track 204 m (669 ft) long and 33.4 m (110 ft) wide (Figure 1.2). It is believed that the stadium had an original seating capacity of 80 000 people.

Stadium of Delphi, Greece

Here it begins – on the starting line (Figure 1.3) of the Stadium of Delphi, at the sanctuary of Apollo. This is where the Greeks held the Pythian Games, one of the four Panhellenic Games of Ancient Greece, from 582 BC. The athletes set their feet in the parallel grooves in the starting line stonework, adjacent to which are round holes where the starting posts were located. Originally only musical contests were held at the Pythian Games, at the Theatre of Delphi, illustrating the amazing longevity of the relationship between stadiums and concerts.

Tailteann Games of Ancient Ireland, County Meath, Ireland

While the archaeological remains of the Greek games are seductive, evidence of site selection and site clearance for assemblies of athletes goes much, much further back in time. Tailté, daughter of a Spanish king, was married to Ireland's Eochaidh, King of the Firbolgs, who was slain at the Battle of Moytura in 1897 BC. Sometime before her husband's death, Queen Tailté had selected her own burial place on the slopes of Caill Cuain, a sunlit location covered in rolling forest. It took 'a host of stalwart men nearly a year to accomplish the task [of clearing the timber]'. Here Tailté wanted her Leath (tomb), Guba (public lamentations) and Nosad (funeral and games) to take place. The funeral games included all types of equestrian, athletic and gymnastic events, and incorporated three events unique to Ireland at the time: horse swimming, hurling and the high jump. The Tailteann Games inspired the content and protocols of the Olympic Games established in Ancient Greece in 1222 BC. They also outlasted the Greek Games, taking place annually, with very few interruptions, until 1169 AD.

Juego de Pelota, Monte Albán, Valley of Oaxaca, Mexico

And so we come to the beginnings of team games played with spherical objects. The Oaxaca Valley in southern Mexico has been populated since around 2000 BC, and between 800 and 500 BC was subject to an influx of Zapotec people. From about 500 BC the Zapotecs began levelling the top of a mountain on which they would establish Monte Albán, with its central Gran Plaza offering spectacular views over the valley. On the eastern side of the Gran Plaza is an I-shaped ball court – Juego de

Figure 1.1 The Panathenaic Stadium, Athens. (© Underwood & Underwood, 1907)

Petola (Figure 1.4). This differs from other Mesoamerican ball courts (Mayan and Toltec) in that there are no goal rings and the sides of the court slope. The game was played by controlling the ball using the hips, shoulders, knees and elbows. The losing team was usually put to death as an offering to the gods.

Figure 1.2 The track at the Panathenaic Stadium. (© Mentor Association Inc, c. 1914)

Figure 1.3 Stadium of Delphi. (© David Monniaux, 2004)

Figure 1.4 Juego de Pelota, Monte Albάn. (© Tjeerd Wiersma)

Figure 1.5 El Djem, Tunisia. (© Octopus at Slovenian Wikipedia) (This figure is reproduced in the colour-plate section as Plate I)

El Djem Amphitheatre, Tunisia

The might that was Ancient Rome was manifested in its stadium-type structures erected in countries around the Mediterranean and through Europe. The largest such structures were the 189 m (615 ft) × 156 m (510 ft), 50 000 capacity Flavian Amphitheatre (Colosseum, Coliseum) in Rome, the ruined 170 m (558 ft) × 140 m (460 ft) Theatre of Capua in southern Italy and the Circus Maximus in Rome, the Empire's principal chariot-racing venue, which measured 621 m (2037 ft) × 118 m (387 ft) and could accommodate approximately 150 000 spectators (its site is now a public park).

The grandiose building that best exemplifies the outreach of the Roman Empire is, however, the 148 m (485 ft) × 122 m (400 ft), 35 000 capacity amphitheatre at El Djem (Figure 1.5). El Djem is the third-largest such building after the Colosseum and the Theatre at Capua. It was built around 238 AD, and may have been the final amphitheatre built in the Empire. Its structure is of large stone blocks with carved incisions to which clamps were fitted to facilitate lifting. Some of these incisions are still visible. The arena was big enough to host chariot racing as well as gladiator shows. Beneath the structure run two passageways in which animals, prisoners and gladiators were held until the time came for them to be brought up into the sunlight to perform – usually, the final acts of their lives.

Arena Gianni Brera, Milan, Italy

Because of their size and functionality, stadiums are likely to contrast with rather than complement their environments. Among the exceptions is the Arena Civica (now known as the Arena Gianni Brera). This arena (Figure 1.6) is a principal neo-classical architectural treasure of Milan along with buildings such as the Villa Reale (Villa del Belgiojoso), Palazzo Belgiojoso, Palazzo Tarsis, Porta Venezia and Arco della Pace (Arco Sempione). The stadium opened on 18 August 1807, and has hosted all sorts of events, including naval battle reconstructions and William (Buffalo Bill) Cody's Wild West Show. Since 2010 it has been the home ground of Amatori Rugby Milano, and is the host venue for an international athletics meeting, the Notturna di Milano. Its facilities include an eight-lane, 400 m track and a soccer/rugby pitch (100 m × 86 m).

Polo Grounds, New York, USA

The geographical outreach of Ancient Rome was a driving force in stadium construction. There would not be another such driving force until Alexander Cartwright wrote down the rules to baseball in 1846, leading to wooden fences and timber grand-stands being erected in ballpark developments throughout North America. Polo and then baseball too (from 1880) were played on the Polo Grounds, where the ballpark's original timber grandstand was demolished in 1889. Polo Grounds II

Figure 1.6 Arena Gianni Brera, Milan, *c.* 1860–1890

(Figure 1.7) was built in 1890, and was also of the then prevalent timber construction. However, after 30 years of destructive fire incidents at ballparks, the construction of major league parks was, from 1909, turning to concrete and steel design alternatives. Polo Grounds II was destroyed by fire in April 1911 and was replaced by a 55 987 capacity horseshoe-shaped concrete and steel stadium in the same year.

Stockholms Olympiastadia, Sweden

The 14 500 capacity (33 000 for concerts) Stockholms Stadion, or simply Stadion, (Figure 1.8) was opened in 1912 to host that summer's Olympic Games. It is one of the smallest athletics stadiums to host the modern Olympics. However, there is perhaps something about a smaller stadium, because this one holds the world record for the number of world records (83) broken at its hosted events. At big venues, the spectators may not know, until it has happened, that a new world record has been achieved – at Stockholm and other smaller stadiums the crowd seems to will the competitors onwards, heightening the aspiration that comes from within the athlete. The atmosphere at Stockholm also finds favour with musicians, notably Bruce Springsteen, who has performed there eight times.

Balboa Stadium, San Diego, CA, USA

Besides hosting sports, concert and other entertainment events, stadiums have long been venues of appropriate size for big political and religious assemblies. Balboa was described as the first true stadium on the Pacific Coast and the largest municipal structure of its type in the world. Here, on 19 September 1919, President Woodrow Wilson spoke from a glass enclosure

Figure 1.7 Polo Grounds II, New York, 1910. (© Pictorial News Co.)

Figure 1.8 Stockholms Olympiastadia, *c.* 1910–1915. (© Bain News Service)

(bottom right in Figure 1.9) in favour of his proposed League of Nations, to 50 000 people (at a time when the population of San Diego was 73 000). The challenge of addressing such large assemblies means that stadiums have been a driving force in the development of communications technologies. The voice amplification system that Wilson used that day was the Magnavox ('great voice') moving coil device. Microphones had previously been used with the system but it was considered undignified to expect the President to take hold of such a device. So a new type of receiver was used, attached to a large horn suspended just over the President's head. A circle was drawn on the floor, within which the President was asked to stand in order to direct his voice to the microphone. The co-inventor of the Magnavox, Edwin S. Pridham, was in the glass enclosure with the President and described the event in his autobiography:

> When we supplied the apparatus for President Wilson's address in the San Diego stadium we were using our latest model power amplifier, a three-stage affair with two five hundred volt tubes in parallel in each stage. Everything was going fine for this test until just before President Wilson started to speak. I had told him of the circle marked on the floor and to direct his speech forward so that it would best affect the microphone. I left the glass enclosure to close the switch on the control box. I no sooner had done this than I saw smoke coming out of the amplifier box. What a fix to be in! Seventy-five thousand [*sic*] people in the stadium, the President inside the glass enclosure – you couldn't have heard him three feet away without the Magnavox. I looked closely at the tubes and saw one was red hot. I yanked it out without disturbing the circuit. [There were two tubes in parallel in each stage.] Everything was okay.

The President's voice was heard for a mile beyond the stadium. There was a problem with tonal quality, and Pridham said this was because he had done the testing in the open stadium. It was the glass enclosure, introduced later, which caused a hollowness in Wilson's voice due to voice vibrations from the glass sides. A solution to this effect was a long time coming. It was never to have surfaces near a microphone that would permit echoes. Today, every studio is designed to minimise echo but not completely eliminate it.

Figure 1.9 Balboa Stadium, San Diego, 19 September 1919. (© O. A. Tunnell)

Figure 1.10 University of Washington Stadium, Seattle. (© Asahel Curtis)

Notes: Pridham perhaps confused the size of the stadium audience with the 75 000 people of San Francisco who heard opera singer Luisa Tetrazinni's voice amplified in the first test of the Magnavox system on Christmas Eve 1915. A blacksmith, Ray Galbrath, reputedly suggested in 1915 that Pridham and his co-inventor Peter Jensen convert their phone apparatus, which they could not sell to the telephone companies, into a loudspeaker that could be used at football matches. It is also believed that Jensen was inspired by James (Foghorn) Murphy, a big Irishman hired to ride around the streets of San Francisco in the baseball season, announcing when the next match would take place. Because Pridham and Jensen amplified gramophone records across residential areas of San Francisco, they are regarded as the world's first disc jockeys.

University of Washington Stadium, Seattle, USA

Tight construction timetables are characteristic of stadium developments and have brought about the introduction of many innovative building techniques. Puget Sound Bridge and Dredging Company signed the contract to build this 30 000 capacity stadium on 7 May 1920, and had to complete by 27 November – a timetable of 6 months and 20 days. A unique method of excavation known as 'sluicing', displacing earth with high-pressure water, was used to clear the area. Approximately 230 000 yd^3 (176 000 m^3) of earth were moved using 687 000 000 gallons (31 231 651 hl) of water. The stadium was completed on time, despite 46 days of rain during the construction period. Work finished 12 hours before the stadium's inaugural match kicked off. The photograph (Figure 1.10) was taken on 13 November 1920, just 2 weeks before that match. The stadium is now known as the Husky Stadium and has a much enhanced seating capacity of 72 500, making it the biggest stadium – college or professional – in the Pacific Northwest. Annually voted the most scenic football structure in the USA, an unusual feature is the boat mooring available to fans arriving by water from across Lake Washington.

Yankee Stadium, New York, USA, 1923

The Yankees played from 1913 on the Polo Grounds, sharing a venue with the New York Giants. On 6 February 1921, the Yankees announced their purchase of 10 acres at West Bronx, across the river from the Polo Grounds. Here, the Yankee Stadium (Figure 1.11) was opened on 18 April 1923 – the first baseball park to be called a 'stadium'. The construction materials used were: 45 000 yd^3 of earth (for grading), 1 million feet of Pacific Coast fir lumber (shipped through the Panama Canal), 800 tons of reinforcing steel, 2200 tons of structural steel, 135 000 steel casings, 400 000 pieces of maple lumber and 1 million brass screws. The completed stadium had an immediately beneficial effect on its neighbourhood and on property values. Small restaurants, parking lots, bars and a theatre opened. Drug stores installed lunch counters. In 1923, the Concourse Plaza Hotel opened nearby and catered for visiting ball clubs. At the first match, 60 000 spectators watched Babe Ruth hit a game-winning homer. This stadium was the home

Figure 1.11 Yankee Stadium, New York, 1923. (© Bain News Service)

of the Yankees from 1923 to 1973, and from 1976 to 2008. In 2006 a new stadium was developed on adjacent land, and the site of the 1923 structure is now parkland known as Heritage Field.

Los Angeles Memorial Coliseum, CA, USA

The Coliseum (Figure 1.12) was commissioned in 1921 as a memorial to veterans of World War I. The 75 144 capacity stadium was completed in May 1923, with construction having taken just 16 months. In 1930 the stadium was extended upwards (new capacity 101 574) in preparation to host the 1932 Olympics. This is the only stadium to have hosted the Olympic Games twice (1932 and 1984) and the only Olympic Stadium to have also hosted the Super Bowl (NFL American Football) and World Series (Major League Baseball). Other events accommodated include motocross, speedway and concerts. On 27 July 1984 the venue was declared a National Historic Landmark. It is now the home of the USC Trojans football team.

Olympiastadion, Berlin, Germany

This stadium (Figure 1.13) was immortalised in Leni Riefenstahl's film *Olympia*. Its 71 rows of seats extended from a depth of 13 m (43 ft) below ground to a height of 17 m (56 ft) above ground. This enabled all the spectators to enter the stadium at ground level and, once inside, be directed to the upper or lower levels of seating. This system reduced the need for steps and stairways. The unusual depth of groundwater at the Reich Sport Field, approximately 30 m (100 ft) below ground level, made possible the construction of the arena at low level.

Zarzuela Hippodrome, Madrid, Spain

Eduardo Torroja designed the 12.67 m (41 ft 6 in.) span cantilever roofs (Figure 1.14) of the Zarzuela Hippodrome as a series of

Figure 1.12 Los Angeles Memorial Coliseum. (© Organizing Committee of the Games of the Xth Olympiad)

Figure 1.13 Opening Ceremony, Olympiastadion, Berlin, 1936

reinforced concrete hyperboloid shells or lobes. Over the main supports each lobe has a thickness of 140 mm (5.5 in.), a rise of 1.37 m (4ft 6 in.) and a radius of curvature of 2.74 m (9 ft). At the free edge the rise of the arch is reduced to 0.48 m (1 ft 7 in.), the radius of curvature increases to 6.7 m (22 ft) and the thickness is 50 mm (2 in.). These proportions give the shell an appearance of intrinsic lightness. The vaults are counterbalanced by vertical tie rods located behind the main supports. Torroja wrote:

> Of the many basic types [of roof considered], the resulting surface could well have been a conoid but for the objection that the conoid is not very attractive. It seemed preferable to choose some other form of curvature. Among the better known ones, none seemed more adaptable than the hyperboloid; hence these cantilevered vaults have the shape of hyperboloid sectors. And the question now arises: is the invention of an especially adapted form to solve a specific problem strictly an imaginative process, or is it the result of logical reasoning based on technical training? I do not think it is either of the two, but rather both together. The imagination alone could not have reached such a design unaided by reason, nor could a process of deduction, advancing by successive cycles of refinement, have been so logical and determinate as to lead inevitably to it.

Figure 1.14 Zarzuela Hippodrome, Madrid, 1935

Note: In 1936, during the Spanish Civil War, the stands were close to the battle line and were subjected to bombardment. They suffered 26 piercings and repeated cracking due to vibration effects. The holes were concreted and stiffeners were added to the tips of the end lobes. The structure continued to perform well.

Campo de Les Corts, Barcelona, Spain

The world confronted much bigger issues than stadium construction in the 1940s. Approximately 30 million soldiers and 50 million civilians (some 4% of the world's population) died in World War II and millions more people died of hunger and cold after 1945. President Truman asked Charles Luckman, the 'boy wonder' of American business (who features later in this chapter), to organise the Friendship Train, which delivered 120 million bushels of grain to the Allies in Europe at the close

Figure 1.15 Campo de Les Corts, Barcelona, 1943

of World War II. Sport helped bring back some normality into people's lives, and the so-called 'austere Olympics' of 1948 were held in London, using the existing Wembley Stadium and buildings erected for the 1924 British Empire Exhibition.

Some countries, including Spain, were neutral during the war years. In 1943, Eduardo Torroja designed an elegant football stadium redevelopment at Campo de Les Corts (Figure 1.15), which continued Barcelona FC's expansion from its original stadium capacity (20 000 in 1922) to its final capacity (60 000 in 1957) when the club relocated to the Nou Camp (now 99 354 capacity – Europe's largest stadium).

Torroja said of Les Corts:

> It was difficult to find alternatives for the structural materials and methods to be used in this great cantilevered roof. Calling for an overhang of 83 ft over the spectator stands, it had to be made in metal in order to reduce weight and cost. It had to be anchored rigidly into a vertical structure strong in bending. This structure could be made of reinforced concrete, as the heavier weight of concrete would not add to the bending moments and the cost would be cheaper.

Torroja would also write, in 1958, a synopsis of his philosophy, which has more in common with today's multidisciplinary team working approach than the separatist norm of his own times:

> In each of these [my] projects I have tried to understand as completely as possible all the factors involved and to apply my ingenuity to achieve a satisfactory solution, both structurally and economically. My final aim has always been for the functional, structural, and aesthetic aspects of a project to present an integrated whole, both in essence and appearance.

Palazzeto dello Sport, Rome, Italy

During the 1940s, Pier Luigi Nervi developed ideas for innovative uses of reinforced concrete which helped in the rebuilding of many factories and other structures throughout Western Europe – he also borrowed from Roman and Renaissance architecture, applying ribbing and vaulting to improve strength and eliminate columns. Nervi designed the 5000 seat Palazzeto dello Sport (Figure 1.16) to host boxing, basketball and other events at the 1960 Summer Olympics. The stadium was actually opened in 1957. The structure is a ribbed concrete shell dome 61 m (200 ft) in diameter, braced by concrete flying buttresses. The flute-edged roof shell is of prefabricated, diamond-shaped sections joined by poured-in-place concrete connecting ribs, creating a webbed ceiling network of flower-like elegance. A

Figure 1.16 Palazzeto dello Sport, Rome, 1957

Figure 1.17 Roads in Akasaka-Mitsuke Ward, near the Olympic Stadium, Tokyo, 1964. (Keystone Press)

compression ring in the centre forms a cupola, providing a central source of natural light. The forces carried through the ribs are gathered in prefabricated, triangular sections, which transfer the load to exterior Y-shaped buttresses and vertical supports. The dome was erected in just 40 days. Nervi designed a larger, 16 000 seat version of the Palazzeto (1958–1959), together with the 50 000 seat Flaminio Stadium (1957–1959), both also for the Rome Olympics. Many years earlier (1930–1932) he had designed the extraordinarily economic Municipal Stadium, Florence, a 35 000-seat facility, with an average cost per seat of $2.90.

Games of the XVIII Olympiad, Tokyo, Japan

Erected on the site of the Meiji Shrine Outer Park Stadium, the 48 000 seat National Olympic Stadium hosted the 1958 Asian Games and the 1964 Summer Olympics. Tokyo's construction achievement for the 1964 Olympics was monumental, as were many of its newly completed concrete buildings. But it was in terms of infrastructure development that Japan really showed the way forward. Train and subway lines were completed, a new highway (Figure 1.17) was built through the metropolitan area, and Haneda International Airport was modernised to accommodate the new generation of jet airliners. Perhaps most dramatically of all, the bullet train concept that dated from the 1930s was revived. The world's fastest train network was completed in time to play its part in getting people to and from the Games.

Tokyo 1964 is also significant for information, communications and broadcasting technologies being developed around big stadium events (Figure 1.18). These were the first Games to be telecast internationally – to the USA using Syncom 3 (the first geostationary communication satellite) and from there to Europe using Relay 1. However, to illustrate electromechanical

Figure 1.18 Press boxes at the National Olympic Stadium, Tokyo, 1964. (Keystone Press)

technologies, I can't resist a trackside image (Figure 1.19) which shows timing-device testing at the Olympic Stadium on 28 September, during torrential rain which followed Typhoon Wilda (the Games commenced on 10 October). Because this book is more about construction than communication technologies, I would like – while thinking about timing – to interject an idea put to me by Brian Piggott, an electromechanical engineering colleague of mine. Brian has proposed placing rail-mounted robots beyond the outside lane of the track, which would circumscribe longer non-championship races at world and other record pace (e.g. European, Asian, national, stadium), as indicated by different-coloured device-mounted light-emitting diodes (LEDs). This would eliminate the inefficiencies and inequalities of pacemakers, give competitors inspiration from racing against the world record (if not the world record holder!) and give spectators a dramatic new dimension. It would provide a kind of image-driven communication to stadium spectators who cannot benefit from the commentaries or visual devices

Figure 1.19 Testing of timing devices, National Olympic Stadium, Tokyo, 1964. (Keystone Press)

Figure 1.20 Madison Square Garden, New York, 1967. (© Charles Luckman Associates)

(e.g. swimming's red-line technology) available only to TV audiences.

Madison Square Garden, New York, USA

In the 1960s/early 1970s I stood watching soccer on the frozen terraces of, first, Occupation Road, Corby, and then St James' Park, Newcastle upon Tyne. My father in those years liked to stay up into the early hours of the morning to watch, in relative comfort, boxing broadcast live by satellite from Madison Square Garden (MSG) into homes in the UK. MSG (built 1963–1967) seemed to me the most glamorous sports venue in the world. The 29-storey Two Penn Plaza Building and 22 000 seat (boxing mode) MSG Sports Center (Figure 1.20) were constructed over terminal facilities for the Penn Central and Long Island Railroads. To accommodate the daily movement of 650 trains and 250 000 commuters without disruption, each building phase was held under constant and critical control: demolition of the Penn Station superstructure; column and slab placement through, above and below operating terminal facilities; modernisation of underground terminal facilities; and construction of superstructures. The MSG's 126.5 m (415 ft) diameter roof (Figure 1.21) covers approximately 4 acres (1.8 ha) and is entirely supported by high-strength suspension cables. In addition to the main arena (Figure 1.22), which seats 20 000 spectators, the centre includes a 5000 seat cinema, a 48-lane bowling centre, public and private clubs, and a sports museum.

To accommodate the new construction, the existing Pennsylvania Station (opened in 1910) had to be demolished. The station had been designed at the turn of the century by Stanford White, based on the Roman Baths of Caracalla, which were constructed in about 217 AD. MSG architect Charles Luckman (1909–1999) said:

> We had to revert to some of the same construction methods employed in building the original Baths of Caracalla in Roma. We used people. Two hundred twenty men used picks and shovels to dig the foundation holes between the tracks while the trains continued to run. The men then put the dirt in old-fashioned buckets and passed the buckets by hand, from man to man, up three flights of stairs to the street level. Each time I visited the site and saw this 'bucket brigade' in operation I shook my head in disbelief. The concrete for the foundations made the trip from street to track level the same way, hand-to-hand. In the twentieth century it was a sight to behold!

Luckman noted that:

> To connect the Garden with the new office building, we created what we believed was the first all-glass-walled-and-ceilinged 'atrium'. It was a hundred feet long, a hundred feet wide, and a hundred feet high, and is used as an exhibit and assembly area for civic and sports events.

He reminisced:

> Another serious problem in designing the new Madison Square Garden was how to get the 22 000 people at sporting events out of the various floors of the arena without interfering with passenger traffic of the new railroad station. I came up with what I thought was the

Figure 1.21 Model of Madison Square Garden showing the roof suspension system. (© Charles Luckman Associates)

Figure 1.22 Completed interior of Madison Square Garden. (© Jack Laxer)

Figure 1.23 The Forum, Inglewood, 1967. (© Charles Luckman Associates)

brilliant idea of constructing four glass-enclosed exterior towers, each containing ten escalators, to serve the four sides of the building. We calculated correctly that the escalators would be able to empty the arena of a capacity crowd in just 22 minutes. By enclosing the outside escalators in glass, people could enjoy the night-time lights of the fabulous New York skyline as they left the Garden. Some ideas, though, work almost *too* well. People were so entranced with the view that they forgot to step off the escalators. Some stumbled or tripped and a few fell. So much for me and my creative ideas! A crisis meeting was called immediately. The result was the installation of a sound system for music with interspersed verbal warnings: 'Please look down. It is time for you to step off the escalator'. It worked. No one was injured; no one sued. But my face was beet red.

The Forum Sports Arena, Inglewood, CA, USA

Madison Square Garden and Inglewood (Figure 1.23) are examples of the introduction of higher levels of comfort beyond the comprehension of earlier generations of stadium and arena users. At Inglewood, early occupancy was a stringent client need. The time elapsed from initial planning through working drawings, construction and the grand opening (December 1967) of the 17 000 seat multiple-use sports arena was just 15.5 months – approximately 9 months less than a normal design and construction period. Precast, post-tensioned concrete arches, each 21.3 m (70 ft) tall (Figure 1.24), were produced on site.

Figure 1.24 Lifting the arches at the Forum. (© Sheedy & Long)

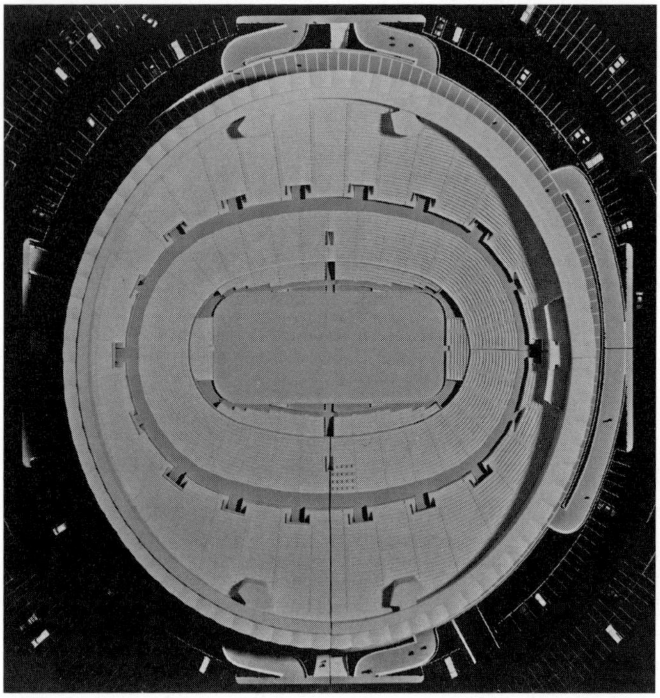

Figure 1.25 Plan of the interior of the Forum. (© Charles Luckman Associates)

They surround the building and form a compression ring that supports a cable suspension roof, creating a completely column-free interior (Figure 1.25) for basketball, hockey, boxing, concerts, circus shows, ice shows, rodeos, track meets, horse shows and conventions. Acoustical design accommodates diverse performances and, at the same time, achieves decibel reduction for jet noise. Parking for 3500 cars was provided.

Charles Luckman was again the architect and, again, he articulates the challenges of the project best:

> The Forum's Romanesque arches were so huge and heavy that it was impossible to have them made at the factory and transported through the city streets. We had to build our own forms on the site and pour the reinforced-concrete arches in place. As each arch was 'cured', two gigantic hoisting cranes lifted it out of the form and placed it in position on the podium. When it came to install the huge air-conditioning units on the curved-dome roof, we used helicopters to lift them and gently place them in position. More than 400 construction workers were employed on day and night shifts. The roof, 400 feet in diameter, was suspended on cables between outside and inside concrete tension rings, this arrangement supplying the enormous strength needed to carry the weight of the roof, the air conditioning, and the electric light grids. This unique concept eliminated all columns within the building and permitted us to place more than 70% of the seats between the goals.

Figure 1.26 Olympiastadion, Munich, 17 May 1972. (Keystone Press)

Olympiastadion, Munich, Germany

This pioneering lightweight structure was considered revolutionary when built and still looks revolutionary today. The 80 000 capacity Olympiastadion was built between 1968 and 1972, to host the 1972 Olympics. Construction was simplified by locating the stadium in a crater caused by Allied bombing during World War II. Large areas of acrylic glass canopy (Figure 1.26) were used, stabilised by steel cables, to symbolise a new transparent, democratic and optimistic Germany. The stadium has gone on to host the 1974 World Cup Final (football) and many other international sporting events, as well as major rock concerts (Michael Jackson performed there five times). It was also home to Bayern Munich and TSV 1860 Munich before these football clubs transferred to the Allianz Arena on its completion in 2005.

Birmingham International Arena, National Exhibition Centre (NEC), Birmingham, UK

I want to include a space-frame project and I was British Steel's marketing planner for Nodus (1974–1976), so it has to be a Nodus (Figure 1.27) project. Steelwork erection of the NEC initial development of six halls (1, 2, 3, 3a, 4, 5), total area 72 400 m^2, commenced on site in late 1973 and was completed in 1975. The Nodus space-frame system was used for the 93 identical roof structures, each measuring 27.9 m × 27.9 m. These contain 45 384 rectangular and circular steel hollow sections (488 members in each space frame). Subsequently, the 10 125 m^2 Birmingham International Arena was built on site in 18 months, with Nodus roof structures suspended from masts

Figure 1.27 Nodus space-frame system, 1980. (© British Steel)

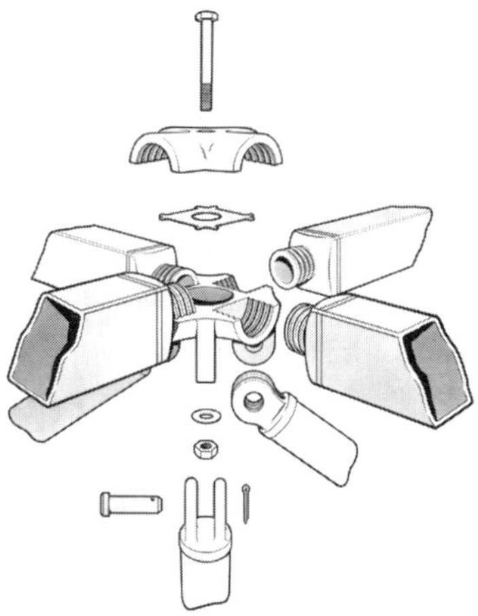

to create a clear span multi-purpose arena 60 m × 60 m. It was opened with a concert by rock band Queen in 1980.

The Nodus system made possible the large uninterrupted spans required and offered the capacity to carry extensive overhead services. The small-size components used in the space-frame construction enabled fast workshop production, easy transportation to site and quick erection. Richard Haryott, the Arup director leading the multidisciplinary NEC engineering team, was still in his thirties when the first halls were built. Years later we were colleagues. I asked him whether, with hindsight, he would have done NEC differently and he said no. Nodus was the real deal aesthetically, technically and logistically.

Mound Stand redevelopment, Lord's Cricket Ground, St John's Wood, London, UK

This highly acclaimed collaboration (Figure 1.28) between architect (Michael Hopkins) and engineer (John Thornton) defies description – it just looks so right. In 1996, John Thornton wrote about his Arup multidisciplinary engineering design team: 'We are a design group which is managed, not a management system which does design ... Excellence arises from conditions and not procedures.' By 'conditions' John meant things like pulling in the same direction, knowledge of the market, skills and experience, focus and keeping responsibility. I could add another condition relating specifically to sports facilities. John and I were keen runners in the 1980s and used to swap ideas on racing, training, injuries and diet. I do not think it's necessary to be a sports enthusiast to engineer a great sports facility – but I think it makes that outcome more likely.

Rogers Centre (Skydome), Toronto, Canada

The Rogers Centre (Figure 1.29) is the world's first stadium to have a retractable motorised roof. It also has a 348-room hotel attached to it, with 70 rooms overlooking the field of play. The idea of building an enclosed sports venue came from a football match held at Toronto's Exhibition Stadium in 1982, at which the unsheltered crowd – including the match patrons – were drenched and the washrooms overflowed. The Skydome was built between October 1986 and May 1989. Because this was new design territory, the architects and engineers kept things simple. They adopted a robust dome form, created using steel trusses supporting profiled steel cladding and a single-ply PVC membrane. Proven technologies were used to move the

Figure 1.28 Mound Stand redevelopment, Lord's, 1987. (© Arup) (This figure is reproduced in the colour-plate section as Plate II)

Figure 1.29 Skydome, Toronto, 2007. (© Marcus Obal) (This figure is reproduced in the colour-plate section as Plate III)

four-panel roof. The panel on the north end is fixed in position, and the other three are moved by electrically powered 'train' engines running on high-strength rails. Opening the roof takes 20 minutes.

Camden Yards Stadium, Baltimore, MD, USA

Before this stadium (Figure 1.30), big league ballparks tended to be symmetrical in form and multipurpose in function. Camden Yards replaced Memorial Stadium in 1992. It is of the traditional asymmetrical ballpark form. Match attendances soared from an average 25 722 over the final 10 years' tenure of Memorial Stadium to an average of 43 490 over the first 10 years of Camden Yards. Because of this success, many other cities began building traditional-style ballparks with modern amenities (such as skyboxes) in downtown settings. Some of these incorporate 'retro' features in the stadium exteriors and interiors. Others, dubbed 'retro-modern' parks, combine retro interiors with modern exteriors. Camden Yards incorporates the historic B&O Warehouse in its architecture, reused as offices, service spaces and a private club. Stadium renovations have included a new HD video display and scoreboard above the right field bleachers (2008), a new high-fidelity sound reinforcement system around the park (2009), replacement seats in the lower seating bowl, drink rails at the club level, and new party suites and casual luxury boxes in place of some skyboxes (2011).

Court Suzanne Lenglen, Stade Roland Garros, Paris, France

You could not dream up a more unique combination than 'La Divine' Suzanne Lenglen (1899–1938), the world's first sports 'star', and pioneer aviator, engineer and war hero Roland Garros (1888–1918), the world's first flying 'ace'. The brutalist concrete Court Suzanne Lenglen (Figure 1.31) is unique too, not so much for what you see as for what you do not see. Stade Roland Garros is the world's premier clay court tennis championship venue but it does not actually contain any clay. The courts have approximately 250 mm (10 in.) of stone surmounted by 150 mm (6 in.) of clinker and 100 mm (4 in.) of limestone, then 3 mm of pulverised brick (which supplies the brown colour to diving players' 'whites'). Because Court Suzanne Lenglen is built on concrete slabs above a car park, it was designed with an underground irrigation system – the first of its kind – to control moisture levels within the court surface. This system features two large vats beneath the court, for rainwater recovery.

Dunc Gray Olympic Velodrome, Sydney, Australia

This (Figure 1.32) is one of the world's largest and most advanced facilities of its type. The unique 130 m × 100 m (426.5 ft × 328 ft) single-layer steel grid-shell roof was developed specifically for the year 2000 Olympics. All the loads on the roof are dealt with by stiffness in the plane of the roof, like an eggshell. The cathedral-like ceiling is the outcome of there being no elements of significant depth and no trusses to clutter the space. The final steel weight for the roof span was 40 kg/m^2 (8.2 lb/ft^2) and 50 kg/m^2 (10.2 lb/ft^2), including the ring beam and columns. The alternative solution of using curved trusses would have been more expensive. By capturing all the roof forces in the ring beam, at eaves level, the structural engineer and architect achieved a dramatic internal–external relationship, with the roof appearing to hover overhead and natural light penetrating the full perimeter. A range of grids and patterns was investigated to form the doubly curved shell surface, including orthogonal and geodesic arrangements. The solution was based on a torus, or donut, with a circular arc of radius 110 m (360 ft) across the venue rotated about a longitudinal radius of 142 m (466 ft). This gives a transverse rise of 13 m (43 ft) and a longitudinal rise of 16 m (53 ft). The elliptical plan shape is then cut out from the surface, leading to a slight but visually pleasing rise and fall around the perimeter. Connections and site erection were simplified by using prefabricated box-

Figure 1.30 Camden Yards Stadium, Baltimore, 1992. (© Carol M. Highsmith)

section arches on the circular set-out lines spanning across the hall at approximately 11 m (36 ft) centres. Pairs of back-to-back cold-formed 'C' purlins at 1.8 m (6 ft) centres span the 11 m from arch to arch. In a world first, the purlins are curved to achieve a uniform curved external surface rather than an awkward faceted surface. The tapered profiled metal deck roofing is the roof cladding.

Utah Olympic Oval, Kearns, Salt Lake City, UT, USA

The challenges here (Figure 1.33) were to create and maintain the fastest sheet of ice in the world, provide a bright and pleasant indoor training facility, set an example of energy efficiency, and accomplish all of this to a limited budget and – not least – complete the facility a year before its hosting of the

Figure 1.31 Court Suzanne Lenglen, Stade Roland Garros. (© Poudou99)

Figure 1.32 Dunc Gray Olympic Velodrome, Sydney, 2000. (© Arup)

2002 Winter Olympics events. The roof was designed as a suspended steel girder system, which reduced the enclosed air volume by 22% over the next best option, a steel truss roof system. This reduction in enclosed volume made the space easier to manage. The preferred temperature for skater comfort is 15.6°C (60.1°F), but racing conditions require that the space temperature be lowered to 10°C (50°F), with the ice held at −6.6°C (20.1°F). These requirements were resolved in an innovative solution that made the building one of the first 12 to be given Leadership in Energy and Environmental Design (LEED) certification for sustainable design. In the 2002 Olympics eight world records were set here in ten Olympic events hosted.

Qi Zhong Stadium, Shanghai, China

Qi Zhong (Figure 1.34) is Asia's largest tennis facility. Japanese company Environment Design Institute (EDI) won the design competition with a unique eight-piece steel roof which resembles

Figure 1.33 Utah Olympic Oval, 2002. (© Arup)

Figure 1.34 Qi Zhong Stadium, Shanghai, 2010. (© Pratik Mehta)

a magnolia flower blooming as it opens. Each 2 tonne 'petal' turns on a fixed point and the petals are supported by a circular truss of inverted triangular section. The opening roof enables the stadium to host both indoor and outdoor tennis events. The 30 649 m^2 (330 000 ft^2), 40 m (130 ft) tall stadium is the centre-piece of a 34 ha (84 acre) tennis centre development, of which 85 000 m^2 (915 000 ft^2) is made up of structures. Parking for 993 vehicles is incorporated. The 15 000 seat Centre Court includes VIP seats, deluxe skyboxes and courtside box seats. There is an air-conditioning outlet under each seat, which is used for cooling in summer and heating in winter. The overall development includes a state-of-the-art 1800 m^2 (19 375 ft^2) media centre equipped with 240 individual journalists' work-stations. The engineer for this project was the same Lin Ying Ru who supervised construction of the 80 000 seat Shanghai Stadium in 1993.

Buddh International Circuit, Greater Noida, Uttar Pradesh, India

The final lap of this chapter is about race circuits. For stadiums arranged around arenas, vertical extension is often the only expansion option. I want now to flag up stadium/grandstand technologies used at racecourses to create exciting possibilities for linear developments. Figure 1.35 includes several very different types of stand structure. Buddh is the venue for the Formula 1 Indian Grand Prix, first hosted in 2011. The circuit has an approximate length of 5.14 km (3.2 miles) within an area of 354 ha (874 acres). Initial seating capacity was planned at 110 000, with provision for future increases to 200 000. The circuit will be part of the 1000 ha (2500 acre) Jaypee Greens Sports City, which also includes a planned 100 000 seat cricket

Figure 1.35 Buddh International Circuit, Greater Noida, Uttar Pradesh, 2011. (© Tina Sharma Tiwari)

Figure 1.36 Suzuka International Racing Course, Mie Prefecture, 2011. (© gpnow)

stadium, an 18-hole golf course, 25 000 seat hockey stadium and a sports academy. The circuit was the judges' unanimous winner of the 2011 Motorsport Facility of the Year award at the Professional Motorsport World Expo 2011 held in Cologne, Germany.

Suzuka International Racing Course, Mie Prefecture, Japan

In the early 1960s, Soichiro Honda wanted to race in Formula 1 in his own country. He selected a piece of land close by a Honda plant, and sent to Dutch track designer John Hugenholtz the message: 'I'm building a racetrack. Come to Japan. S Honda.' The 5.8 km (3.6 mile) racecourse, originally named Motor Sportsland, opened in 1962, and went on to become the host venue of the Japanese Grand Prix. It has, unusually, a figure-of-eight layout (with the back straight passing over the front of the track). The 350 m (1150 ft) long, 40 687 m^2, three-storey, steel-framed Circuit Paddock Grandstand (Figure 1.36) was built by Takenake Corporation as part of the circuit's major reformation works in preparation for re-hosting Formula 1 in 2009. The grandstand's lowest level is for repairing engines, and the two floors above provide spectator seating areas, where race goers can eat and drink at tables while watching the action on the track. The existing building was demolished and the new one erected in just 7 months. The grandstand roof was assembled into units on the ground and lifted into position. Because the very long building is straight, there is no impairment to the line of sight from the spectator seats.

Bahrain International Circuit, Sakhir, Bahrain

This Grand Prix motor racing circuit (Figure 1.37) opened in 2004. In 2007 it was awarded the Fédération Internationale de l'Automobile (FIA) Institute Centre of Excellence Award for excellent safety, race marshal and medical facilities, and for the high standards of technology required to maintain these. Sand is kept off the track by spraying an adhesive on adjacent desert. The highly acclaimed surface of the track is made of Graywacke aggregate shipped to Bahrain from Bayston Hill quarry in Shropshire, UK. Principal facilities of the circuit include: an eight-storey VIP tower, the main grandstand (10 500 spectators), a pit and paddock complex, the Oasis Complex (3000 spectators and a second pit building), the race control centre, a technical resource centre, racing team buildings, administration and hospitality buildings, a medical centre, a TV broadcast centre, a media centre, a photographers' centre, under-track tunnels, vending area, dual carriageway road access, and parking for 13 000 cars. There is seating for 45 000 spectators in total, with 47 VIP hospitality suites incorporated in the main and Oasis grandstands).

Figure 1.37 Bahrain International Circuit, Sakhir, 2014

Rockingham Motor Speedway, Corby, Northamptonshire, UK

Our stadium and arena tour is at an end – or at least it is for me. Figure 1.38 shows the view from my office window – well, it would have to be, wouldn't it? Rockingham Motor Speedway was opened by HM The Queen on 26 May 2001. The circuit is uniquely quadrilateral and it is Europe's fastest. Visitors have included Paul Newman (1925–2008), who arrived with the Newman/Haas Racing (NHR) team contesting IndyCar Series. This chapter has been all about change and diversity, and Paul Newman gives me the perfect sign-off: 'Each town has its own allure and it has its own character. This race would be different from Long Beach, it would be different from Edmonton, and you guys will put your own signature on it.'

Figure 1.38 Rockingham Motor Speedway, Corby, 2014. (© Simon Atkinson)

FURTHER READING

Benson M (1989) *Ballparks of North America*. McFarland, Jefferson, NC, USA.

Culley P and Pascoe J (2009) *Sports Facilities and Technologies*. Routledge, London, UK.

Huxtable AL (1960) *Pier Luigi Nervi*. George Braziller, New York, NY, USA.

Luckman C (1988) *Twice in a Lifetime: From Soap to Skyscrapers*. WW Norton, New York, NY, USA.

March W (1936) The Reich Sport Field. *Olympic Games 1936*, No. 12. Publicity Commission for the Eleventh Olympic Games, Berlin, Germany.

Nally TH (*c*. 1922) *The Aonac Tailteann and the Tailteann Games: Their Origin, History and Ancient Associations*. Talbot Press, Dublin, Republic of Ireland.

Shepherd GA (1986) When the President spoke at Balboa stadium. *The Journal of San Diego History* **32(2)**: n.p.

Torroja E (1958) *The Structures of Eduardo Torroja*. FW Dodge Corporation, New York, NY, USA.

Stadium and Arena Design
ISBN 978-0-7277-5790-6

http://dx.doi.org/10.1680/saad.57906.025

Chapter 2

Sustainable design of major sports stadiums: case study analysis of stadiums for the Olympic Games 2000 in Sydney, 2004 in Athens and 2008 in Beijing

Sven Schmedes

Definition of sustainability

The English word 'sustain' is based on the Latin word *sustenare*, meaning to 'keep' or 'hold up'. The modern use of the word can be traced to 17th century German forestry science, in which it was the precursor to today's 'sustainable yield' (a level of resource extraction that can be maintained over time). The meaning of the word 'sustainable' could be defined as 'capable of being maintained at a certain level' (Silverman, 2009). With reference to the environmental point of view the most commonly cited definition is from the United Nations report *Our Common Future*, also known as the Brundtland Report from 1987: 'Sustainable development is development that meets the needs of the present without compromising the ability of future generations to meet their own needs' (United Nations, 1987).

Environmental impact

Sports stadiums have a considerable impact on the urban, environmental and social context. In particular, where several new stadiums are built within the same city for a single mega-event like the Olympic Games, Commonwealth Games or Soccer World Cup, the implications are significantly higher. Therefore, the requirements for the operation of each stadium after the mega-event are of great importance to ensure sustainable integration into the urban, environmental and social context, as well as to maximise their use on a long-term basis.

In the first part of this chapter, a review of the subject is presented. A brief history of the development of stadium design in the Olympic context is given, and the evolving requirements for staging Olympic Summer Games, the structure of organisations involved, existing literature research, certification methods and further detailed background information are summarised. In the second part, the methodology and the development of the bespoke research tool is described. Subsequently, case studies of three different stadium types (Olympic, football and indoor) used for the Olympic Summer Games in Sydney (2000), Athens (2004) and Beijing (2008) are analysed based on literature research, field surveys and interviews. A comparative assessment of each stadium type is conducted in three domains: Urban Category, Environmental Category and Social Category. In each category two indicators with two parameters are evaluated based on a five-point score system. In the third part of this chapter the results are discussed and conclusions are drawn based on comparative assessment. The applicability of the evaluation matrix is demonstrated using an example appraisal of Wembley Stadium for the Olympic Summer Games in London (2012). In the last section, lessons learned and recommendations for future Olympic Games are shared.

In the Urban Category, sports stadiums built on a site previously used for sports venues or adjacent to other existing venues are rated significantly higher, because existing sites are already integrated into the urban context and located in proximity to the city centre as well as other mixed-use areas, resulting in synergy effects, with extended catchment areas and good connectivity.

In the Environmental Category, sports stadiums are rated higher if specific requirements for operation of the sports stadium after the Olympic Games were considered in the design stage to ensure maximum adaptability and flexibility. Dismantling of the overlay (tailor-made structures/installations required for staging the event) after the Olympic Games reduces energy consumption in subsequent operation. An overlay designed for permanent use or reuse at a different venue further increases the level of sustainability. Use of energy-efficient systems with power generation and water conservation preserves resources.

In the Social Category, sports stadiums are rated higher if a balanced proportion of sport usage and other non-sport uses is achieved. The long-term utilisation of a sports stadium correlates with the level of urban integration, urban context, building type and use mix. Sports stadiums used by more than one home team and integration of other non-sport uses (e.g. retail, commercial, recreation) lead to a significantly higher level of utilisation.

Introduction

The Olympic Summer Games are the largest sporting event in the world, with 28 different competitions held almost simultaneously over 16 days at multiple venue locations. In addition to the significant construction costs, there are the major costs of operating and maintaining the sports stadiums after the Games have ended, and in the long term these are significantly higher than the initial construction costs. To cover the operational costs for maintenance and operation, a minimum number of events have to be held at the sports stadium or an alternative use found to generate income.

The aim of this research is the analysis, evaluation and comparative assessment of three different types of sports stadiums (Olympic, football and indoor) used for the most recent modern Olympic Summer Games in order to identify key parameters influencing the level of sustainability after the Games. The three types of stadium are assessed across three domains: Urban Category, Environmental Category and Social Category.

The objective of this research was to develop an evaluation matrix to make a comparative assessment of the indicators influencing the level of sustainability in the three categories and to understand the key differences between the selected sports stadiums for the Olympic Games in Sydney, Athens and Beijing, as well as to determine which stadium of each type is the most sustainable.

With regard to academic relevance, it is assumed that this thesis is meeting the expectations of a relevant academic contribution, as value, utility and importance of information are well accounted for. Furthermore, the subject itself is an 'evergreen' one and one of current interest, as the Olympic Games are a repeated mega-event that is celebrated every 4 years.

In 2002, the International Olympic Committee (IOC) organised the *International Symposium on Legacy of the Olympic Games: 1984–2000*, which in its conclusions calls for more attention to be paid to research into the legacy of the Games:

> There is a clear need for more research of a longitudinal nature into all aspects relating to Olympic legacy, beginning well before the Games and lasting for a sustained period after their completion. (Olympic Studies, 2002)

The analysis and comparative assessment of selected stadiums used for the first three Olympic Games in the 21st century, in 2000, 2004 and 2008, provide new and important results. The author of this research was granted a scholarship from the Olympic Studies Centre (OSC), as the IOC considered the subject to be significant. Accordingly, the literature review of this research is predominantly based on data and references obtained through the unique opportunity to use the archives at the Olympic Studies Centre in Lausanne. In addition, the literature search was complemented by a search of documents available in other libraries and on the internet to find published design reviews.

Review of the subject

In the literature review the existing studies and researches are revised and discussed. The most relevant literature relating to the aims and objectives of this chapter are briefly mentioned in this section. *A Framework for Evaluating Olympic Urban Development for Sustainability* was written by Hanwen Liao in 2006. This thesis includes a systematic collection of data and an assessment of factors that define environmental sustainability. The evaluation framework considers nine themes (e.g. energy consumption, water consumption) and is based on a multicriteria assessment method (Liao, 2006).

Natalie Essig's *Nachhaltigkeit von Olympischen Sportbauten (Sustainability of Olympic Sports Facilities)* is a thesis that was subsequently published as a book in 2010. Essig analysed the practicability and measurability of sustainability aspects with regard to competition facilities for the Olympic Games. She reviewed 12 case studies of the modern Olympic Summer and Winter Games, and concluded with four theses. Furthermore, environmental evaluations, reports and guidelines of different non-governmental organisations as well as of the applicant and candidate cities established by the IOC in the 1990s were investigated.

International assessment methods and certification systems such as the Building Research Establishment's Environmental Assessment Method (BREEAM), Leadership in Energy and Environmental Design (LEED) and the Deutsche Gesellschaft für Nachhaltiges Bauen (DGNB) were evaluated by Essig with regard to relevant criteria and their applicability to Olympic sports facilities (Essig, 2010) (Tables 2.1 to 2.4).

The book *Olympic Cities: City Agendas, Planning and the World's Games, 1896–2012* is a compendium of essays by John R Gold and Margaret M Gold published in 2007. It states that 'several authors note that in the bid process it is common to overestimate the long-term and other benefits that accompany hosting the games, but none of the authors completely answers the question of whether the Olympics are worthwhile for cities in the long run. This demonstrates the need for further research on the experience of cities years and decades after hosting the games' (Parks, 2014).

In addition, the book covers historical background of the Olympic Games, and the outstanding accomplishments of Baron Pierre de Frédy de Coubertin in founding the modern Olympics in 1894 are reviewed and briefly summarised. The requirements for the hosting city for staging the Games, the election process and its historical development, and the introduction of a two-phase system in 1999 with a process duration of about 2 years and the organisations involved are described. Furthermore, the topic of 'Green Olympics' with reference to David Chernushenko's work and the principles of the Olympic Movement's *Agenda 21* 'as a useful reference tool for the sports community at all levels in the protection of the environment and enhancement of sustainable development' (IOC, 1999) are elaborated.

Certification methods

The most important certification methods are briefly mentioned in this chapter. Some criteria used in the Evaluation Matrix (Table 2.5 – Summary of the comparative assessment ratings for each stadium) in this thesis are developed in consideration of similar criteria of the certification methods BREEAM, LEED, DGNB and GOBAS. The BREEAM Tables 2.1, 2.2, 2.3 and 2.4 illustrate the environmental sections and ratings.

The main internationally recognized certification systems mostly build up on each other and were further developed based on the cumulative experience and knowledge. Initially the British BREEAM (Building Research Establishment's Environmental Assessment Method) was published as an evaluation catalogue in 1990. Subsequently the French HQE (Haute Qualité Environnementale) was released in 1996, however it is still mainly in French. LEED (Leadership in Energy & Environmental Design) was published by the US Green Building Council in 1998. The Japanese CASBEE (Comprehensive Assessment System for

Table 2.2 BREEAM rating benchmarks

BREEAM rating	% score
Outstanding	85
Excellent	70
Very good	55
Good	45
Pass	30
Unclassified	<30

BREEAM, BREEAM New Construction, Non-domestic Buildings, Table 3-2: Minimum BREEAM standards by rating level, Technical Manual, SD5073 2.0: 2011, Manual issue: 1, Watford, Hertfordshire, UK, BRE Global Ltd 2011, page 26

Table 2.1 BREEAM 2011 new construction: environmental sections and assessment issues

Energy
Reduction of CO_2 emissions
Energy monitoring
Energy-efficient external lighting
Low- or zero-carbon technologies
Energy-efficient cold storage
Energy-efficient transportation systems
Energy-efficient laboratory systems
Energy-efficient equipment (process)
Drying space

Transport
Public transport accessibility
Proximity to amenities
Cyclist amenities
Maximum car-parking capacity
Travel plan

Land use and ecology
Site selection
Ecological value of site/protection of ecological features
Mitigating ecological impact
Enhancing site ecology
Long-term impact on biodiversity

Health and well-being
Visual comfort
Indoor air quality
Thermal comfort
Water quality
Acoustic performance
Safety and security

Water
Water consumption
Water monitoring
Water leak detection and prevention
Water-efficient equipment (process)

Waste
Construction waste management
Recycled aggregate
Operational waste
Speculative floor and ceiling finishes

Materials
Life-cycle impacts
Hard landscaping and boundary protection
Responsible sourcing of materials
Insulation
Designing for robustness

Pollution
Impact of refrigerants
NO_x emissions from heating/cooling source
Surface water run-off
Reduction of night-time light pollution
Noise attenuation

Management
Sustainable procurement
Responsible construction practices
Construction-site impacts
Stakeholder participation
Service-life planning and costing

Innovation
New technology, process and practices

BREEAM, BREEAM New Construction, Non-domestic Buildings, Table 1-1: BREEAM 2011 New Construction environmental sections and assessment issues, Technical Manual, SD5073 2.0: 2011, Manual issue: 1, Watford, Hertfordshire, UK, BRE Global Ltd 2011, page 14–15

Table 2.3 Minimum BREEAM standards by rating level

BREEAM issue	Pass	Good	Very good	Excellent	Outstanding
Management 01: Sustainable procurement	1 credit	1 credit	1 credit	1 credit	2 credits
Management 02: Responsible construction practices	0	0	0	1 credit	2 credits
Management 04: Stakeholder participation	0	0	0	1 credit	1 credit
Health & wellbeing 01: Visual comfort	1	1	1	1 credit	1 credit
Energy 01: Reduction of CO_2 emissions	0	0	0	6 credits	10 credits
Energy 02: Energy monitoring	0	0	1 credit	1 credit	1 credit
Energy 04: Low- or zero-carbon technologies	0	0	0	1 credit	1 credit
Water 01: Water consumption	0	1 credit	1 credit	1 credit	2 credits
Water 02: Water monitoring	0	1 credit	1 credit	1 credit	1 credit
Materials 03: Responsible sourcing	0	1 credit	1 credit	1 credit	1 credit
Waste 01: Construction waste management	0	0	0	1 credit	1 credit
Waste 03: Operational waste	0	0	0	1 credit	1 credit
Land use & ecology 03: Mitigating ecological impact	0	0	1 credit	1 credit	1 credit

BREEAM, BREEAM New Construction, Non-domestic Buildings, Table 3-2: Minimum BREEAM standards by rating level, Technical Manual, SD5073 2.0: 2011, Manual issue: 1, Watford, Hertfordshire, UK, BRE Global Ltd 2011, page 27

Building Environmental Efficiency) originated from BREEAM and LEED experiences. It is an assessment system which is primarily used in Japan as it is based on local Japanese norms, standards and regulations. The Australian Green Star followed in 2002. These systems are regarded as first generation systems which focused on the assessment of ecological and energy efficiency criteria of a building. The second generation systems like the German DGNB (Deutsche Gesellschaft für Nachhaltiges Bauen) issued in 2008 based on the works of expert groups, or GOBAS, which is a green building assessment system sponsored by the Chinese Ministry for Science and Technology and published in 2003 for the Olympic Games 2008 in Beijing, have an emphasis on a holistic approach including the consideration of the ecological situation as well as energy efficiency of a building but also economical aspects, sociocultural criteria, technical criteria, location and process quality in regard of the whole lifecycle (Ebert *et al.*, 2010).

Table 2.4 BREEAM environmental section weightings

Environmental section	Weighting
Management	12%
Health & well-being	15%
Energy	19%
Transport	8%
Water	6%
Materials	12.5%
Waste	7.5%
Land use & ecology	10%
Pollution	10%
Total	100%
Innovation (additional)	10%

BREEAM, BREEAM New Construction, Non-domestic Buildings, Table 3-3: BREEAM Environmental section weightings, Technical Manual, SD5073 2.0: 2011, Manual issue: 1, Watford, Hertfordshire, UK, BRE Global Ltd 2011, page 28

Stadium design in the Olympic context

The early monostadium model was one mainly of hosting sports of:

> basic physical exertion belonging to the oldest disciplines such as running, jumping, throwing and swimming as well as sports of the so called civilised era with its conflicts and wars including physical exercise in a battlefield manner such as archery, shooting, fencing, boxing, judo and equestrianism. (Liao, 2006, p. 1)

Liao is stating that:

> the infield of the Olympic Stadium was accordingly adapted to the requirements with portable turf for the football tournament, set up with hurdles and obstacles for the equestrian routine, erected with a demountable podium for boxing, gymnastics, wrestling and weightlifting contests. In the event of the London 1908 and the Amsterdam 1928

> Games the Olympic Stadium was even excavated for a swimming pool or ringed by a cycling track. Additionally the example of the 'White City' Olympic Stadium of the London Games 1908 is given, which was developed by James Fulton and is considered as a good example of a monostadium with an emphasis on the design but compromising the sports and safety requirements. The crucial impulse for creating a new kind of sports venue was the time schedule. The Olympic time was restricted to 16 days for the Los Angeles Games 1932 and thereafter, which made a simultaneous performance at different locations inevitable. (Liao, 2006, p. 297)

The development of the multi-purpose stadium took place following the separation of the facilities for swimming, cycling and equestrian events. The Olympic Stadium remained the largest and most important venue, staging the athletic competitions as well as the opening and closing ceremonies. Subsequent to the Olympic Games, the playing field is usually used for soccer, rugby union, American football and other non-sports events such as concerts. The trend to utilise the stadium for additional functions by integrating facilities for recreation, hotels, offices, retail and other purposes started towards the end of the 20th century. The main reason for adding and combining different functions is to maximise the utilisation of the building beyond sports events on a day-to-day basis. One example for stadiums where additional facilities became an integral part of the stadium complex, but are operating independently, is the Olympic Stadium in Berlin (1936). This structure was built for the 1936 Summer Olympic Games in Germany, and was completely refurbished as well as upgraded between 1999 and 2004 for the 2006 Soccer World Cup. It now includes a museum, chapel, fan shops and VIP boxes which can be used as offices.

The stadium design of the 21st century is described with special attention to the key drivers for enhancing the adaptability of stadiums. These are the operational requirements to stage different types of sports events and non-sports events with differing and partly conflicting requirements at the same venue, for example the running track of an Olympic Stadium encircling the turf in a certain dimension and thus limiting the future use. The necessity to modify the playing field for different event types and for flexibility with regard to spectator capacity of sports venues is emphasised. The reason for a reduced spectator capacity is the different operational requirements in the long term after the Games, when there is usually significantly lower demand. Another important reason for the subsequent reduction in spectator capacity and dismantling of the overlay is to minimise operational and maintenance costs. Examples of temporary arenas, venues at the 2000 Olympics in Sydney and the 2004 Olympics in Athens, and their use are given as paradigms for, but are not part of, this research.

Development research tool

The methodology of this chapter is a combined assessment of the obtained knowledge on the subject consisting of the findings of analytical research, which is based on an evaluation of 'library research, including historical research with utilisation of historical sources like documents and remains' (Pattron, 2009). Subsequently, a 'field research including case studies and personal interviews was carried out' (Lingayas Institute of Management and Technology, 2012). The analytical aspects are an integral part of this chapter as 'the researcher had to use facts or information already available and analyse these to make a critical evaluation of the material' (Pattron, 2009, p. 3). In the analytical research, three domains are established: A – Urban Category, B – Environmental Category and C – Social Category. The results were analysed by means of a comparative assessment using a five-point score system and summarised in the evaluation matrix.

In accordance with Pattron's (2009) publication, the applied research methods used are qualitative as well as quantitative. The qualitative method is a survey (personal interviews with the venue operators) and the quantitative method is the case-study research. The data collection can be described as a mixed matrix of categorical data through ordinal observations, which are ranked, and continuous data, which are measured and counted.

In the literature search evidence was found from primary sources supplemented by secondary sources. The majority of literature reviewed for this research was obtained from the Olympic Studies Centre in Lausanne, Switzerland, as part of the scholarship (Postgraduate Research Grant Programme, 2008). In addition, the literature search was complemented by a search of documents available in other libraries and on the internet for published design reviews.

A preliminary evaluation matrix was developed based on the insights gained from the literature research and a technical review of drawings as well as other project-relevant information. The results of the literature research and preliminary evaluation matrix were then reviewed and re-evaluated after each case study.

After completion of each piece of field research the results, consisting of site pictures, drawings, site plans, maps, measurements, etc., were analysed and evaluated. Based on the documentation for each case study a comparative assessment was carried out using the five-point scoring system. The evaluation matrix was refined and simplified during the field research in order to bring out the significant facts.

Analysis of the case studies

In preparation for the field research, an analogical literature research and a technical review of relevant drawings and other project-related information for the case studies were carried out. The research for each of the selected three venues (an Olympic stadium, an indoor stadium and a football stadium used for the 2000, 2004 and 2008 Summer Olympic Games) was summarised in a project brief containing relevant key project information and three pictures (satellite picture, exterior view and interior view). The project brief for each venue contained the following details: selection criteria, architect, location, location type, construction period, year of opening, cost of construction, cost of Olympic overlay, operator after the games, use before the games, use during the games, use after the games, tenants, capacity before the games, capacity during the games and capacity after the games.

The majority of information was obtained from official bidding documents and other reports on the Olympic Games provided by the Olympic Studies Centre in Lausanne, Switzerland, as part of the Postgraduate Research Grant Programme (2007). In addition, other information from documents available in libraries and publications and on the internet as well as documents collected during the field survey (e.g. maps, site plans, site pictures) are presented.

Analysis

Table 2.5 summarises the ratings for each stadium with regard to category, subcategory and related parameter, and gives the overall rating for the host city.

Discussion

In the domain A – Urban Category, which consists of subcategories A.1 – Urban integration and A.2 – Urban context (Givoni, 1998), the venues of the 2004 Olympic Games in Athens were rated the highest, with a score of 43 points out of a maximum of 60, followed by the venues of the 2000 Olympic Games in Sydney (39 points) and the venues of the 2008 Olympic Games in Beijing (34 points).

In the subcategory A.1 – Urban integration, which consists of two parameters, A.1.1– Location type and A.1.2 – Building type, the venues of the 2004 Olympic Games in Athens were rated highest with a score of 23 points out of a maximum 30, followed by the venues of the 2000 Olympic Games in Sydney (19 points) and the venues of the 2008 Olympic Games in Beijing (14 points).

In the subcategory A.2 – Urban context, which consists of two parameters, A.2.1 – Proximity and A.2.2 – Connectivity, the venues of the 2000 Olympic Games in Sydney, 2004 Olympic Games in Athens and 2008 Olympic Games in Beijing were all rated equally with a score of 20 points out of a maximum of 30.

In the domain B – Environmental Category, which consists of subcategories B.1 – Building and B.2 – Building services, the venues of the 2000 Olympic Games in Sydney were rated the highest, with a score of 44 points out of a maximum of 60, followed by the venues of the 2008 Olympic Games in Beijing (27 points) and the venues of the 2008 Olympic Games in Athens (20 points).

In the subcategory B.1 – Building, which consists of two parameters, B.1.1 – Overlay and B.1.2 – Adaptability/flexibility, the venues of the 2000 Olympic Games in Sydney were rated the highest, with a score of 18 points out of a maximum of 30, followed by the venues of the 2004 Olympic Games in Athens (16 points) and the venues of the 2008 Olympic Games in Beijing (6 points).

In the subcategory B.2 – Building services, which consists of two parameters, B.2.1 – Energy efficiency and B.2.2 – Water conservation, the venues of the 2000 Olympic Games in Sydney were rated the highest, with a score of 26 points out of a maximum of 30, followed by the venues of the 2008 Olympic Games in Beijing (21 points) and the venues of the 2004 Olympic Games in Athens (9 points).

In domain C – Social Category, which consists of two subcategories, C.1 – Usage mix and C.2 – Utilisation, the venues of the 2000 Olympic Games in Sydney and the 2004 Olympic Games in Athens were rated equal highest, with a score of 34 points out of a maximum of 60, followed by the venues of the 2008 Olympic Games in Beijing (22 points).

In subcategory C.1 – Usage mix, which consists of two parameters, C.1.1 – Sport usage and C.1.2 – Other usages, the venues of the 2004 Olympic Games in Athens were rated the highest, with a score of 14 points out of a maximum of 30, followed by the venues of the 2000 Olympic Games in Sydney (13 points) and the venues of the 2008 Olympic Games in Beijing (10 points).

In subcategory C.2 – Utilisation, which consists of two parameters, C.2.1 – Average number of events and C.2.2 – Diversity of events, the venues of the 2000 Olympic Games in Sydney were rated the highest, with a score of 21 points out of a maximum of 30, followed by the venues of the 2004 Olympic Games in Athens (20 points) and the venues of the 2008 Olympic Games in Beijing (12 points).

Conclusions

The conclusions for each of the three domains A – Urban Category, B – Environmental Category and C – Social Category are summarised at the beginning of each section below. Subsequently, the conclusions for each of the four parameters in each category are summarised, concluding with a 'statement'. Special emphasis is placed on the findings for those stadiums having the highest score for each parameter, allowing for valuable

Table 2.5 Summary of the comparative assessment ratings for each stadium

		Sydney			Athens			Beijing			Maximum possible points
		Olympic stadium	Indoor stadium	Football stadium	Olympic stadium	Indoor stadium	Football stadium	Olympic stadium	Indoor stadium	Football stadium	
A Urban Category											
A.1	Urban integration (Σ)	5	5	9	8	7	8	3	3	8	10
A.1.1	Location type	3	3	4	4	3	5	1	1	4	5
A.1.2	Building type	2	2	5	4	4	3	2	2	4	5
A.2	Urban context (Σ)	9	8	3	5	5	10	7	7	6	10
A.2.1	Proximity	5	5	1	2	2	5	3	3	2	5
A.2.2	Connectivity	4	3	2	3	3	5	4	4	4	5
	Subtotal	**14**	**13**	**12**	**13**	**12**	**18**	**10**	**10**	**14**	**20**
B Environmental Category											
B.1	Building (Σ)	10	6	2	4	4	8	2	2	2	10
B.1.1	Overlay	5	2	1	3	1	5	1	1	1	5
B.1.2	Adaptability/ flexibility	5	4	1	1	3	3	1	1	1	5
B.2	Building services (Σ)	10	10	5	2	2	4	9	6	6	10
B.2.1	Energy efficiency	5	5	3	1	1	3	5	3	3	5
B.2.2	Water conservation	5	5	2	1	1	1	4	3	3	5
	Subtotal	**20**	**16**	**7**	**6**	**6**	**12**	**11**	**8**	**8**	**20**
C Social Category											
C.1	Usage mix (Σ)	7	2	4	4	3	7	6	2	2	10
C.1.1	Sport usage	5	1	3	3	2	2	1	1	1	5
C.1.2	Other usages	2	1	1	1	1	5	5	1	1	5
C.2	Utilisation (Σ)	6	10	5	6	6	8	8	2	2	10
C.2.1	Average no. of events	4	5	3	4	3	4	4	1	1	5
C.2.2	Diversity of events	2	5	2	2	3	4	4	1	1	5
	Subtotal	**13**	**12**	**9**	**10**	**9**	**15**	**14**	**4**	**4**	**20**
	Total for each venue	**47**	**41**	**28**	**29**	**27**	**45**	**35**	**22**	**26**	**60**
	Total for hosting city	**116**			**101**			**83**			**180**

insights that can be utilised in the planning of future Olympic Games. All conclusions for each category are summarised at the end of the chapter.

A – Urban Category

Athens Football Stadium was rated the highest score in this category (Figure 2.1). For three of the four parameters, Location type, Proximity and Connectivity, the stadium was allocated the maximum of 5 points, and for one parameter, Building type, it was allocated 3 points. Although Athens Football Stadium is a new building, it was constructed on an existing site which had been used for sports venues for more than 100 years. Therefore, the venue is fully integrated into the Urban context and, furthermore, due to additional uses within the Football Stadium, functions for an independent usage have been created. Accordingly, the Connectivity of the stadium is excellent, as the connection between the metro station and the stadium is convenient, with two pedestrian bridges, and direct entry to the upper concourse for access to the spectator stands is provided.

A high sustainability level is achieved, as the Location type is ideal, being an existing site, and the Building type being ideal also, because it is an existing building. This correlates with a very good Proximity within the Urban context and excellent Connectivity. This result can be explained by the synergic effects of pre-existing infrastructure and a site that was already integrated into the Urban context.

Figure 2.2 shows the evaluation results for subcategories A.1 – Urban integration (parameters A.1.1 – Location type and A.1.2 – Building type) and A.2 – Urban context (parameters A.2.1 – Proximity and A.2.2 – Connectivity).

A.1 – Urban integration

A.1.1 – LOCATION TYPE

Athens Football Stadium was rated the highest on this parameter as the stadium is built on an existing site which had been used for sport venues for more than 100 years. The original stadium on this site was constructed in 1895–1896, and it was used as a velodrome for the first Olympic Games of the modern area in 1896. In 1952–1954 the existing stadium was demolished and a new track and field stadium was constructed. In preparation for the 2004 Olympic Games the existing stadium was once again demolished, and the new Athens Football Stadium was constructed in 2003–2004.

Statement: The use of an existing site with existing infrastructure for access and egress (public transport, road network, etc.) and existing utilities (electricity, water, sewage, gas, etc.) results in a higher sustainability level than can be achieved with a brownfield or greenfield site. The longer that venues operate at an existing site the higher the actual level of integration into the Urban context. With reference to Athens Football Stadium and Athens Olympic

Figure 2.1 Athens Football Stadium (Karaiskakis Stadium), Greece – the 24-hour stadium access concept. (© Sven Schmedes)

Figure 2.2 Evaluation results for the subcategories Urban integration and Urban context based on the summary of results for Sydney, Athens and Beijing

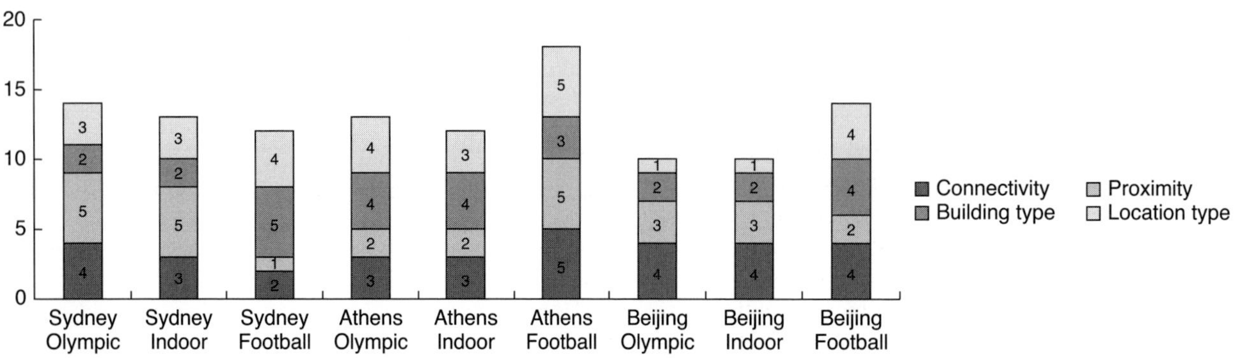

Stadium it was noted that the longer an existing site is used for sport venues the better Urban integration is achieved.

These findings are in accordance with *Agenda 21*, which suggests that the 'Protection of conservation areas and countryside as sports activities, facilities and events must be so arranged as to ensure the protection of conservation areas, the countryside, the cultural heritage and natural resources as a whole' (IOC, 1999, p. 34).

A.1.2 – BUILDING TYPE

Canberra Football Stadium was rated highest on this parameter because the existing stadium was constructed more than 30 years previously and had been upgraded and redeveloped between 1997 and 1999 to meet the requirements of the IOC for the Olympic Games. The redevelopment included lowering of the playing field to build additional spectator seats closer to the playing field, with conversion from an oval to a rectangular shape. In addition, a new spectator stand opposite the existing grandstand was constructed. Other works included the upgrading of the existing grandstand. The total cost was approximately Australian $47 million.

Statement: The upgrade and redevelopment of an existing sport facility was undertaken to meet the requirements for the Olympic Games. Accordingly, modifications and additional elements (a new roof structure, hospitality facilities, athletes' areas, etc.) were necessary to meet the IOC's requirements. Depending on the specific conditions of the existing building, the refurbishment/renovation of an existing sports facility is generally considered to result in a higher sustainability level compared with the construction of a new sport facility, which consumes significantly more resources.

These findings are in accordance with *Agenda 21*, which suggests that 'Special effort shall be undertaken for best possible use of existing sports facilities including to keep them in good condition and to improve them by increasing safety and reducing their environmental impact. The creation of new sports facilities must be confined to cases in which demand cannot be satisfied by using or renovating existing facilities' (IOC, 1999, p. 35).

A.2 – Urban context

A.2.1 – PROXIMITY

Sydney Olympic Stadium (Figure 2.3), Sydney Indoor Stadium and Athens Football Stadium were rated the highest on this parameter because they are located within the immediate urban context of mixed-use areas (e.g. commercial, retail, recreational and residential areas within a linear distance of approximately 0–2 km). The linear distance between Sydney Olympic Stadium/Sydney Indoor Stadium and the city centre, e.g. Central Station or Sydney Opera House is considerable (approximately 17 km (10.6 miles)). However, the stadiums are fully integrated in the

Figure 2.3 Shading elements reducing solar radiation at Sydney Olympic Stadium (ANZ Stadium), Australia (© Christine Goodman) (This figure is reproduced in the colour-plate section as Plate IV)

Olympic Park, with an area of 260 ha (640 acres) providing a significant number of different options for recreation and entertainment. The Olympic Park is surrounded by residential areas that were developed in parallel with and also subsequent to completion of the Olympic Games. The linear distance between Athens Football Stadium and the city centre (e.g. Central Station) is approximately 12 km (7.5 miles) and the distance to the Acropolis is approximately 8 km. The additional uses integrated within the football stadium are creating an independent usage area. Residents of the surrounding areas use the integrated supermarket and gym on a daily basis. Visitors to the adjacent velodrome regularly use the bars, cafés and restaurants in the football stadium before and after events. There are adjacent residential areas, and the Metropolitan Hospital is in the vicinity of this sport facility.

Statement: Commercial areas such as the city centre and mixed-use, residential areas or other areas of attraction are catchment areas for potential users, positively affecting the operation of the sports facility. These areas act as a catalyst to attract more visitors to the stadium and, conversely, visitors to the stadium utilise the areas before or after an event for shopping, recreation, etc., resulting in a higher sustainability level. Proximity to residential areas results in positive effects, allowing users convenient access, but could also lead to negative effects due to traffic congestion, parking issues, noise and other disturbance during events.

A.2.2 – CONNECTIVITY

Athens Football Stadium was rated highest on this parameter because of its outstanding connectivity via the metro system, the linear distance to the metro station being 62 m (203 ft). Access to the stadium is very good, as it is connected via the metro and bus systems. The connection between the metro station and the stadium is convenient, with two pedestrian bridges and direct access to the upper concourse for access to the spectator stands. All circulation areas on the ground floor around the stadium have been allocated as public areas. This system has the benefit that these areas can be used by the public during periods of non-operation of the stadium. Although a large number of security checkpoints is required (there are 35 entrances) to access the spectator stands, this is still considered to be an efficient design, allowing maximum usage of the circulation areas during non-events. There are approximately 1600 parking spaces available close to the stadium. An empty plot of land (owned by a third party) reserved for the construction of an office building is currently used for additional parking. Once the new office building is constructed the available parking sites will be reduced by approximately 40%. Park-and-ride facilities are generally not provided.

Statement: The availability of sufficient public transport (e.g. there are metro stations close to the stadium, within a linear distance of less than 500 m (1640 ft) results in a higher sustainability level compared with transportation by bus, car or other modes of transport. In addition, the use of public transport systems is more environmentally friendly compared with cars.

These findings are in accordance with *Agenda 21*, which emphasises that: 'Non-polluting means of transport and public transport shall be promoted as it contributes beneficially to various environmental problems, including air pollution, the consumption of non-renewable energy and excessive use of land for highways and parking areas' (IOC, 1999, pp. 36–37).

B – Environmental Category

Sydney Olympic Stadium was rated the highest in this category. For the four parameters Overlay, Adaptability/flexibility, Energy efficiency and Water conservation the stadium was allocated the maximum of 5 points. The stadium overlay was removed after the Olympic Games in order to meet the requirements of the nine home teams for day-to-day use and to reduce energy consumption. The stadium was well planned with regard to Adaptability/flexibility. It allows for reconfiguration of the playing field and provides retractable seating and removable turf.

Sydney Olympic Stadium utilises state-of-the-art renewable-energy (Figure 2.4), energy-efficient air-conditioning, energy-efficient lighting and environmentally friendly building materials. Furthermore, the stadium has systems for rainwater harvesting and water recycling, and has water-efficient appliances. The general awareness among Australian society about the shortage and preciousness of resources in this part of the world is considerably high, and was subsequently a co-factor in such an outstanding achievement.

Figure 2.4 'Towers of power' generating solar power energy fed into the electric grid at the Sydney Olympic Stadium (ANZ Stadium), Australia. (© Christine Goodman)

An important factor in the sustainability of a stadium is the overlay, which ideally is removed after the Olympic Games in order to meet the requirements of home teams and the day-to-day use of venues. At the same time, in terms of Adaptability/flexibility a venue will ideally have provision for reconfiguration of the playing field, retractable seating, and removable flooring or modular turf, resulting in a more frequent usage of the sports facility for different types of events. These factors, combined with technologically state-of-the-art Energy efficiency and Water conservation, which ideally will include usage of energy-efficient systems for power generation and conservation, environmentally friendly building materials and systems for water conservation, as well as a reduction of water wastage through water recycling and water-efficient appliances, result in a higher sustainability level.

Figure 2.5 shows the evaluation results for subcategories B.1 – Building (parameters B.1.1 – Overlay and B.1.2 – Adaptability/flexibility) and B.2 – Building services (parameters B.2.1 – Energy efficiency and B.2.2 – Water conservation).

B.1 – Building

B.1.1 – OVERLAY

Sydney Olympic Stadium and Athens Football Stadium were rated equal highest on this parameter, but for different reasons.

Figure 2.5 Evaluation results for the subcategories Building and Building Services based on the summary of results for Sydney, Athens and Beijing

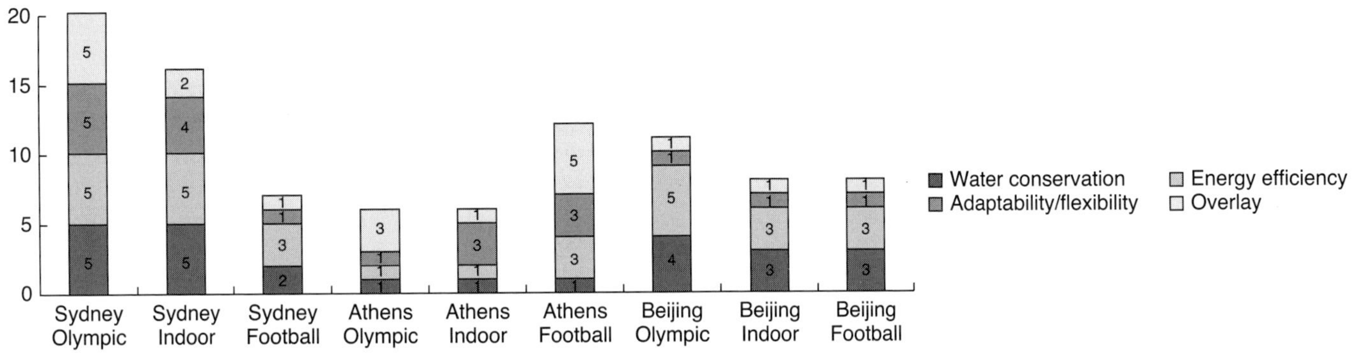

It was decided at an early design stage of the Sydney Olympic Stadium that, to facilitate post-Games operational requirements, the running tracks should be removed and the spectator capacity would be reduced from 115 600 to 83 500 (82 500 in oval mode) after the Games (Sydney Organising Committee for the Olympic Games, 2001). For this purpose the north and south spectator stands (seating capacity 15 000 spectators each) were constructed as temporary seating without a roof. These 30 000 temporary seats were only leased from the contractor, and after their removal from the stadium in October 2001 the seats were installed at another stadium in New Zealand. Subsequently, roof structures (steel members covered by polycarbonate panels with UV protection) were constructed to cover the remaining seats on the north and south spectator stands. After removal of the running track, the lower tiers of the spectator stands were extended and partially converted into retractable seating, significantly reducing the distance to the playing field (to approximately 15 m (49 ft)) (ANZ Stadium, 2012).

The overlay was removed after the Olympic Games over a period of approximately 2.5 years, the removal being completed in 2003, at an estimated overall cost of approximately Australian $80 million (approximately €62 million) (ANZ Stadium, 2012). During the bidding stage the consortium for the ANZ Stadium consisted of the architect, consultants, investors and operators (Octagon). Accordingly, the operation of the stadium after the Olympic Games was the main focus. The consortium was selected to design, build and operate the stadium on a 30-year lease. The agreement with the operator Octagon was limited to 10 years, and after 9 years the personnel of the operator were taken over by the Australian government.

One of the key reasons for the successful conversion of Sydney Olympic Stadium after the Games was the long-term approach taken by the expert bidding team and the direct legitimate interest of the operator.

Athens Football Stadium was constructed as per the requirements of the IOC but at the same time with full consideration of its post-Games use for football by the home team and its day-to-day use by integration of other usages. No temporary overlay was considered, and hence the stadium remained largely unchanged after the Olympic Games in 2004. In April 2003, Olympiacos CFP undertook the responsibility to build a new football-only ground for the 2004 Olympic Games. In return, Olympiacos was granted exclusive use of the stadium until 2052. The club has to cover all maintenance costs and also pay 15% of revenue to the Greek state (Stadia GR, 2013). A private consortium was appointed for the design, construction and operation of the new stadium. This was the first stadium project in Greece financed, designed, built and operated by a private consortium.

Statement: Venues with an overlay that is removed after the Olympic Games in order to meet the requirements of home teams and of day-to-day uses, or venues that are constructed as a perfect match for post-Games use have a higher sustainability level than do venues having a permanent overlay for limited post-Games utilisation or venues having an overlay that results in disadvantages for post-Games utilisation of the venue.

B.1.2 – ADAPTABILITY/FLEXIBILITY

Sydney Olympic Stadium was rated the highest on this parameter because of the consideration of the differing and partially contradicting requirements of the nine teams that would use the stadium as their home venue (Qantas Wallabies (rugby), HSBC Warathas (rugby), NWS State of Origin Blues (rugby), Sydney Swans (rugby), Bulldogs (rugby), South Sydney Rabbitohs (rugby), Pirtek Parramatta Eels (rugby), Qantas Socceroos (football) and Sydney Thunder (cricket)). The lower spectator stands are retractable in order to meet the requirements of the five different professional sports played at the stadium: rugby league, rugby union, Australian rules football, football and cricket.

For rugby league and football (FIFA regulations), a rectangular playing field is required, while the regulations of the American Football League (AFL), rugby and cricket require an oval-shaped playing field. In order to meet these variable requirements, the east and west spectator stands are retractable.

Statement: Provisions for reconfiguring the playing field, retractable seating and removable flooring or modular turf result in a more frequent usage through accommodating different requirements, and therefore in a higher sustainability level.

The operator of Sydney Olympic Park said that the lesson learnt from the Sydney Olympic Stadium was that: 'Focus in the design should be more on the operation after the Olympic Games rather than on the requirements for the Olympic Games. Costs for modification of the stadium for operation after the Olympic Games should be more intensively considered at bidding stage.'

The operator said that the lessons learnt from the Sydney Indoor Stadium were that: 'Interiors of the indoor hall including ceiling, walls and seating should be in black to create a better ambient. Catwalks should have connection points for lighting (in the original design for the Olympic Games no power was available for the catwalks). Furthermore no curtain-system and mother grid should be installed.'

B.2 – Building services

B.2.1 – ENERGY EFFICIENCY

Sydney Olympic Stadium, Sydney Indoor Stadium and Beijing Olympic Stadium were rated equally on this parameter.

At Sydney Olympic Stadium: 'air-conditioning and passive ventilation has been integrated into the design of the stadium. By minimising the extent of air-conditioning provided, valuable energy is conserved' (ANZ Stadium, 2008). 'Natural ventilation was maximised by the use of over-sized lift shafts, stairwells and escalator voids. The use of air-conditioning was minimised by about 40 percent' (Sydney Organising Committee for the Olympic Games, 2001, p. 357).

With regard to lighting, 'a computerised Clipsal C-BUS lighting system with programmable timing is used throughout ANZ Stadium [Sydney Olympic Stadium] – lighting in rooms with no motion or programmed activities automatically switches off after a set period. The need for artificial lighting has been reduced, as the design of the stadium allows maximum intake of daylight through specially constructed light voids. Overall, natural lighting reduced energy needs by 20 percent' (Sydney Organising Committee for the Olympic Games, 2001, p. 357). With regard to power, 'environmentally friendly gas fired co-generators serve as a backup to the main supply of electricity' (ANZ Stadium, 2008). 'The two gas co-generation units achieved an estimated saving of 5000 tonnes of carbon dioxide emissions each year from when the venue was completed in 1999. With regard to building materials, no polychlorinated biphenyls (PCBs) were used in the developments and the use of polychlorinated anisoles (PCAs) was minimised, and at the same time extensive use of recycled building materials was introduced. These initiatives have resulted in major environmental achievements such as a reduction in greenhouse gas emissions by about 10 000 tonnes a year compared to similar facilities' (Sydney Organising Committee for the Olympic Games, 2001, p. 357).

Sydney Indoor Stadium was awarded the prestigious national Banksia Award for Construction Practices in 1999 (Allphones Arena, 2013). With regard to air-conditioning, 'the system has been designed to cool only the air immediately surrounding the spectators, not the entire airspace in the venue, resulting in lower energy use for cooling. With regard to power, the stadium featured Australia's largest rooftop solar power system. Photovoltaic solar cells were installed on the 19 Olympic Plaza lighting towers on the Olympic Boulevard, providing light for major events at the Olympic Stadium and Sydney Indoor Stadium, both in the lead up to, and during, the Sydney 2000 Games and a permanent clean energy legacy for the site. With regard to building materials, no PCBs were used in the developments and the use of PCA was minimised. At the same time extensive use of recycled building materials was introduced as well as maximum use of natural light and natural ventilation in the building design was implemented.

Beijing Olympic Stadium 'was constructed with regard to air-conditioning, lighting and power in accordance with an advanced energy-saving design and such environment-friendly features as natural ventilation and lighting, utilisation of renewable geothermal energy resources and application of solar photovoltaic power generation technologies. Under the football field of Bird's Nest there are 312 ground source heat pump (GSHP) wells, which heat the stadium in winters by absorbing thermal energy from the soil through embedded heat exchange pipes, and cool the place in summers by absorbing cold energy from the soil, saving considerable amount of electric power' (National Stadium – Bird's Nest, 2011). 'Waste water recycling was achieved in all the venues, and the goal of cutting energy consumption by 50 percent was realised' (Beijing Organising Committee for the Games of the XXIX Olympiad, 2011).

Statement: Use of energy-efficient systems for power generation and conservation (e.g. solar power, photovoltaic power, heating system with heat recovery, heat pumps, heat storage), air-conditioning/ventilation (e.g. natural ventilation), lighting (e.g. natural lighting, automatic control for lighting devices, light fittings) and environmentally friendly building materials (recycled, technically advanced) result in a higher sustainability level.

These findings are in accordance with *Agenda 21*, which emphasises: 'Reducing energy consumption and use of renewable non-polluting sources shall be promoted as access to energy is a motor of development, but its excessive, pollutive or uncontrolled use mortgages the long-term future of that development and is contrary to the principles of sustainable development' (IOC, 1999).

B.2.2 – WATER CONSERVATION

Sydney Olympic Stadium and Sydney Indoor Stadium were rated the highest on this parameter. At both stadiums: 'water is conserved throughout Homebush Bay through the use of water conservation and recycling practices. The most important water conservation initiative is the Water Reclamation and Management Scheme (WRAMS). Sewage generated from Olympic venues and facilities is collected, treated and cleansed for non-potable reuse. Reclaimed water is used for toilet flushing in venues and facilities as well as in the Olympic Village and for irrigation across Sydney Olympic Park. WRAMS is designed to save 850 million litres (187 million gallons) of water annually that would otherwise be drawn from Sydney's main water supply, providing a permanent environmental and economic benefit and a useful model for other developments. As well, Olympic venues conserve water through water-efficient appliances, including water flow reduction valves and shower roses, roof-fed rainwater tanks and drip irrigation systems' (Sydney Organising Committee for the Olympic Games, 2001, p. 357).

'At Sydney Olympic Stadium rainwater from the roof is collected in 4 tanks. The total capacity of rainwater harvesting is 3.2 million litres' (ANZ Stadium, 2008). 'Australian native trees and shrubs have been planted across Sydney Olympic Park to reduce water needs' (Sydney Organising Committee for the Olympic Games, 2001, p. 357).

Statement: Usage of systems for Water Conservation (e.g. rainwater harvesting) and reduction of water wastage (e.g. water recycling, separate systems for potable water and grey water supply for toilet flushing, usage of grey water for irrigation of the landscape) as well as water-efficient appliances result in a higher sustainability level.

These findings are in accordance with *Agenda 21*, which emphasises: 'Careful water management and encouraging and supporting world-wide and local activities intended to protect water reserves and preserve the quality of natural waters as water reserves are limited and form part of a closed system. They are essential for farming activities, drinking and hygiene. In an increasing number of regions of the world, the quantity and quality of water reserves are under threat' (IOC, 1999, pp. 38–39).

C – Social category

Athens Football Stadium was rated the highest in this category. For the parameter Other usages, the stadium was allocated the maximum of 5 points, for two further parameters, Average number of events and Diversity of events, it was awarded 4 points and for the parameter Sport usage it was awarded 2 points. The stadium is used by two home teams for the same sport. The stadium integrates eight retail/commercial facilities and various other temporary attractions such as a cinema, a children's playing field and a permanent museum, resulting in a very good level of Other usages. Furthermore, it hosts three events per month and has 23% non-sports events.

Statement: A higher sustainability level is achieved at sports facilities when they are used by home teams, resulting in regular sports events. The integration of retail/commercial concepts significantly enhances the utilisation of the sports facility during non-events. The incorporation of tourism and cultural aspects, with sightseeing tours, adds value, enabling a sports facility to also be an independent tourist attraction. Furthermore, it is noted that the higher the number of events hosted by a venue and, where applicable, a high percentage of non-sport events, indicating a wider variety of usage options, and different usage alternatives for venues with fewer sports events, the better the post-Games usage achieved.

Figure 2.6 shows the evaluation results for the subcategories C.1 – Usage mix (parameters C.1.1 – Sport usage and C.1.2 – Other usages) and C.2 – Utilisation (parameters C.2.1 – Average number of events and C.2.2 – Diversity of events).

C.1 – Usage mix

C.1.1 – SPORT USAGE

Sydney Olympic Stadium was rated the highest on this parameter. The stadium has nine different home teams, which use the stadium for three different sports types (Qantas Wallabies (rugby), HSBC Warathas (rugby), NWS State of Origin Blues (rugby), Sydney Swans (rugby), Bulldogs (rugby), South Sydney Rabbitohs (rugby), Pirtek Parramatta Eels (rugby), Qantas Socceroos (football) and Sydney Thunder (cricket)).

Statement: Sports facilities that are used by different home teams have more regular sports events. Options for different types of sport result in a higher number of events, and therefore a higher sustainability level.

C.1.2 – OTHER USAGES

Beijing Olympic Stadium and Athens Football Stadium were rated equal highest on this parameter.

Figure 2.6 Evaluation results for the subcategories Usage mix and Utilisation based on the summary of results for Sydney, Athens and Beijing

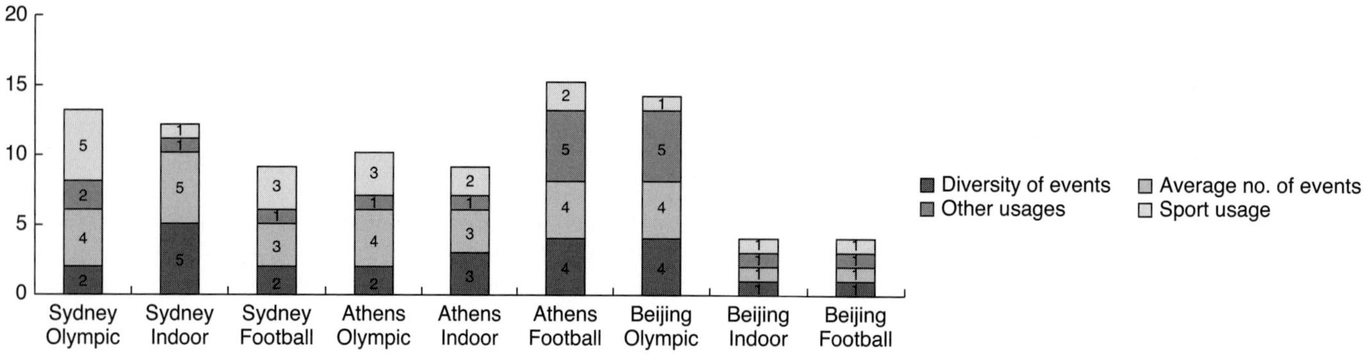

At Beijing Olympic Stadium the exhibition area was added after the Olympic Games. The operator mentioned that the museum would be extended soon. During the visit to the venue it was noted that the exhibition itself is relatively basic. The operator said that feasibility studies were being undertaken to integrate a hotel within the Beijing Olympic Stadium in the future. Currently the following facilities are provided: one restaurant, four food and beverage stores, three shops, four souvenir shops, one Walt Disney shop and one money-exchange facility. With regard to the day-to-day use by citizens, the precinct of the Olympic Stadium is used by the local community (e.g. for Taijiquan, skating and kite flying). The stadium itself is only accessible by payment of an entrance fee, and therefore it is not used on a daily basis by the local community.

At Athens Football Stadium the revenue generated by guided tours is limited. Accordingly, the operator is not promoting guided tours. On average there are 2000 visitors per year.

The museum of the local team Olympiacos is located below the spectator stands and is accessible to the public. The museum is open 6 days per week (closed on Mondays) and has approximately 5000 visitors per year with an entrance fee of €2. The museum mainly relates to the local football club Olympiacos. Accordingly, schools are not fully utilising the museum because the level of general information about the history of sport and the location of the stadium is limited. Various other functions are integrated within the stadium. The fan shop, three coffee shops, two bars, one supermarket and one gymnasium, which are located below the spectator stands, are open 7 days per week and are fully accessible to the public. In 2010 the operator converted some parking areas adjacent to the stadium into a playing field for children. Permission for a change of other land use is pending.

Statement: The integration of retail/commercial concepts (e.g. fan shop, tourist shop, church, coffee shop, restaurants, offices, hotel, spa) significantly enhances the utilisation of a sports facility during non-events. Also, the incorporation of tourism and cultural aspects result in a transfer of knowledge and insight about local culture through the sports facility (e.g. Olympic museum, sports museum). The provision of sightseeing tours for national and international tourists adds value and results in a higher sustainability level.

The operator at Athens Football Stadium confirmed as lessons learnt that: ‘Additional VIP boxes and areas for VIP services, as well as independent access to the restaurant, should be considered. A revised approach for fencing of the stadium and a building management system to control individual areas would be of great value.’

C.2 – Utilisation

C.2.1 – AVERAGE NUMBER OF EVENTS

Sydney Indoor Stadium was rated the highest on this parameter. The operator stated that, on average, 85–110 events are held at the stadium each year (the exact number is confidential). Subsequently, the operator stated that in previous years approximately 75 events were held per year. This would amount to six events per month.

Statement: The higher the number of events hosted at a venue the better the post-Games usage is achieved, resulting in a high sustainability level.

C.2.2 – DIVERSITY OF EVENTS

Sydney Indoor Stadium was rated the highest on this parameter. The operator stated that in previous years there were approximately 75 events per year, with five of these being sports events, amounting to 93% non-sports events and 7% sports events. The non-sports events are mainly concerts and music award functions. The Hillsong Conference has been held at the stadium annually since 2001, and *Disney on Ice* was staged once.

Statement: A balance of sports events and non-sports events indicates greater variety of usage mix and utilisation alternatives for venues having fewer sport events, resulting in a higher sustainability level.

The operator of Sydney Olympic Park confirmed as lessons learnt that: 'Sports culture should be enhanced for diverse cultural understanding. Extraordinary events should include also ordinary daily events for more variety and single-use of facilities should be converted to multi-use.'

These findings are in accordance with the *Olympic Charter* (IOC, 2013, p. 11), which states that 'Olympism is a philosophy of life, exalting and combining in a balanced whole the qualities of body, will and mind. Blending sports with culture and education, Olympism seeks to create a way of life based on the joy found in effort, the educational value of good example and respect for universal fundamental ethical principles. The practice of sport is a human right. Every individual must have the right of practising sport without discrimination in any kind and in the Olympic spirit, which requires mutual understanding with a spirit of friendship, solidarity and fair play. The goal of Olympism is to place sport at the service of the harmonious development of humankind, with a view to promoting a peaceful society concerned with the preservation of human dignity.'

Summary conclusions

In this section the high-level findings are summarised separately for each of the three domains: Urban Category, Environmental Category and Social Category.

Urban Category

The level of sustainability of sports facilities built on a site previously used for sports venues (A.1.1 – Athens Football Stadium) or adjacent to other existing venues is significantly higher because pre-existing infrastructure (e.g. roads, public transport, utilities) can be utilised and developed further.

Utilisation, redevelopment and modernisation of an existing sports facility (A.1.2 – Sydney Football Stadium) are generally more sustainable because they preserve resources and reduce material consumption.

Existing sports facilities sites are more sustainable because such sites are already integrated into the urban context, and therefore in close Proximity to mixed-use areas (e.g. retail, commercial, recreation), resulting in synergy effects and extended catchment areas (A.2.1 – Athens Football Stadium). The level of Connectivity between an existing site and the city centre as well as other mixed-use areas is higher for previous sports sites than new sites, unless the new site is part of a city redevelopment, like at the Sydney Olympic Park where the urban context and infrastructure were created as part of an overall master plan.

Environmental Category

The level of sustainability is higher when the specific requirements for the operation of a sports facility after the Olympic Games are considered during the design stage at the beginning of the project. Dismantling of the Overlay (temporary structures, elements and equipment to fulfil the requirements for staging special events) after the Olympic Games reduces energy and water consumption in the subsequent operation of the stadium. Major structures for the overlay, such as a new roof structure designed for permanent usage (B.1.1 – Athens Olympic Stadium) or a roof reused at a different venue (B.1.1 – Sydney Olympic Stadium) increase the level of sustainability. Adaptability/Flexibility maximises utilisation of the venue (B.1.2 – Sydney Indoor Stadium), reducing the need to use new resources to construct additional sport facilities.

The level of sustainability is significantly enhanced by use of systems for Energy efficiency (e.g. natural ventilation, natural lighting), with systems for power generation (e.g. photovoltaic, heat pumps), Water conservation (e.g. rainwater harvesting, water recycling) and environmentally friendly building materials (B.2.1 – Sydney Indoor Stadium).

Social Category

Sustainability results from a balanced Usage mix consisting of Sport usage and Other usages, providing a broad mix of activities for the community and public. Sports facilities used by more than one home team and integration of various other usages (e.g. retail, commercial, recreation) achieve a significantly higher level of social sustainability (C.1.2 – Athens Football Stadium). The exception is Beijing Olympic Stadium, where the usage is mainly sightseeing rather than sports or other events.

The utilisation of a sports facility correlates with the level of Urban integration (Location type, Building type), Urban context (Proximity, Connectivity), Building (Overlay, Adaptability/flexibility) and Usage mix (Sport usage, Other usages). A high rating in these subcategories results in a high Average number of events and high Diversity of events. Sports facilities where a private operator was involved at the beginning of the project and specific operational requirements were integrated at the project brief and design stages (C.2.1 – Sydney Indoor Stadium) achieved a significantly higher sustainability.

Lessons learnt

During the election process of the host city for the Olympic Games the impact of permanent sports stadiums on the urban, environmental and social context tends to be underestimated, while the opportunities for utilisation of the venues after the Olympic Games are overestimated by the applicant/candidate cities. Therefore, it is recommended to establish more specific guidelines for the preparation of the project brief and feasibility studies in order to verify and assess the proposed level of sustainability of each sports stadium in the urban, environmental and social contexts. In addition to the feasibility study, it is recommended that applicant/candidate cities prepare a business plan, with reserve funds for maintenance and modernisation of the stadium and entire precinct in order to meet the evolving requirements for the long-term operation of facilities. Considering the diverse operational requirements for each sports stadium, it is recommended that applicant/candidate cities involve an operator in the preparation of the project brief and when undertaking feasibility studies. This is to ensure that the focus is not only on the fulfilment of the short-term requirements for staging the Olympic Games or other mega-events but also, and more importantly, on the long-term utilisation of each sports stadium after the Games are over.

Closing comments

This chapter draws on the substantial research done by Sven Schmedes which, in itself, was developed based on an enormous amount of source material. Within the space available in this chapter it has only been possible to cover selected parts of the subject area and to make selected points from the ongoing research. The editors have done their best to present the ideas and information on sustainability that they think may be most useful. Further information related to this chapter may be obtained from the research published under the title *Sustainable Design of Sports Stadiums*.

REFERENCES

Allphones Arena (2013) See http://www.allphonesarena.com.au/Venue_Information/History.aspx (accessed 11/02/2015).

ANZ Stadium (2008) *Recycling First for ANZ Stadium*, 27.02.2008. See http://www.anzstadium.com.au/News/LatestNews/NewsArticles.aspx?NewsContentId = 1d3ac6e9-df48–44da-af3f-0 aa128f37126 (accessed 16/05/2013).

ANZ Stadium (2012) *ANZ Stadium: Creating History Every Day*. See http://www.anzstadium. com.au/AboutUs/History.aspx (accessed 22/12/2012).

Beijing Organising Committee for the Games of the XXIX Olympiad (2011) *Preparation for the Games: New Beijing Great Olympics. Official Report of the Beijing 2008 Olympic Games*, Vol. III, LA84 Foundation, p. 170. See http://olympic-museum.de/o-reports/report2008.htm (accessed 11/02/2015).

Ebert T, Essig N and Hauser G (2010) *Zertifizierungssysteme fuer Gebäude*, München, APPL, pp. 26, 56

Essig N (2010) *Nachhaltigkeit von Olympischen Sportbauten*. Frauenhofer Verlag, Stuttgart, Germany, Band 5, p. 19.

Givoni B (1998) *Climate Considerations in Building and Urban Design*. Wiley, New York, NY, USA.

Gold JR and Gold MM (eds) (2007) *Olympic Cities: City Agendas, Planning, and the World's Games, 1896–2012*, 2nd edn. Routledge, London, UK.

IOC (International Olympic Committee) (1999) *Olympic Movement's Agenda 21. Sport for Sustainable Development*. Foreword by Klaus Töpfer, p. 10. IOC Sports and Environment Commission. See http://www.olympic.org/Documents/Reports/EN/en_report_300.pdf (accessed 27/12/2014).

IOC (2013) *Olympic Charter*. See http://www.olympic.org/Documents/olympic_charter_en.pdf (accessed 11/02/2015).

Liao H (2006) *A Framework for Evaluating Olympic Urban Development for Sustainability*, PhD Thesis, University of Sheffield, Sheffield, UK, pp. 1 and 297.

Lingaya's Institute of Management and Technology (2012) *Research Methodology: An Introduction*, p. 7 (table). See http://www.limat.org/data/research/Research%20Methodology.pdf (accessed 27/12/2014).

National Stadium – Bird's Nest (2011) Overview of the National Stadium. See http://www.n-s.cn/en/test/n214619929.shtml (accessed 20/05/2013).

Olympic Studies (2002) *Provisional Remarks, Conclusions and Recommendation. International Symposium on Legacy of the Olympic Games, 1984–2000*. See http://olympicstudies.uab.es/olympiclegacy/web/conclu/legacy_conclusions.pdf (accessed 16/12/2014).

Parks J (2014) *Urban Planning and the Olympic Games*. See http://www.h-net.org/reviews/showrev.php?id = 30515 (accessed 27/12/2014).

Pattron DD (2009) *Research Methodology*. See http://www.authorstream.com/Presentation/drpattron68–138583-Research-Methodology-CONTENTSConstitut-es-Topic-Select-Limitations-method-Entertainment-ppt-powerpoint (accessed 13/11/2012).

Postgraduate Research Grant Programme (2007) International Olympic Committee (IOC) Olympic Studies Centre created by the IOC Olympic Studies Centre (OSC)/Information Management Department.

Silverman H (2009) Sustainability: The S-word. *People and Place* **1(3)**: n.p.

Stadia GR (2013) *New Karaiskakis Stadium*. See http://www.stadia.gr/karaiskaki/karaiskaki.html (accessed 13/06/2013).

Sydney Organising Committee for the Olympic Games (2001) *Official Report of the XXVII Olympiad. Volume One: Preparing for the Games*. Paragon, Fyshwick, Australia.

United Nations (1987) *Our Common Future*. Report of the World Commission on Environment and Development. See http://www.un-documents.net/our-common-future.pdf (accessed 12/02/2015).

Stadium and Arena Design
ISBN 978-0-7277-5790-6

http://dx.doi.org/10.1680/saad.57906.041

Chapter 3
Symbiosis between nature and advanced technologies

Kisho Kurokawa (1934–2007)
Founder, Kisho Kurokawa Architect & Associates (KKAA), Tokyo, Japan

Kisho Kurokawa determined to become an architect in 1945 when his father took him to see what little remained of his home city of Nagoya. After graduating from Kyoto University, he became a postgraduate student at Tokyo University under Professor Kenzo Tange, designer of the twin 'sweeping-roof' gymnasiums for the Tokyo Olympics. He founded Kisho Kurokawa + Associates in Tokyo in 1962, and became the leader of the Metabolist Movement – a movement away from 'machine principle' towards 'life principle'. His major works in this style include the Kuala Lumpur International Airport, Malaysia, a beautiful fusion of nature and technology located at the edge of a rain forest. As a boy, Kisho Kurokawa had been educated by Buddhist monks, and would develop a philosophy of living together that he articulated in his book *The Philosophy of Symbiosis* – he gave me a signed advance copy. Altogether, Kisho Kurokawa designed some 100 buildings, and still found time for writing, translating, philosophy, politics, teaching, print-making and driving his speedboat around Tokyo Bay. It was an honour to have him on our 'stadium engineering' team and, as a tribute, we are here publishing for the first time his original chapter draft for that book. It was one of his final contributions to the construction industry, because he subsequently went into politics, standing for Mayor of Tokyo and then for a seat in the Japanese House of Councillors. Kisho Kurokawa is survived by his second wife, the actress Ayako Wakao – whom he famously met on a TV programme – and by his son Mikio and daughter Kako (Matsuura) from his first marriage.

John Pascoe

Two World Cup stadiums featuring advanced technologies

Both Toyota Stadium and Oita Stadium were designed and built using state-of-the-art technologies. They comply with Fédération Internationale de Football Association (FIFA) criteria and were built to host World Cup games at the beginning of the 21st century.

Core facilities with movable roofs, inside sports parks

The two stadiums come with retractable roofs that enable games to be played under optimal conditions, even in the wettest weather. The stadiums are also designed to hold concerts, exhibitions and other events. Even as the 2002 World Cup ended, numerous games and other events that take advantage of the retractable roof were, and are, being planned and held. The design of the Toyota Stadium project, moreover, includes the Toyota Bridge (Figure 3.1). Because of this, the stadium has become the core facility in the riverbed sports park that also serves as an enforced dike or bank. Oita Stadium, meanwhile, has track and field facilities, and has been approved by the International Association of Athletics Federation as an athletics stadium with a retractable roof (Figure 3.2). Like Toyota Stadium, Oita Stadium is also the core facility inside a sports park. There are plans to build an indoor athletics arena and pool nearby in the future.

Harmony with the surrounding landscape

Toyota Stadium was planned to match the Toyota Bridge, which I had previously designed. Together, they create a magnificent landscape that harmonises with the riverine environment. In the

Figure 3.1 The Toyota Bridge links the Toyota Stadium and a playing field on the bank of the River Yahagi. (© KKAA)

Figure 3.2 View of Oita Stadium from the north. (© KKAA) (This figure is reproduced in the colour-plate section as Plate V)

Figure 3.3 Model showing an integrated landscape design for Oita Stadium and nearby sporting facilities. (© KKAA)

case of Oita Stadium, I had intended to create an integrated landscape by designing sporting facilities in the shape of a simple sphere that was planned to be built nearby (Figure 3.3).

Structural system

To accommodate 45 000 seats on a narrow tract of land, which stretches from north to south, Toyota Stadium has a structure in which the roof is hung from four masts (Figure 3.4). The main structure comprises a steel frame and precast concrete panels. This approach helped increase the proportion of factory-built modules, reduced the time spent on construction work, and succeeded in drastically reducing costs. I took advantage of Oita Stadium's large land space and adopted an arch featuring a trussed girder made from steel structural hollow sections. This

Figure 3.4 Drawings of Toyota Stadium: (top left) plan drawing; (top right) axonometric drawing; (bottom left) north elevation; (bottom right) west elevation. (© KKAA)

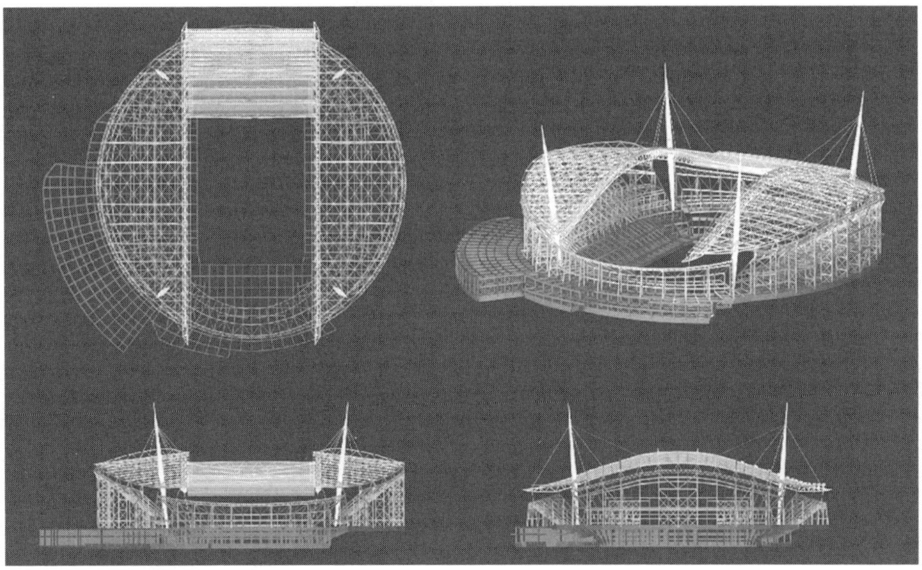

Figure 3.5 East–west section of Oita Stadium. (© KKAA)

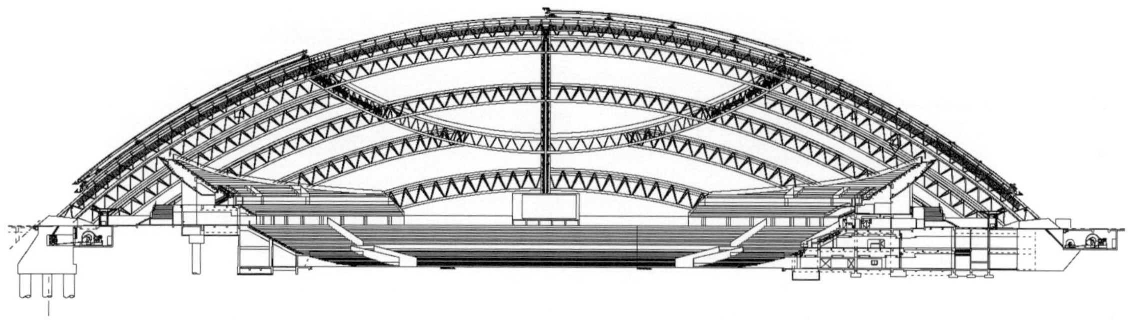

enabled a simple retractable roof structure that slides over the surface of the sphere (Figures 3.5 and 3.6).

A sense of oneness between the players and the spectators

To allow Toyota Stadium spectators to come as close to the field as possible and experience a feeling of 'seeing live action take place right before their eyes', I made the rake of the seats as steep as was realistically possible. For Oita Stadium, on the other hand, I developed a system that makes it possible to move the spectator seats easily over the track. This allows spectators to watch a soccer game even more closely to the field than before.

A stadium as a 'broadcasting studio'

I believe that the number of people who watch sports will continue to grow, and that stadiums will increasingly serve as 'studios' for the benefit of people who watch sports on TV. My two stadiums not only meet FIFA's requirements; they are also designed such that a variety of events can be broadcast to the rest of the world, using the latest acoustic and lighting technologies. At Toyota Stadium, a large-scale visual device is hung from an arch truss (Figure 3.7). It can be moved to any location inside the field, and acts to dramatise and produce events. At Oita Stadium, I developed a system whereby a high-speed moving TV camera is set up on a rail underneath the main arch (Figure 3.8). The camera shoots the players in action and broadcasts the images from directly above the field. I have contrived these mechanisms in the knowledge that stadiums will sooner or later act as studios.

Symbiosis with the environment

At Toyota Stadium, I placed the seats facing east–west, drastically lower than those facing north–south. This was to enable the spectators to enjoy the beautiful scenery of the nearby riverbed from their seats. Meanwhile, I designed Oita Stadium so that as large a space as possible was left between the upper-row stands and the lower-row stands. The result is an arena space that creates a feeling of oneness with the surrounding environment. Thanks to designs such as these, the air moves freely through the seats, creating an environment that is not only pleasant and comfortable for the spectators but also keeps the grass on the field healthy.

Development of technology for opening and closing a huge roof

At Toyota Stadium, I designed a system whereby a roof with an air-film structure is set up between an arched truss (250 m long) at the tip of the fixed roof. The roof moves, using a self-propelled

Figure 3.6 North–south section of Oita Stadium. (© KKAA)

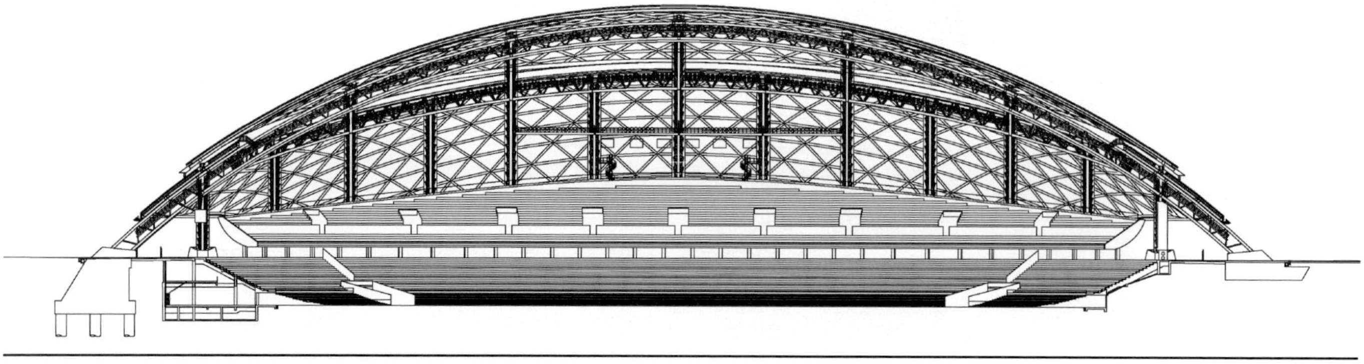

Figure 3.7 Detail drawings of Toyota Stadium's large video LED display screen. (© KKAA)

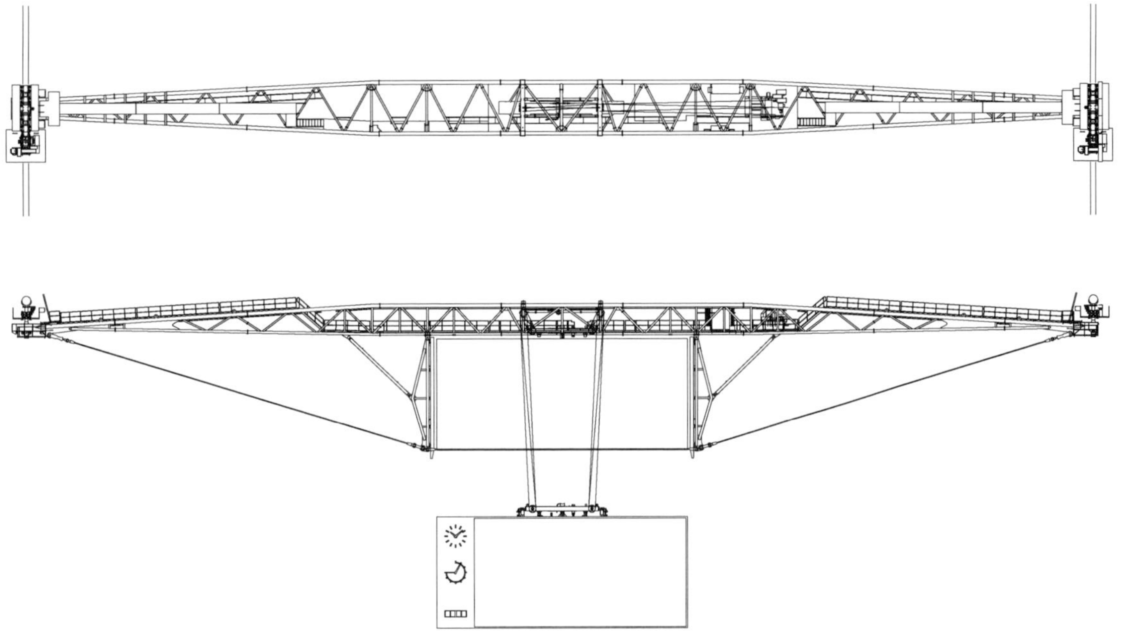

drive dolly, and opens and closes by folding (Figures 3.9 and 3.10). Vinyl-coated polyester cloth is used for the air film, in consideration of its folding performance when the roof opens and closes. This also makes it possible to cope flexibly with the repeated folding movement (Figure 3.11). At Oita Stadium, I developed a mechanism whereby two sections of spherical movable roof are supported by pin points on 25 sliding dollies on one side, which move along a rail installed at the upper part of the main arch to open and close the roof (Figure 3.12). Teflon™, a film with outstanding translucency that has double the transparency of its alternatives, was used to allow natural light to pour over the field (Figures 3.13 and 3.14). Both stadiums incorporate many advanced technologies, and demonstrate successful, reliable roof opening/closing operations.

Figure 3.8 Oita Stadium's roof structure with camera. (© KKAA)

Development of grass-growing technology

For both stadiums, the size of the movable roof was determined, based on the results of a computer simulation, to secure sufficient hours of sunshine during winter periods for the grass in the field to remain green. To enable spectators to watch sporting events against a lush green background, even during winter, I adopted a soil-heating system that monitors the soil temperature and maintains it at the optimum level. The system automatically measures at all times the temperature close to the roots of the grass and controls the ground temperature by sending warm water through pipes buried underground. It promotes the growth of grass during the winter months, making it into an 'evergreen carpet'.

Development of disaster-prevention technology

In both stadiums, I conducted simulations to study evacuation procedures during times of maximum capacity, when spectators

Figure 3.9 Toyota Stadium: the retractable roof glides along the fixed roof's keel trusses. (© KKAA)

Figure 3.10 Toyota Stadium: the closed roof and the movable LED display screen. (© KKAA) (This figure is reproduced in the colour-plate section as Plate VI)

Figure 3.11 Toyota Stadium: (a) the retractable roof comprises a steel frame truss and an air mat; (b) the air mat is folded in half towards the bottom and packed away; (c) the movable truss moves along a rail installed on top of the fixed roof; (d) the movable truss moves and slides over a driving dolly installed at the edge of the truss; (e) at the edge of the fixed roof is a sliding parapet. (© KKAA)

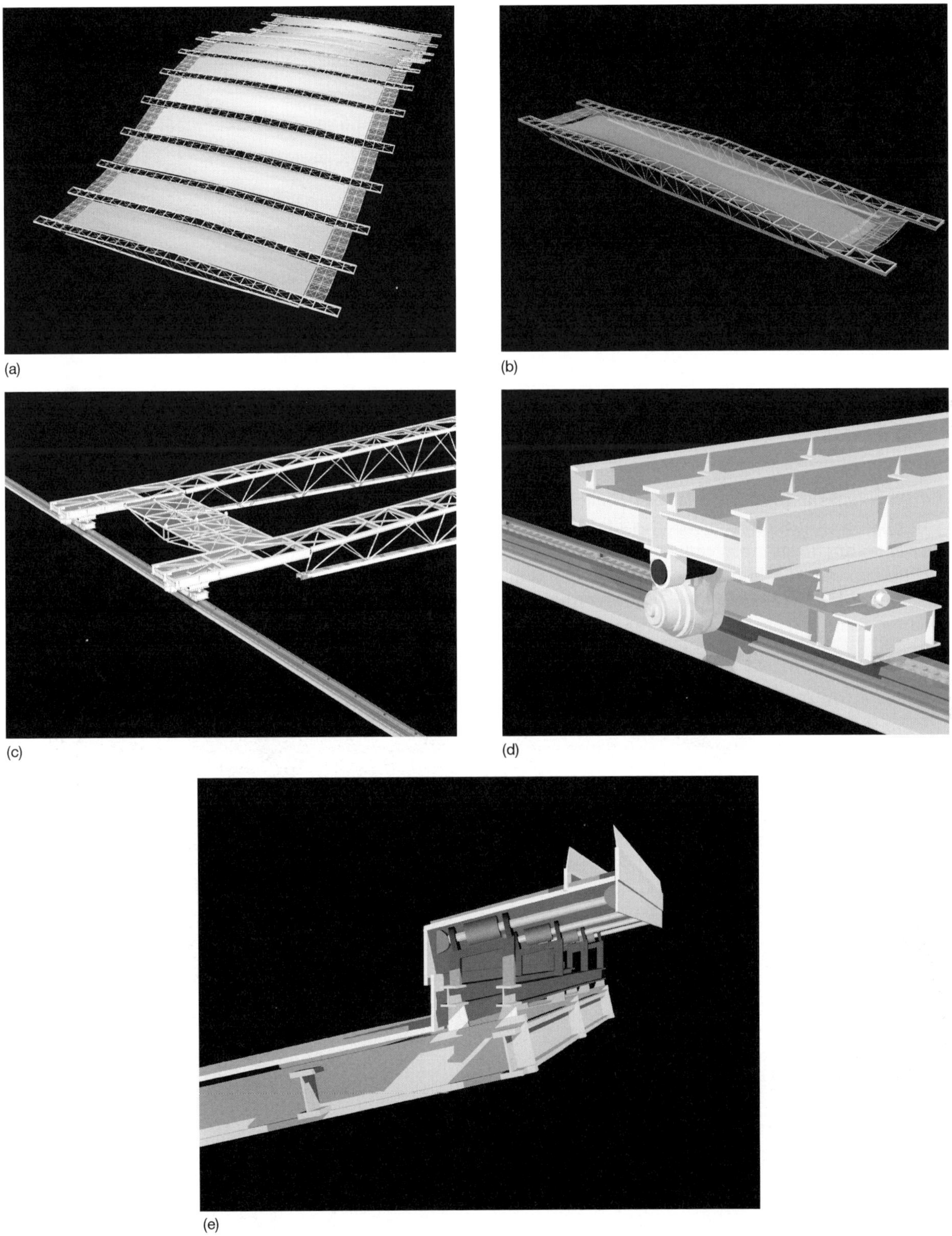

Figure 3.12 Detailed section drawing of Oita Stadium, showing sliding roof and motor. (© KKAA)

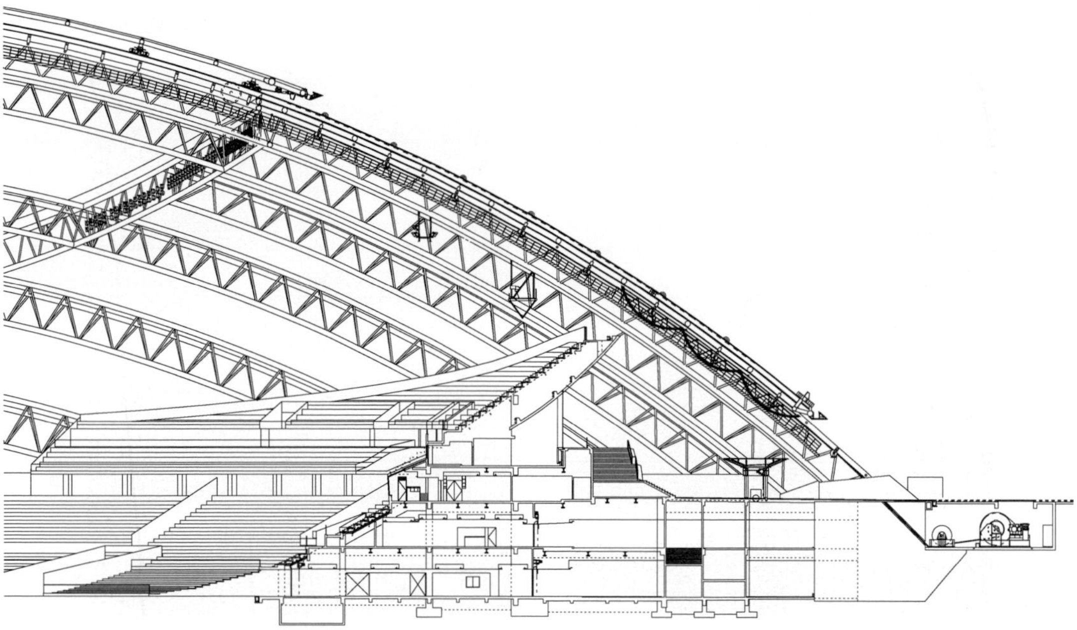

fill both the seats and the field. The plan is to enable all the spectators to evacuate to an open place near the stadium in 5 minutes. The seating areas come equipped with a state-of-the-art mechanism featuring four water cannons that come with a special embedded fire sensor (designed exclusively for dome-shaped stadiums). This system is capable of automatically identifying and extinguishing fires.

Collaborations with various specialists

Many specialists took part in building the Toyota and Oita stadiums: they helped with structural designs and with developing an array of technologies, including body construction, the retractable roof, acoustic/lighting systems, disaster prevention, healthy grass growth, environmental symbiosis, materials and maintenance. Their knowledge, intellect and expertise were

Figure 3.13 Interior view of Oita Stadium with open roof. (© KKAA)

Figure 3.14 Interior view of Oita Stadium with closed roof. (© KKAA) (This figure is reproduced in the colour-plate section as Plate VII)

brought together and used to best effect. In designing the structure for Toyota Stadium, in particular, I was fortunate to have the support and assistance of Ove Arup & Partners. To create an even better stadium, designers require the knowledge and expertise of all relevant experts so as to carry the latest technologies to new levels and to work in a spirit of international cooperation. I believe that numerous architectural technologies can be developed through collaborations with professionals such as these.

Three years [*sic*] have elapsed since the two stadiums were opened. The advanced technologies used in them have been fully utilised and, as the core facilities of sports parks, both stadiums are used and enjoyed fully by their local populations.

Part II

Specialist skills

Stadium and Arena Design
ISBN 978-0-7277-5790-6

http://dx.doi.org/10.1680/saad.57906.053

Chapter 4
Stadium site appraisal

Tim Chapman, Chris Barrett and Sarah Glover Arup, London, UK

Reducing ground risks

Introduction

For urban regeneration and ease of transport issues, most new stadiums tend to be sited on brownfield sites in large urban areas. Because stadiums need large areas of land, particularly if other facilities or parking are to be provided alongside, they tend to be located in former industrial districts, where sufficient derelict land is available. This chapter makes frequent reference to the *ICE Manual of Geotechnical Engineering* (Burland *et al.*, 2012), which acts as an encyclopaedia for geotechnical design, construction and risk mitigation, and contains amplification of much of the guidance given here.

In common with all building and infrastructure projects, many of the major risks lie in the ground. This is especially true on sites where the soil and groundwater have been heavily contaminated, and costs can escalate enormously in unexpected ways if the site appraisal has not been properly executed.

Stadium foundations are distinct from those for other structures:

- because some columns tend to be exceptionally heavily loaded
- due to large cantilever roofs (and sometimes supporting arches), the foundation system is subject to high horizontal loads and large overturning moments
- a high proportion of the loads are live, either induced by crowds or by wind. These dynamic effects have to be considered in the design of the structure and its foundations. A related issue is that the consequences of structural failure in a stadium can be horrendous, with the potential to kill hundreds of people. Therefore, a higher margin against failure is appropriate than for most other structures.

These effects are illustrated in Figure 4.1.

Other site appraisal considerations specific to stadium projects at the initial stage include:

- The presence of utilities and whether they contain sufficient capacity for the new facility (e.g. power or sewerage capacity especially). If not, then reinforcement of the network will be needed, which may extend for some distance away from the site to provide sufficient capacity. Often, new stadiums put far higher demand on utility provision than the facilities they replace. Sometimes the augmentation of utility provision takes several years, so it is an important consideration.
- The clearance and re-engineering of the area around the stadium, with landscaping, car parking and ancillary structures all need careful design to be compatible with the overall vision and to be completed on time and to budget alongside the main stadium building. Sometimes they are neglected until too late, leading to delayed final completion, commissioning and use of the stadium they serve.
- The capacity of the local transport network to support the very high flows of people and vehicles, especially at game or competition time. This needs careful consideration as the efficacy of a stadium project depends hugely on its reputation for easy and simple access and egress by fans.

Typical foundation issues

Perhaps even more so than for other structures, it is important for stadiums that the geotechnical risk reduction process should address the particular types of foundation under consideration. This is because the types of foundation cannot easily be guessed at unless the architectural and structural forms planned for the stadium are known.

First, the extent of earthworks needs to be considered. The chosen pitch level dictates the requirements for excavation and filling operations. In turn, this influences the need for retaining walls and whether gravity or embedded retaining walls are appropriate.

Next, the approximate foundation locations and the loads to which each major foundation type will be subjected need to be

Figure 4.1 Different types of load applied to a typical stadium foundation

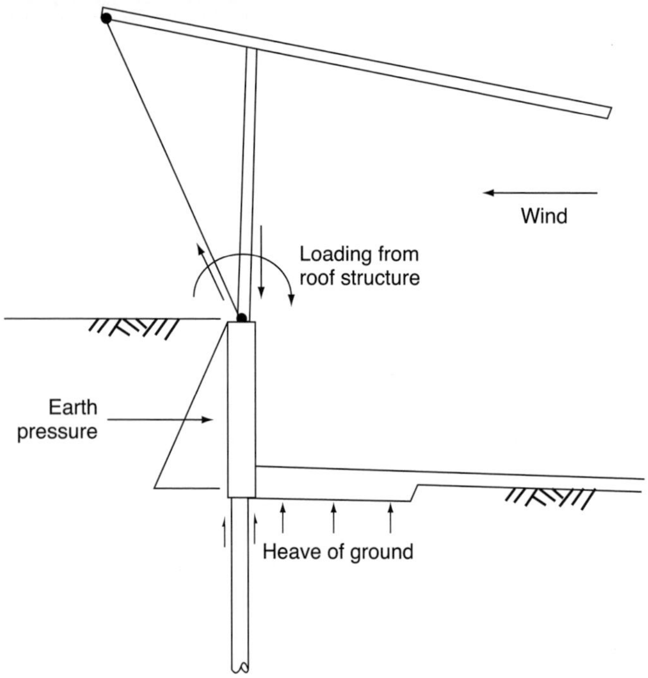

assessed. For heavy horizontal forces, large thrust blocks may be required. These would need the passive foundation capacity of the shallow soils to be assessed, and may involve the excavation of significant depths of soil, which may be significantly contaminated and costly to dispose of.

Geotechnical risk reduction

Various authors have emphasised the high risks that lurk in the ground and the disproportionately small expenditure typically devoted to investigating them (see Burland *et al.*, 2012, chapter 7). The best way to control risks is for them to be recorded in a risk register. This allows the risks to be mitigated throughout the site investigation, design and construction process, even if the companies and individuals involved change as the development of the stadium progresses. This process is explained well by Clayton (2001).

In the UK, according to analyses of projects by organisations such as the National Economic Development Office, the Public Accounts Committee and the Transport and Road Research Laboratory, approximately 20% of construction projects are delayed by more than a month due to unforeseen ground conditions. For most buildings, the average costs of these delays is normally greater than the costs of investigating the hazards in the first place. For stadiums, which are often purpose built to serve a particular event, the consequences of a significant construction delay can be even more extreme. Slipping behind a tight programme may mean that a structure never fulfils the function for which it was intended. Delays in such landmark projects are often in full media glare, and can even be seen as a source of national embarrassment. It is therefore vital that the ground risks are properly investigated at the outset, and their effects mitigated as the design process progresses.

Geotechnical design process

The most effective way to control geotechnical risks is to recognise that they exist, investigate them in a timely manner, and design to mitigate them wherever possible. The typical geotechnical intervention stages, based on the traditional Royal Institute of British Architects (RIBA) work stages, are shown in Table 4.1 (RIBA, 2007). To deal with all the ground risks in a coherent manner, it is essential to gather all the possible information at a sufficiently early stage. Mitigation by maintaining a financial contingency while not addressing the risks explicitly is not recommended for the management of geotechnical risks, because the consequences can be disproportionate to the cause, as they invariably lie on the project's critical path.

The site

Hazards and their mitigation

Some of the site's constraints and hazards may be obvious – most of the obvious ones will be above the ground surface. The more difficult ones will generally be those buried beneath the ground surface. Without a dedicated programme of investigation, the buried hazards are likely to remain hidden until they are found during site construction.

Hazards should be first identified in the desk study. This is the best opportunity to establish the risk register, so that the hazards can then be tracked right through the project. The hazards are then investigated in a targeted programme of fieldworks, known as the 'ground investigation'. The design should address the more significant risks.

Getting it right and avoiding problems should not be difficult. Most problems tend to be due to the same factors, which are easily countered. These include:

- Starting with the wrong designer: the organisation must be suitably experienced and competent to undertake the work, and the right people from that organisation must be assigned to the project.
- Confused responsibilities: the design and procurement process must be coherent, with clearly allocated responsibilities at every step of the way. Using a single design organisation is easiest, but using a combination of several can also prove effective, provided interface responsibilities are clear.

Table 4.1 Traditional RIBA work stages

Project phase	RIBA work stage	Corresponding geotechnical activities
Preparation	A Appraisal	■ Preliminary desk study
	B Design brief	■ Full desk study ■ Establish geotechnical risk register
Design	C Concept	■ Ground investigation ■ Geotechnical interpretative report ■ Initial designs for various options ■ Advice on preferred options ■ Update risk register
	D Design development	■ Design development ■ Detailed calculations for preferred option ■ Agree need for site supervision if design relies on it ■ Final calculation package and meet Building Control ■ Highlight risks that have not been mitigated
	E Technical design	■ Start on tender documents for comment ■ Decide on risk allocation between contractor and client
Pre-construction	F Production information	
	G Tender documentation	■ Complete tender documents ■ Identify preferred foundation contractors
	H Tender action	■ Tender interviews ■ Ensure that contractor understands risks that are being carried, and how they will be dealt with
Construction	J Mobilisation	■ Brief resident engineers on outstanding issues
	K Construction to practical completion	■ Resident engineer supervision ■ Resolve construction issues ■ Complete foundation close-out report
Use	L Post practical completion	

- Poor checking/reviewing: the key to avoiding problems is to anticipate what might go wrong and then ensure that the design can cater for those problems. This involves learning from the experience of designers who have been involved in similar projects. When design fees are tight, it is usually input by more experienced designers that is sacrificed first, thereby increasing the likelihood of problems later.
- Unclear procurement: communicating the complete design intent to the construction team is of paramount importance, particularly on more complex projects like large stadiums. Specifications must be clear, and the various contractors' understanding of the design should be checked for compliance by careful review of their method statements – a contractor that does not appreciate the special needs of the project at the outset is unlikely to demonstrate better understanding as the project progresses. The procurement route should also ensure that subcontractor interfaces are well managed in order to minimise subsequent claims and delays.
- Inadequate supervision: particularly where work will be covered over or become inaccessible, there is only one opportunity for appropriate and knowledgeable checks to be made. If these are not made diligently, then the client team has no easy way of resolving discrepancies, understanding the potential severity of problems or countering claims.

The ground

The major hazards and constraints that need to be investigated, particularly on formerly industrial sites on which many new stadiums are sited, include:

- ground conditions, including geological and geomorphological hazards, and especially groundwater levels and pressures
- buried obstructions
- buried services, which can also include, in some places, large tunnel systems for utilities and transport networks
- remnants of vegetation, such as high suction in clay soils caused by trees, or roots of noxious or non-native invasive plants, such as Japanese knotweed
- the presence of valuable historical or archaeological remains
- mining
- contaminated ground and groundwater
- seismic risks, which are generally considered low for most structures in the UK but will vary from country to country
- the presence of protected species (flora and fauna) or designated ecological receptors
- unexploded ordnance.

Buried services are expensive to divert but are even more expensive to replace or repair after being damaged by construction plant. Some, such as high-voltage cables or gas mains, are also exceptionally dangerous to site workers. Plans showing services are notoriously inaccurate, and do not even purport to show domestic connections from main services. It is also important to consider the effects on these buried services of ground movements induced by the stadium.

Early investigation of the ground is recommended because the project can normally be adapted for particularly adverse conditions at an early stage and potential costs minimised, or at least properly recognised, in the initial cost plans. For instance, excavation for a basement will be relatively expensive on sites with high groundwater levels or when a structure is being built on land affected by contamination. As the project progresses, the opportunities to adapt reduce, and the discovery of problems after the start of construction are usually very expensive and time consuming to correct. This can present real funding problems if a suitable allocation of funds was not made in the original cost plan.

It is preferable that some form of intrusive investigation should be available, or have been undertaken, before the site has been bought, or at least before the development proposals are finalised.

Ground contamination

Contamination of the ground and groundwater is potentially a very significant factor during the planning and construction stages of a development. A wide range of solid, liquid and gaseous contaminants can arise on sites, especially those that have a previous industrial use. In addition, there can be problems of natural contaminants in certain parts of the country, such as metals, gases and radon. If a seemingly innocuous site turns out to be heavily contaminated the effects on the project cost and programme can be severe. The potential presence of ground contamination at the site could give rise to a range of effects and issues as follows:

1 Construction-related issues, including:
   (a) health and safety (on and off site), with associated additional mitigation
   (b) impact on environment (waters) during construction activity
   (c) increased costs, particularly for materials management and waste disposal or soil treatment
   (d) possible delays due to unforeseen contamination
   (e) additional planning conditions and requirements to discharge (before, during and after development)
   (f) requirements for verification records and reporting.
2 Design issues, including:
   (g) requirements for investigation and assessment
   (h) design and agreement of remediation measures
   (i) ground gas/vapour protection measures;
   (j) aggressive ground conditions (materials and services)
   (k) foundation design (limiting pathways to groundwater)
   (l) design of capping/cover layers, in ground barriers.
3 A potential requirement for statutory remediation, for the current use (e.g. Part 2A of the Environmental Protection Act 1990, UK).
4 The requirement for remediation under planning consent for a change of use.
5 The requirement for an environmental impact assessment and associated additional assessment and mitigation.

Land contamination is regulated under several regimes, including environmental protection, pollution prevention and control, waste management, planning and development control, and health and safety. Remediation of historic land contamination has been, and will continue to be, principally managed through the planning regime. National planning guidance on assessing risks to human health and the environment under planning and development control (i.e. the Town and Country Planning Act, UK, 1947, 1990) is provided by the National Planning Policy Framework. The framework states that, where a site is affected by contamination, responsibility for securing a safe development rests with the developer and/or landowner. It states that local planning policies and decisions should also ensure that the site is suitable for its new use, taking account of ground conditions, and including natural hazards or former activities such as mining, pollution arising from previous uses and any proposals for mitigation, including land remediation or impacts on the natural environment arising from that remediation. At the time of a proposed redevelopment, the local authority may require remediation works to be undertaken as part of the redevelopment of the site and for the new land use. These works usually encompass site investigation, risk assessment, consultation, remediation and verification works.

The UK framework for the assessment of contaminated land endorses the principle of risk assessment and a 'suitable for use' approach to contaminated land. Remedial action is only required if there are unacceptable risks to human health or the environment, taking into account the use of the land and its environmental setting. The methodology of risk assessment is normally set out in terms of plausible pollutant linkages within a source–pathway–receptor model of the site. All three of these elements must be present for a site, or area of a site, to be determined as contaminated. The plausible pollutant linkages are set out in a conceptual model initially developed during the preliminary risk-assessment stage (desk study).

A site conceptual model describes the scenario of the risks to human health and the environment. It describes the ground and surface conditions, and the activities performed on the site in terms of the ground works and the final form of the development. In particular, the model identifies and describes the sources of the potential contamination, and the behaviour of the contamination in the environmental media such as soil and groundwater, surface water and air. It also identifies and characterises potential human health and environmental receptors, and the plausible pathways linking the sources and receptors. It is imperative that the conceptual model informs the scope of the ground investigation and underpins the subsequent quantitative risk assessment and remediation strategy. Finally, once remediation and development are complete, the verification report confirms that the pollutant linkages set out in the conceptual model have been adequately managed or removed.

Guidance on the process of investigation, assessment and management of risks on potentially contaminated sites, relevant to redevelopment projects, is set out in a wide range of documents – too many to list here. A few essential references in this respect for construction in the UK are listed below (the list is not exhaustive):

- *Model Procedures for the Management of Land Contamination* (Environment Agency, 2004)
- *Guiding Principles for Land Contamination* (Environment Agency, 2010)
- *Piling and Penetrative Ground Improvement Methods on Land Affected by Contamination: Guidance on Pollution Prevention* (Environment Agency, 2001)
- BS 10175:2011. Code of practice for investigation of potentially contaminated sites (BSI, 2011)
- *Contaminated Land Risk Assessment: A Guide to Good Practice* (CIRIA, 2001), (C552).
- *Assessing Risks Posed by Hazardous Ground Gases to Buildings (revised)* (CIRIA, 2007)
- BS 8485:2007. Code of practice for the characterisation and remediation from ground gas in affected developments (BSI, 2007).

If a risk assessment determines that land is contaminated, there is a wide range of options available to manage or break the pollutant linkages, including:

- treating the sources (either in situ or ex situ) using a range of different physical, chemical or biological technologies
- managing the pathways through barriers (e.g. in-ground barriers such as membranes, vertical barriers to prevent migration, engineered break layers, soil cover layers or simply carefully designing slabs, basement walls and hardstanding public realm to act as barriers)
- removing/changing the receptor (e.g. changing the layout of the development to place a landscaped area in an uncontaminated part of the site, or, for instance, to locate a nursery in a different area)
- removing the source (e.g. excavating soil having elevated concentrations of contaminants) – quite often these soils may then be moved around the site, treated on site, treated off-site or sent for recovery/recycling rather than disposal.

Examples of some particular effects of ground contamination that have to be addressed include the following:

- *Effects on site workers*. Under the Construction, Design and Management (CDM) (2007) Regulations (UK), designers need to consider the risks to site workers. Designers and contractors therefore need to reduce exposure to hazardous ground and groundwater insofar as is practicable. This is most often accomplished through the use of appropriate health and safety procedures along with personal protective equipment, although the designer should consider whether aspects of the layout or construction (such as basement details and services) could be altered. In addition, the designer should consider the suitability of the foundation design (including piling techniques) and dewatering options.
- *Effects on future users and visitors after completion*. As an example, a thick reinforced concrete raft with a vapour barrier may cut the pollutant linkage to future users of the building. Conversely, a playing pitch or landscaped area may result in a direct route for human contact with contaminated soil following capillary rise of contaminated groundwater through the imported formation soils. The assessment process must take account of potential ground gases and vapours which may penetrate building fabric.

Current UK assessment protocols assume soils from landscaped areas may be tracked into buildings, resulting in human exposure to dusts in buildings.

- *Effects on vegetation.* Contamination (solid, liquid and gas) can have phytotoxic effects (i.e. cause harm to vegetation). For stadiums, the growth of the grass playing surface is of obvious importance for the success of the facility. Soft landscaping and trees planted around the development also need to be considered.
- *Effects on stadium construction materials.* The stadium structure and all its service connections, particularly those buried in the ground, must meet their planned design life. Materials appropriate for the particular exposure conditions should be used. The selection of materials for potable water services is particularly important.
- *Risks to groundwater and surface water.* Piling and construction through contaminated soils may create new pathways that allow the migration of mobile contaminants to protected groundwater or surface water resources.

Surplus soil or groundwater requires appropriate disposal. Reductions in the number of landfills for hazardous waste, coupled with the requirement to treat all waste before disposal and increases in landfill tax in the UK, have resulted in on-site and off-site remediation being more economically viable. Having a coherent strategy for material classification and reusing as much suitable fill materials on site as possible is likely to provide an economical solution. There are aspects of environmental permitting that may apply when reusing materials arising from development, especially if those materials require treatment to make them suitable for use. Key guidance on the definition and use of materials arising from development (both on site and the transfer of materials and waste between sites) is set out in *Definition of Waste: Development Industry Code of Practice* (CL:AIRE, 2011). The approach provides a clear, consistent and streamlined process that enables the legitimate reuse of materials on site or their movement between sites with a significantly reduced regulatory burden. In many instances, the code can provide an alternative to environmental permits or exemptions from permitting when seeking to reuse or transfer excavated materials (both contaminated and uncontaminated).

Site investigation

A site investigation is composed of two phases – the desk study and the ground investigation. For building projects, site investigation costs seldom exceed 0.3% of the building cost and, although statistics are not available for stadium projects, it is thought that ground investigations typically represent a smaller proportion of the final building cost. The extent of investigation should sensibly address the site's specific hazards and the requirements for the new structure. It should not be limited by arbitrary budget constraints.

Site investigations are too often inadequate. It is not unusual that they are considered as a commodity, to be bought in at minimum cost, rather than an important risk-reduction element, to be incorporated in the flow of the design process. This can lead to inappropriate site investigations, where the total amount expended might have been sufficient but the information obtained does not address the specific problems of the development on the site. (Section 4 of the *ICE Manual of Geotechnical Engineering* (Burland *et al.*, 2012) discusses the principal aspects of site investigation planning, procurement and delivery.)

Inevitably, even properly planned site investigations sometimes fail to detect features in the ground that will lead to problems during construction. It is seldom practicable to sample more than 0.001% of the ground beneath a site. However, problems that arise following an appropriate ground investigation are usually easier to solve, and blame for the problem is often easier to apportion, which speeds up problem resolution.

Desk study

The first stage of a site investigation should be the desk study, allowing the costs of the subsequent ground investigation to be optimised (a desk study is also a requirement of all relevant codes in the UK). BS 10175:2011 identifies the requirement for a preliminary investigation, which includes desk study, site reconnaissance and conceptual model.

The desk study should investigate at least the following areas:

- topography – to reveal risks such as incipient slope instability
- geology – to reveal risks such as major geological features (e.g. scour hollows or alluvium-filled former river courses)
- local stratigraphy and ground conditions – to investigate the particular soils beneath site and to warn of problems for their use in foundations (e.g. swelling clays, soils of very low strength and stiffness, or permeable gravels that could allow water to inundate excavations)
- site history – to identify former industries and past site uses on or in the vicinity of the site which could have contaminated the ground and groundwater, or other historical features such as back-filled ponds, brick pits or old structures which could affect foundation construction or capacity
- buried services – to detect services that could be affected by the ground investigation or by the building foundations
- consultation – with the local authority and Environment Agency regarding information held on their files with respect to ground conditions and contamination on or in the vicinity of the site.

A site walkover should be carried out at the desk-study stage to identify current-site and local conditions, and early identification of site constraints and opportunities for investigation. This may also include a phase 1 ecological survey to identify current ecological constraints on the investigation and subsequent development.

The desk study can also provide data of benefit to the developer, say by warning of surviving archaeology that could affect the granting of planning permission or by identifying a source of good-quality soil that might allow relatively cheap and sustainable filling operations.

A good desk study will:

- provide data for foundation design, allowing more economical foundation types to be identified
- identify possible materials for use in earthworks operations, allowing early consideration of the potential for using site-won materials
- give useful background information on the site and allow a complete scheme design to be developed, including the need for ancillary structures such as retaining walls
- identify the ground-related hazards that need to be investigated in greater detail during the ground investigation. (Without a desk study, the ground investigation becomes a hit-and-miss affair. What are the chances of identifying a former pond 10 m across when your boreholes and trial pits are at 20–30 m centres?)
- set out a conceptual model and preliminary risk assessment for contaminated land and define the steps necessary for further investigation and assessment. (The conceptual model informs the scope of the ground contamination investigation.)
- identify other site constraints associated with unexploded ordinance, archaeology/heritage and ecological receptors.

Procuring a ground investigation

There are many different techniques for investigating the ground. Choosing which ones to use on a particular site, how many exploration points to investigate and how deep to explore are all matters of judgement. *Without Site Investigation Ground is a Hazard*, commissioned by the Site Investigation Steering Group (SISG, 1993a), gives outline guidance on why a ground investigation is required and examples of what could otherwise go wrong. A second SISG booklet, *Planning, Procurement and Quality Management* (SISG, 1993b), gives detailed advice on which professionals should be engaged, how the ground investigation and its interpretation should be incorporated in the design process, and how the ground investigation should be procured. These documents identify the following roles:

- The client's principal technical advisor (PTA), who will often be the architect for landmark stadium projects. The PTA will be in charge of the design team and control the design process, and is the most important decision-maker with regard to the quality of site investigation. The PTA will normally recommend the appointment of a geotechnical advisor, unless the structural design team already includes such a person.
- The geotechnical advisor should have sufficient knowledge and experience to plan and oversee the site investigation process.
- The geotechnical contractor is the body engaged to carry out the ground investigation.

Most ground investigations should use the *UK Specification for Ground Investigation* (SISG, 2012), as this has been broadly adopted as the national UK specification and people throughout the industry are familiar with its content.

BS 10175:2011 (BSI, 2011) sets out the ground investigation process with regard to characterising land potentially affected by contamination. The process of investigation may involve one or a number of phases of investigation, and should begin with setting the objectives based on the conceptual model set out the desk study (preliminary investigation). At the end of each phase of ground investigation, the information obtained should be reviewed as to whether the objectives have been met or whether there is a need for further investigation of potential contamination at the site. Where further investigation is necessary, the design of the next phase is based on the information previously obtained and the updated conceptual model.

After completion of the ground investigation, someone with appropriate and sufficient geotechnical experience should interpret the data. Often, some of the recovered information is contradictory, and needs careful examination for the truth to come out. Overreliance on statistical techniques can be misleading in solving geotechnical problems; often it is the apparently spurious data that turn out to be the most significant.

The ground investigation should determine the following factors as appropriate for the particular set of development proposals:

- *Stratigraphy and its variation across the site*. The depth of investigation should be appropriate for all likely foundation options, and the spatial frequency of exploration points should ensure that likely variations in key layers are catered for.
- *Those ground properties and parameters required for the design of the temporary and permanent works*. Less testing usually means less certainty, which, in turn, leads to a more conservative foundation design, so economies in the extent of testing are normally dwarfed by consequentially increased construction costs.

- *Groundwater conditions and their variation across the site, over depth and time.* The only reliable way to establish groundwater conditions is to measure water levels in standpipes and piezometers. All too frequently this is not done, which explains why misunderstood water conditions are one of the main causes of problems and failures during construction, and also for the completed structures.
- *The detection of any significant levels of contamination in the ground or groundwater.* This is carried out by a process of risk-based assessment for human heath, controlled waters and, sometimes, ecological receptors. The information informs decisions to ensure that the structure and connecting services can be made sufficiently durable and that spoil disposal costs can be assessed at an early stage. The results of the ground investigation, if properly specified, can be used to provide a preliminary classification of potential waste classification and material management strategy so as to maximise the reuse of suitable soils as part of the cut/fill balance design and to minimise disposal off-site.

Foundations

Purpose

The most heavily loaded parts of any structure are the foundations. They are also the only part which is virtually impossible to uncover to inspect after installation, and are one of the parts of the structure most prone to defects. It is, therefore, most important that knowledgeable people design the foundations following a coherent design process.

Foundations have two functions:

- to support the applied loads without significant risk of failure (referred to as 'ultimate limit state')
- to support the applied loads with sufficiently small movements so that no function of the building is impaired (i.e. no cracking, sticking of doors, cutting of service connections, etc.) (referred to as 'serviceability limit state').

The second function is obviously more onerous than the first but, as the consequences are less severe, it can have a lower margin against exceedence. This approach to design is now embodied in most structural and geotechnical design codes in the UK and is termed 'limit state design'. Its applicability to geotechnical problems is obvious, dealing directly with the two issues most likely to cause concern.

Stadium foundations can receive vertical loads, horizontal thrusts and applied moments, and must resist these with a factor of safety (normally greater than 2 for vertical loads) so that deflections are within acceptable bounds. There is a skill in selecting the correct deflection limits – the greater they can be the bigger the range of acceptable foundations for the particular project and the cheaper they are likely to be. Often limits are selected that are too tight because such a choice limits subsequent risks. Conversely, the effects for a project of allowing movements that are too big can be expensive to repair, and the resulting delays can be even more costly – the balance to be struck is a fine one and in many cases it is appropriate for the client to be involved in the choice, so he or she can balance his or her potential savings against increased risks.

Generally, for foundation installation, as the completed products are buried and cannot be seen again, the available methods to check quality using non-destructive methods are not reliable and, given the potentially severe consequences arising from the relatively high incidence of foundation problems, it is good practice to employ knowledgeable independent supervision during the installation process. This should ensure that the random testing can be targeted at the foundations that are most likely to be defective, and that any problems are detected earlier, when they are easier to deal with.

On completion of foundation installation, it is good practice to complete a 'foundation close-out report'. This report should summarise all the key issues that arose during the foundation installation contract, as well as the key design information, the as-built drawings and all test results (Burland *et al.*, 2012, chapter 101). The report serves three functions:

- It serves as a record demonstrating that the foundations were installed competently, that all construction issues were properly closed out, and that the foundations can safely fulfil their design function, thereby giving reassurance to the project promoter and funders.
- It provides an indication of possible extra capacity available in the foundation system for future modifications made to the stadium over its design life.
- It acts as a record of the completed initial foundation system, so that when the stadium is eventually demolished or dismantled the location of these obstructions is known, and thereby maximises the opportunity to reuse the foundations in the future. This is especially important for the relatively dense foundation systems that are needed under major stadiums, where the available ground remaining for the installation of a second set of foundations to facilitate future redevelopment is very limited.

Shallow foundations

Generally, shallow foundations will be the most economical option, especially if loads are relatively light, the near-surface ground is strong and the water table is deep. They are, therefore, most suitable for smaller stadium stands.

Options include:

- trenchfill – for the lightest loads
- pad footings
- strip footings
- raft – for heavier loads.

If rock or granular soils are close to the surface, allowable bearing pressures will be higher (say 250 kN/m^2). For stiff clays (like London or Oxford clays), allowable bearing pressures will typically be lower (say 150 kN/m^2) and settlements may be considerably higher. Stiff clays, when unloaded, for instance at the bottom of basement excavation, can result in large heave movements, and these can complicate foundation design. Shallow foundations on soft clays should only be contemplated if the movement tolerances for the new building are larger than normal and expert geotechnical advice is involved in assessing the foundation design.

All shallow foundations should be installed to a minimum depth. In frost-susceptible ground such as chalk, the National House Building Council (NHBC) recommends a minimum depth of 0.45 m. In shrinkable clays, where trees do not already exist, minimum depths of between 0.75 m and 1.0 m are recommended (NHBC, 2011), although it should be noted that these depths do not allow for future tree planting.

As shallow foundations will normally be most efficient, techniques for ground improvement have been developed to stiffen soil that would otherwise be too soft. These ground-improvement techniques are most effective for lighter structures on areas of deep loose-fill deposits, such as smaller stadiums on areas reclaimed from former industrial use. While ground improvement is reliable when appropriate controls are applied, it generally will not be as reliable as a piled solution. It should, therefore, only be used when it results in significant cost or programme savings over piles (Burland *et al.*, 2012, chapters 84 and 90).

Piled foundations

For weaker ground, or heavier or unusual loads, or where tighter control on foundation movements is required, piles are needed. Piles come in many varieties to deal with different circumstances.

Typical features of different pile types are described in Table 4.2 and more detailed descriptions are given in chapter 81 of the *ICE Manual of Geotechnical Engineering* (Burland *et al.*, 2012).

Bored piles are most versatile and can be used for high column loads. In favourable ground conditions, such as the Thanet sand underlying London's Docklands, their capacity is limited by the crushing strength of the concrete, which corresponds to about 18 MN load for a 1.5 m diameter pile. In central London, a similar pile founded in London clay would have a capacity of about 6 MN. To carry large loads in London clay, single piles of 2.1 m diameter with a 6.3 m diameter under-reamed bell can carry loads of up to about 23 MN. A bored piling rig installing piles for the Wembley project is shown in Figure 4.2 and rigs working for the Manchester Stadium project are shown in Figure 4.3. Bored piles are often used instead of driven piles in built-up urban areas because of concerns about the effects of noise and vibration on nearby third parties.

Often the refurbishment or extension of existing stadiums necessitates an increase in loadings which the existing structure cannot readily accommodate. Provided no damage has been noted in the original building, an increase in applied loads on

Table 4.2 Pile types

Type	Bored	Driven
Common types in the UK	■ Large-diameter auger ■ Under-reams ■ Continuous flight auger ■ Mini-piles	■ Precast concrete ■ Steel circular and H-sections ■ Driven cast in situ
Normal diameter range	■ 0.15–2.4 m for large-diameter auger ■ 1.2–2.4 m shaft and 3.6–6.3 m base diameters for under-reams ■ 0.3–1.2 m for continuous flight auger ■ Up to 0.3 m for mini-piles	■ 0.15–0.4 m^2 for precast concrete ■ 0.3–2.0 m for steel circular ■ 0.34–0.6 m diameter for driven cast in situ
Normal maximum depth	■ 65 m for large-diameter auger ■ 28 m for continuous flight auger ■ 25 m for mini-piles	■ 20 m and above for precast concrete

Figure 4.2 Piling rig at the New Wembley Stadium. (© Stent Foundations Ltd)

the foundations of up to 10% is often proposed by the designers and accepted by Building Control. Larger increases in applied load frequently require new foundations. The new foundations need to provide a very stiff response under load to minimise differential movements with the rest of the structure, which will not settle further. Very small diameter mini-piles installed by a rig capable of working in less than 3 m headroom are often the most effective solution in this situation. As these piles can cost up to three times the amount per tonne carried for normal piles, they are mainly used when other options are not possible.

Figure 4.3 Start of piling in a confined area at City of Manchester Stadium. (© Stent Foundations Ltd)

Driven precast concrete piles can be very economical for new stadiums being built away from people or structures that would be adversely affected by noise and vibration.

Steel piles tend only to be used to support stadiums where part of the new structure is being built over docks, harbours or rivers – although, for pile positions within 5 m of land, bored piles installed through a steel casing can offset the high mobilisation costs of marine working.

There are various types of test that can be applied to piles. During construction the materials are sampled to make sure that they have sufficient strength and durability, integrity tests to verify that the piles are substantially free of defects, and load tests to confirm that the piles perform as expected. All these tests contribute by some extent to reducing risks – judgement is needed to choose the correct testing strategy so that risks are reduced without incurring excessive cost.

Substructure

Basements are increasingly used to provide additional space beneath portions of stadiums, especially when the new stadium complex incorporates extensive leisure centres, hotels or shops. They can be used for the provision of delivery areas, car parking, storage, water tanks and plant space.

For all basements, a perimeter retaining wall will need to be dug to hold back the soil and groundwater. A discussion of retaining wall types and design methods can be found in Section 6 of the *ICE Manual of Geotechnical Engineering* (Burland *et al.*, 2012). The following factors frequently dictate the choice of wall type:

- proximity to neighbouring buildings and their sensitivity
- water-table level and need for water retention in both the temporary and permanent conditions
- depth of basement
- ability to use final floor plates for horizontal propping, dictated by the floor-to-floor heights and the number of holes in the slabs
- vertical loading from the superstructure, whether it is helpful to carry loads on the perimeter wall, and whether shallow or piled foundations for neighbouring loads will be most economical
- compatibility with construction sequence for the basement
- ease of overcoming obstructions
- programme
- cost.

Normally, it is most economical if the temporary works measures needed to enable the basement to be built can be combined with the equivalent measures for the completed permanent structure.

Table 4.3 Grades of basement space[a]

Grade of basement	Example use of structure[b]	Performance level
Grade 1	Car parking; plant rooms (excluding electrical equipment); workshops	Some seepage and damp patches tolerable, depending on the intended use[c] Local drainage may be necessary to deal with seepage
Grade 2	Plant rooms and workshops requiring a drier environment (than Grade 1); storage areas	No water penetration acceptable Damp areas tolerable; ventilation might be required
Grade 3	Ventilated residential and commercial areas, including offices, restaurants, etc.; leisure centres	No water penetration acceptable Ventilation, dehumidification or air-conditioning necessary, appropriate to the intended use

[a] From BS 8102:2009 (BSI, 2009)
[b] The previous edition of BS 8102:2009 referred to Grade 4 environments. However, this grade has not been retained in the current standard as its only difference from Grade 3 is the performance level related to ventilation, dehumidification or air-conditioning (see BS 5454:2000 (BSI, 2000) for recommendations for the storage and exhibition of archival documents)
[c] Seepage and damp areas for some forms of construction can be quantified by reference to industry standards, such as the ICE's *Specification for Piling and Embedded Retaining Walls* (ICE, 2007)

To dig basements within the retaining walls, two main types of construction sequence are commonly used – 'bottom up' or 'top down'. The bottom-up sequence is most commonly used on open sites and for shallower excavations. In this sequence the basement walls are initially temporarily propped by raking or flying struts until the base slab is built, then the rest of the substructure is built, until the floor slabs that permanently prop the retaining walls are installed. Conversely, in the top-down sequence the permanent propping slabs are installed first and are then progressively undermined until the base slab's formation level is reached.

A top-down sequence has the advantage of providing a stiffer propping system and is more appropriate for very constrained sites surrounded by neighbouring properties that are susceptible to damage.

In urban areas a need to keep ground movements within acceptable bounds normally dominates the design. In greenfield areas there is usually little need to control ground movements, so economy and programme drive the choice.

A principal function of retaining walls is to cut off the flow of water into the completed basement structure. However, no embedded retaining wall should be assumed to be completely watertight. Guidance on watertightness is given in BS 8102: 2009, Code of practice for protection of below ground structures against water from the ground (BSI, 2009) and CIRIA Report 139, *Water-resisting Basement Construction – A Guide* (CIRIA, 1995). These documents give guidance on the means by which particular grades of basement space can be achieved. The qualities and requirements for these grades are explained in Table 4.3.

Hard–hard secant pile walls or diaphragm walls alone may directly provide an acceptable level of water retention if a low grade of basement space is all that is required. For higher grades of space, structural facing walls designed to Eurocode 2 Part 1-2 (BSI, 2004) (in front of contiguous or hard–soft secant pile retaining walls for Grade 1 space) or the more onerous Eurocode 2 Part 3 (for Grade 2 space) (BSI, 2006) or drained cavities (for Grade 3) should also be provided. An example of a contiguous bored pile wall is shown in Figure 4.4.

Drained cavities are designed to be kept essentially free of water to prevent penetration of water vapour into the substructure. A means of removing any water that flows into the cavity is required, and ventilation may also be required for this purpose. Water authorities normally have concerns about allowing groundwater to be discharged into their sewers, and permission to discharge is normally only given after assurance that the flows will be very low.

Problems of poor waterproofing are, unfortunately, common, and are usually due to the following (avoidable) factors:

- failure to anticipate water levels over the design life of the basement
- failure to understand the stadium operator's requirements for the basement space or to explain the possible consequences of the designed performance on the selected grade of use

Figure 4.4 Completed contiguous bored pile wall to retain soil at the New Wembley Stadium. (© Stent Foundations Ltd)

- poor design or detailing
- bad workmanship.

Acknowledgement

Stent Foundations Ltd is thanked for providing images of the construction of major UK stadiums.

REFERENCES

BSI (British Standards Institution) (2000) BS 5454:2000. Recommendations for the storage and exhibition of archival documents. BSI, London, UK.

BSI (2004) BS EN 1992-1-2:2004. Eurocode 2: Design of concrete structures. Part 1-2: General rules – Structural fire design. BSI, London, UK.

BSI (2006) BS EN 1992-3:2006. Eurocode 2: Design of concrete structures. Part 3: Liquid retaining and containment structures. BSI, London, UK.

BSI (2007) BS 8485:2007. Code of practice for the characterization and remediation from ground gas in affected developments. BSI, London, UK.

BSI (2009) BS 8102:2009. Code of practice for protection of below ground structures against water from the ground. BSI, London, UK.

BSI (2011) BS 10175:2011. Code of practice for investigation of potentially contaminated sites. BSI, London, UK.

Burland J, Chapman T, Skinner H and Brown M (2012) *ICE Manual of Geotechnical Engineering. Volume 1: Geotechnical Engineering Principles, Problematic Soils and Site Investigation.* ICE, London, UK.

Burland J, Chapman T, Skinner H and Brown M (2012) *ICE Manual of Geotechnical Engineering. Volume 2: Geotechnical Design, Construction and Verification.* ICE, London, UK.

CIRIA (Construction Industry Research and Information Association) (1995) *Water-resisting Basement Construction – A Guide.* CIRIA, London, UK, Report 139.

CIRIA (2001) *Contaminated Land Risk Assessment: A Guide to Good Practice.* CIRIA, London, UK, C552.

CIRIA (2007) *Assessing Risks Posed by Hazardous Ground Gases to Buildings (revised).* CIRIA, London, UK, C665.

CL:AIRE (Contaminated Land: Application in Real Environments) (2011) *Definition of Waste: Development Industry Code of Practice,* version 2. See http://www.claire.co.uk/index.php?option = com_phocadownload&view = file&id = 212:initiatives&Itemid = 230 (accessed 27/12/2014).

Clayton RRI (2001) *Managing Geotechnical Risk.* Thomas Telford, London, UK.

Environment Agency (2001) *Piling and Penetrative Ground Improvement Methods on Land Affected by Contamination: Guidance on Pollution Prevention.* Environment Agency, Rotheram, UK, Report NC/99/73.

Environment Agency (2004) *Model Procedures for the Management of Land Contamination.* Environment Agency, Rotheram, UK, Contaminated Land Report 11 (CLR11).

Environment Agency (2010) *Guiding Principles for Land Contamination.* Environment Agency, Rotheram, UK, GPLC1, 2 and 3.

ICE (Institution of Civil Engineers) (2007) *Specification for Piling and Embedded Retaining Walls.* Thomas Telford, London, UK.

NHBC (National House Building Council) (2011) *NHBC Standards*. Chapter 4.1 Land quality – managing ground conditions; Chapter 4.4 Strip and trench foundations. National House Building Council, Amersham, UK.

RIBA (Royal Institute of British Architects) (2007) *Plan of Work*. RIBA, London, UK.

SISG (Site Investigation Steering Group) (1993a) *Without Site Investigation Ground is a Hazard*. Thomas Telford, London, UK.

SISG (1993b) *Planning, Procurement and Quality Management*. Thomas Telford, London, UK.

SISG (2012) *UK Specification for Ground Investigation*, 2nd edn. ICE, London, UK.

FURTHER READING

BSI (British Standards Institution) (1999) BS 5930:1999. Code of practice for site investigations. BSI, London, UK.

BSI (2004) BS EN 1997-1:2004. Eurocode 7: Geotechnical design – Part 1: General rules. BSI, London, UK.

BSI (2007) BS EN 1997-2:2007. Eurocode 7: Geotechnical design – Part 2: Ground investigation and testing. BSI, London, UK.

Stadium and Arena Design
ISBN 978-0-7277-5790-6

http://dx.doi.org/10.1680/saad.57906.067

Chapter 5
Stadium circulation: transport planning and crowd control

Will Durden Momentum Transport Planning, London, UK

At sports venues the field of play is defined by sport. Everything else – the seating bowl, circulation areas, food and drink outlets and toilets, turnstiles and exit gates – is defined by the needs and demands of the spectators that visit the building.

In planning a new stadium, getting the pedestrian circulation right is of paramount importance, and the consequences of getting it wrong can be severe. At best, stadium management may not be able to operate the stadium efficiently, and, at worst, the safety of spectators could be jeopardised.

Crowd modelling underpins the whole stadium design process; developing an understanding of how and why spectators will move through the building and shaping the infrastructure to ensure they can move safely and effectively.

The demands of a major stadium do not end at the edge of the building, and the understanding of the needs of spectators extends beyond into the wider area and the transport facilities that bring them to the site. Crowd and transport strategies should be developed in conjunction to provide a coordinated approach, ensuring the safety and comfort of spectators from home to their seats and back again.

The delivery of every major stadium event requires a significant management operation, with many venues employing thousands of staff on an event day. The design of a venue should not be developed in isolation of the eventual management strategy. A well-designed stadium should minimise the amount of management required and provide an environment that makes the life of the stadium manager as easy as possible.

Crowd modelling

Crowd modelling is the study of how spectators will use a stadium and then using that information to shape the venue design. How crowds arrive and enter the stadium, circulate around the concourses before entering the seating bowl, and finally arrive at their designated seats is factored into the design from the outset to ensure that spectators move in an orderly fashion and that control is maintained.

Protecting the safety of spectators is at the heart of the crowd-modelling process, and much of the guidance used in the industry has been developed in the aftermath of serious crowd disasters to improve safety. Alongside consideration of the normal movement of spectators (in arrivals, departures or mid-event), the simulation of an emergency evacuation is one of the most important, and most onerous, tests of a stadium design.

Once the pattern of spectator movement is understood the process of crowd modelling uses the basic principles of static pedestrian modelling, utilising a set of agreed numerical modelling parameters that define how spectators will move.

These include *flow rates*; the rate at which spectators can move through a particular space, often defined as the number of spectators that can pass through a metre width within one minute. Flow rates will vary depending on the type of surface in question (stairs, ramps, flat) and the type of flow (normal movement or evacuation).

Other parameters utilised might include: *maximum permissible egress times* from any area in an evacuation; *queuing densities*, which define how many people can queue within a given space; and *travel speeds*, which define how quickly people can walk around a venue.

These parameters are then used to study the capacity of each area of the building. For example, to calculate the width of an exit route from a seating bowl (assuming the exit route is a stairway) the following calculation would typically be carried out:

$$\text{Exit route width} = \frac{\text{population of seating bowl}}{\text{maximum permissible egress time}} \Big/ \text{flow rate on stairs}$$

This logic is then extended across the entire stadium to create a system of interdependent flow calculations, determining queuing patterns, spectator movement profiles and sizing key infrastructure. The modelling can be used to determine how quickly and effectively people can move between each zone of the stadium, and the suitability of areas to accommodate queuing.

Although the logic is simple in isolation, the complex patterns and huge accumulation of individual movements and interdependencies within a stadium mean significant work is required to model a venue fully. Sports venue design teams will typically incorporate a crowd-modelling specialist to focus on this discipline and carry out discussions with the licensing and safety authorities. While there are standard practices, many venues will throw up unique challenges that require bespoke design solutions.

Different countries will often have their own crowd-modelling guidance; in the UK it is the *Guide to Safety at Sports Grounds* (Department for Culture Media and Sport, 2008), often called the 'Green Guide' (this has also been used in South Africa and Brazil as part of their FIFA World Cup projects). Event owners may also have their own standards, such as *Football Stadiums: Technical Recommendations and Requirements* (FIFA, 2011). However, the principals are always the same and aim to factor the movement and behaviour of spectators into the design of the building.

Increasingly, crowd modelling and the complex understanding of crowd behaviours is used to define the safety case and improve the comfort of spectators and the functionality of a stadium.

Dynamic modelling

In addition to the process of static crowd flow modelling, the use of state-of-the-art pedestrian modelling software to simulate spectator movement and shape stadium designs is increasingly popular. These *micro-simulation* programmes work by simulating individual spectators within the stadium environment and studying the performance of the building designs as they move.

Simulating pedestrian movement in this way allows the design team to understand the effects of relative minor building geometries on the flow of people. The outputs from this sort of modelling are both quantitative (egress times, maximum crowd densities) (Figure 5.1) and qualitative (video outputs that allow the user to study how crowds are moving and establish why).

Seating bowl

As well as being the core of any stadium – defined by the viewing distances and angles to the field of play – the seating bowl is also the area of the venue that works the hardest in terms of crowd movement.

Movement within a seating bowl is typically studied in terms of space available for spectators to move between their seats and the seating bowl exit or vomitory. This is important during normal operations as people move between their seats and the stadium facilities, but is essential were there to be an emergency evacuation.

Typically, a seating bowl will be modelled to ensure that all the seated spectators can evacuate the area within a given period of time. This time will be dictated by the local guidance and the 'fire risk factor' of the particular seating bowl. For example, an outdoor, predominantly concrete seating bowl might be considered low risk and be designed to allow spectators to evacuate within a given period of time (in the UK this would be 8 minutes), while an indoor venue where the construction materials might represent a greater fire risk and ventilation is not as good might be considered medium or high risk and be designed to evacuate spectators more quickly.

The simplest assessment of seating-bowl egress would take into account the width of the vomitory or exit route and use flow rates to determine the length of time it would take for every spectator to leave. However, a more comprehensive assessment would take account of the population of each individual area of the seating accommodation and the effect of individual gangway geometries on egress.

In the past, detailed modelling of seating bowls has highlighted the vertical and lateral gangways immediately around the head and above the vomitory and the width of the vertical gangway as greatest restrictions to crowd movement.

Figure 5.1 A crowd density heat map taken from the London 2012 Olympic Stadium micro-simulation crowd model. (This figure is reproduced in the colour-plate section as Plate VIII)

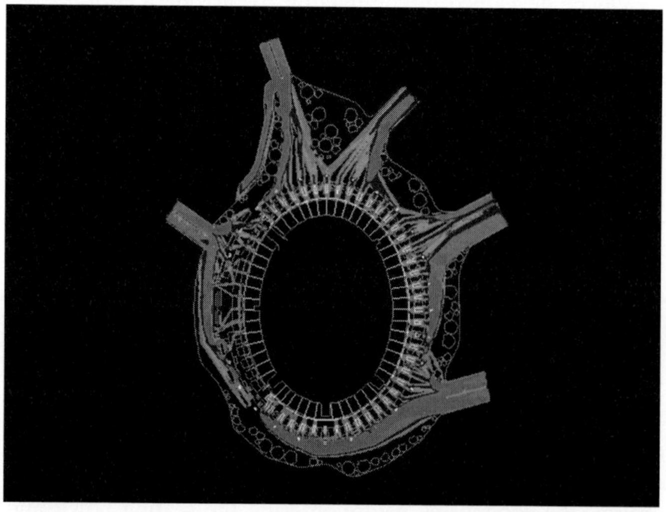

Concourses

The concourses and lounges that surround the seating bowl are also critical areas for crowd movement. They provide the circulation zone between the seating bowl and stadium exit routes but also accommodate most of the stadium facilities such as toilets and concessions.

Careful consideration should be given to the pedestrian densities on the concourse and the impact of queues at facilities. At sports venues where there are distinct breaks in play, such as half-time in football or rugby, the effect of significant peaks in demand for facilities should be considered. Concourses should be sufficient to allow guests to leave the seating area, enjoy the facilities offered and then return to their seats before the end of the interval. This is of particular importance for hospitality seating areas, where half-time hospitality can be an important part of the package sold to premium guests.

The standing space provided on concourses will vary by type of event. For example, while a significant proportion of a UK football crowd will leave their seats at half-time, putting pressure on the concourse area, in many US sports there might be more numerous breaks in play and a culture of selling drinks and snacks within the seating bowl, meaning concourses are used less intensely.

At the end of an event, or during an evacuation, concourses will often have to accommodate large queues that will form at the head of exit stairs or gates. Crowd modelling will be used to determine the size of those queues and ensure they can be safely accommodated within the space available.

Vertical circulation

Continuing through the building, the various levels of a stadium will be linked together by different types of vertical circulation elements, defined by the requirements of crowd flow.

Traditional venue designs will use large stairs or ramps to bring spectators to and from the upper levels, and these will be studied to ensure they have sufficient capacity to accommodate the required numbers of people.

External concourses

Once beyond the building line, spectators will move onto the external concourse or podium. This large outdoor area provides a critical space for the circulation of much of the stadium population, either inside or outside of the turnstile line.

Typically, a clear zone of around 15 m should be provided to accommodate circulating crowds. However, where podium exits or onward routes to transport nodes are not evenly dispersed, the external circulation areas should respond accordingly.

For example, where there is a significant desire line towards a particular transport node, the external podium should ensure that there is increased width to enable circulation movements. This might increase the width of a podium in some places to more than 30 m.

An estimate of where spectators will look to travel after an event is essential in determining whether the podium will be sufficient to accommodate them. This is where the process of crowd modelling within the stadium begins to segue with the spectator transport strategy.

For example, at the London 2012 Olympic Stadium, the venue sits on an island between two rivers, and the podium can only be accessed using significant bridge structures. The width of these bridges was defined by crowd modelling to ensure they respond to the requirements of the stadium. This includes a significant desire line towards nearby transport facilities placing disproportionate demand on one route.

Arrivals

Typically, turnstiles will provide the key arrival point to a stadium, providing a secure line between the *paid* interior of the venue and the surrounding area.

There are a number of different approaches to entry facilities to suit different requirements, ranging from manual ticket-rip facilities and temporary fencing, to low-impact half-height turnstiles, to full-height secure rotary turnstiles.

The chosen design will be dictated by the security and ticketing strategies, and will take account of the demographic of the crowd. While an Olympic Games might employ simple ticket-rip facilities, a UK football ground is likely to employ the most secure full-height turnstiles. Increasingly, entry facilities will utilise technology for the checking of tickets, including smart-card readers and mobile-phone technology.

The differing approaches to access control will dictate the rate at which spectators can pass into the stadium, typically this is between 600 and 1200 spectators per hour per turnstile. One of the most important crowd-modelling processes is to ensure that the number, design and location of entry facilities is sufficient to admit everyone in the required time.

A standard crowd flow assessment of turnstile numbers would consider the need to admit all the spectators within a given period of time, typically an hour. However, experience has shown that local conditions and spectator behaviour will affect the arrival profile of spectators at a stadium.

At many football venues in Europe and South America, where the sale of alcohol might be restricted, there is a culture of late

Figure 5.2 Three-dimensional visualisation taken from a micro-simulation crowd model of the security screening area proposed for Arena Cuiaba, a venue for the 2014 FIFA World Cup in Brazil. (This figure is reproduced in the colour-plate section as Plate IX)

arrivals – spectators using facilities in the local area before arriving at the stadium shortly before the beginning of an event. This puts tremendous pressure on the stadium facilities, particularly the turnstiles.

Crowd modelling should look to understand the behaviour of local spectators and ensure there are sufficient turnstiles in the right places to deal with these late surges without compromising security or safety.

For major events such as an Olympic Games or World Cup, spectators typically arrive over a much longer period. However, for these types of events, there are typically requirements for additional security and accreditation that create their own design pressures.

Security-screening areas provided at many major sporting events, similar to facilities used at airports, restrict the throughput of spectators more significantly than any turnstile. The number of screening units provided and the design of queuing areas should be carefully modelled and considered (Figure 5.2).

Onward routes

The consideration of major crowd movements does not end when the crowd reaches the edge of the building, and there will often be considerable challenges faced in the areas immediately surrounding the venue leading up to local transport nodes.

This requires close cooperation with adjacent land owners and local authorities to ensure a joined-up approach – designing the local area as an extension to the design of the stadium. It is, however, often outside of the immediate control of stadium design teams.

It is important that the design of surrounding spaces is consistent with the principals of stadium crowd movement – providing clear open spaces, designed to provide sufficient width to accommodate major flows, free of trip hazards or sudden changes in level or direction.

The most significant areas of crowd movement beyond the stadium will be the routes to transport modes, either private car parks or public transport hubs. Understanding the movement of crowds towards these facilities requires a broader understanding of the travel behaviour of spectators.

In order to ease the impact on the surrounding pedestrian network and transport facilities, measures are often employed to retain spectators within the venue or local area as long as possible. These measures might include showing replays after the match, providing additional entertainment such as double-headers, or putting on drinks and food promotions.

At venues that do not employ retention measures, crowds will often leave within 15 minutes of the end of an event, putting increased pressures on the surrounding pedestrian routes and transport facilities. Sports such as cricket and athletics might have 'softer' finishes, with some spectators leaving the venue before play has been completed, easing demands on the external infrastructure.

Once they have left the stadium, spectators will often need to queue nearby for public transport facilities. The size of queues

Figure 5.3 The crowd management operation on Olympic Way leading up to Wembley Stadium

at local stations will be determined by the frequency of services and station capacity, and could require spectators to wait for up to an hour. The management of these queues is one of the key considerations in the design of the areas surrounding a stadium.

At Wembley National Stadium the pattern of pedestrian movement is dominated by the 18 m wide Olympic Way route linking the stadium with Wembley Park Station. The significant post-event demand for this route requires a significant crowd-management operation developed in coordination between the Metropolitan Police, British Transport Police, London Underground and the Stadium Management Team (Figure 5.3).

Transport strategy

A comprehensive transport strategy is one of the key factors in the delivery of a successful new stadium, and will be unique for every venue.

Much like the crowd-modelling process within the venue, the process of defining a successful transport strategy starts with understanding how spectators will behave. This will often include the study of the origins of spectators, using databases of season-ticket holders and members to determine where they will be travelling from on event days, as well as spectator travel surveys to determine how they will travel.

Many new venues will look to maximise the use of sustainable public transport, increasingly moving away from the use of private cars. Stadium projects will often provide significant upgrades or new public transport facilities as part of their design.

Stadium developers might contribute towards the upgrade of existing facilities, or develop bespoke event-day transport services, such as dedicated bus services or park-and-ride facilities.

Where spectators do use private cars, car parks should be located and designed and managed effectively to reduce excessive delays on exit. This will often be combined with efforts to prevent spectators parking on-street in the local area and thus affecting residents and businesses.

Shaping behaviour

While ensuring that there is sufficient transport capacity available to accommodate demands, a modern stadium transport strategy should also seek to shape travel behaviours through engagement with spectators.

Communicating with regular spectators through the use of event-day programmes, websites and ticketing, stadium managers can encourage changes in spectator behaviour that will have a beneficial impact on event-day operations. These packages of measures are often described as 'travel-demand management'. For example, information might be provided to spectators on effective ways of avoiding excessive queuing after an event by utilising a different transport mode or changing the time of their journey home.

The transport strategy at Arsenal's Emirates Stadium has been highly successful in implementing such measures. When it moved from the nearby Highbury Stadium in 2006, the club was able to reduce the proportion of its spectators using private car to access its new stadium by 50%, significantly reducing congestion in the local area.

Integrated thinking

Successful stadium design projects have shown the importance of ensuring that the work to understand spectator behaviour, from moving around the seating bowl to travelling to the event, is undertaken as early as possible. This gives the best chance of this aspect being integrated into the fabric of the building, ensuring that the safety and comfort of spectators is safeguarded and that the venue can be managed efficiently.

Where this sort of thinking has been left until late in the design process or, worse still, until after opening of the venue, the results are predictably dire: venues that cannot operate at full capacity due to licensing constraints; transport strategies that result in heavy congestion and long delays; and designs that put the safety or spectators at risk.

REFERENCES

Department for Culture Media and Sport (2008) *Guide to Safety at Sports Grounds*, 5th edn. HMSO, London, UK.

FIFA (Fédération Internationale de Football Association) (2011) *Football Stadiums: Technical Recommendations and Requirements*, 5th edn. FIFA, Zurich, Switzerland.

FURTHER READING

Nixdorf S (2007) *Technical Recommendations for Grandstands in Modern Stadia*. Ernst & Sohn, Berlin, Germany.

Stadium and Arena Design
ISBN 978-0-7277-5790-6

http://dx.doi.org/10.1680/saad.57906.073

ice
Institution of Civil Engineers
publishing

Chapter 6
Engineering secure stadiums

Andrew Sieradzki Buro Happold, Bath, UK
Andrew Kelly Buro Happold, Abu Dhabi

Background

As the market for live events, especially regular sports such as football, horse racing, cricket and athletics, moves increasingly away from the hobbyist and enthusiast market into mainstream entertainment, the appetite for high-capacity stadiums to support large fan bases and maximise ticket revenue is growing.

The boost in revenue and the ability of venues to host high-profile events has obvious positives, but this chapter focuses on how to engineer out some of the negative aspects, particularly the potential attractiveness of modern stadiums and arenas to criminals and terrorists.

Core aims

Modern stadiums are key commercial assets to the organisations that own and run them. It is thought that the Staples Center in Los Angeles has an annual revenue of over $13 000 per seat (Van Riper, 2010), and Real Madrid, one of the world's richest football clubs, earns over £100 million in annual ticket revenue (Deloitte, 2012) from its 85 000 seat Bernabéu Stadium. Once broadcast revenue is factored in, stadiums and their facilities can represent a huge percentage of a club's income. In conjunction with a stadium's 'mainstream' sports venue use, there are also other modes of use that bring in significant revenue streams, such as the venue-naming rights, conferences, concerts and other events (Figure 6.1).

This emerging 'multi-mode' use puts a significant emphasis on the venue design requirements to ensure that a broad set of uses is catered for. Indeed, in modern stadiums there is a greater emphasis on ensuring that all parties involved with the design, construction and operation are engaged early on in the design process. This is not only to provide a 'safe' place that is both compliant with local codes/standards and licensing requirements, but also a 'secure' space that is integrated within its surroundings and environment.

The overall objectives of security measures are thus to facilitate the protection of the public and building users, and to ensure that appropriate physical and technical measures are put in place to protect the assets and associated business continuity.

Risk management

Given such a wide pattern of usage and commercialism, modern stadiums now face a diverse range of threats. These can range from low-level crime such as petty theft, pick-pocketing and anti-social behaviour, through to the more potentially catastrophic issues such as serious organised crime, terrorism (e.g. vehicle- or person-borne improvised explosive devices (VBIEDs and PBIEDs), chemical, biological or radiological (CBR) agents, control of telemetry systems (e.g. IT and building control systems)), reputational damage and direct action protest.

This diverse threat landscape means it would be virtually impossible for modern stadiums to be '100% protected' against 'all threats', as well as maintaining their ability to be open and accessible public spaces within an operational environment. The costs associated with the provision of such security would also be prohibitive and disproportionate in the context of an individual stadium.

Therefore, in order to best align the security provision with the business need, a process of reasoning the threats against their likelihood and impact if they happen is undertaken early on in the design process. This is known as a 'risk-management approach', and at its core is the risk assessment (Figure 6.2). Once this assessment has been conducted a balanced view can be taken by all project stakeholders. Typically, this would involve the client/developer, design team, operator, licensing and security authorities discussing what issues could be reasonably mitigated through intelligent design and use of physical and technical measures, combined with the operations strategy (e.g. stewarding). This should be documented in the risk

Figure 6.1 Graph showing the significance of match-day and broadcasting revenue to Real Madrid football club. (Data from Deloitte, 2012)

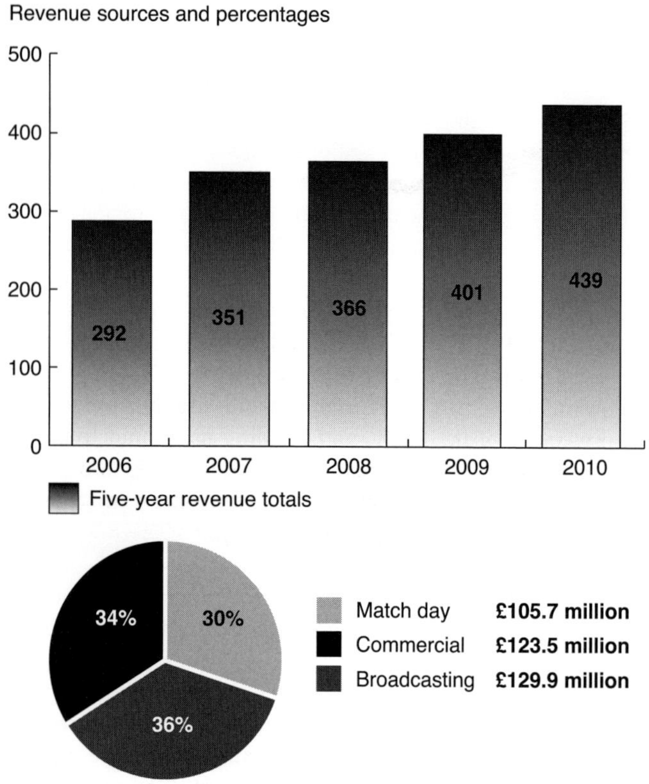

treatment plan or security strategy. Finally, there may be a number of issues that cannot be dealt with at the design stage; these will typically become 'residual risks' that will need to be managed over the course of the venue's lifetime.

Provision within the risk-assessment process should be made to ensure that the security strategy is developed in conjunction with the client/operator from the outset, so that all risks identified are considered throughout the design, construction and handover phases. This should remain a 'living' document that can be used to assess the impact of existing and emerging threats as they become known.

Planning and design principles

It should be noted that this chapter is concerned with the key underpinning principles behind security engineering design for stadiums. Generally, this is a specialised subject and expert design level input is essential throughout the design and construction process. This needs to be coupled with regular consultation with the relevant authorities, other designers and stakeholders, including police, licensing authorities and emergency services.

Furthermore, information regarding the project itself may be of interest to those with hostile or malicious intent, and the adoption of designs incorporating digital built environments and Building Information Modelling (BIM) tools, emphasises the need to develop an effective information management strategy from the outset. The PAS 1192-5:2015, 'Specification for

Figure 6.2 Risks (typical examples are provided for reference) can be summarised graphically using a risk matrix that highlights the relative consequences/impacts versus the likelihood of the risk occurring

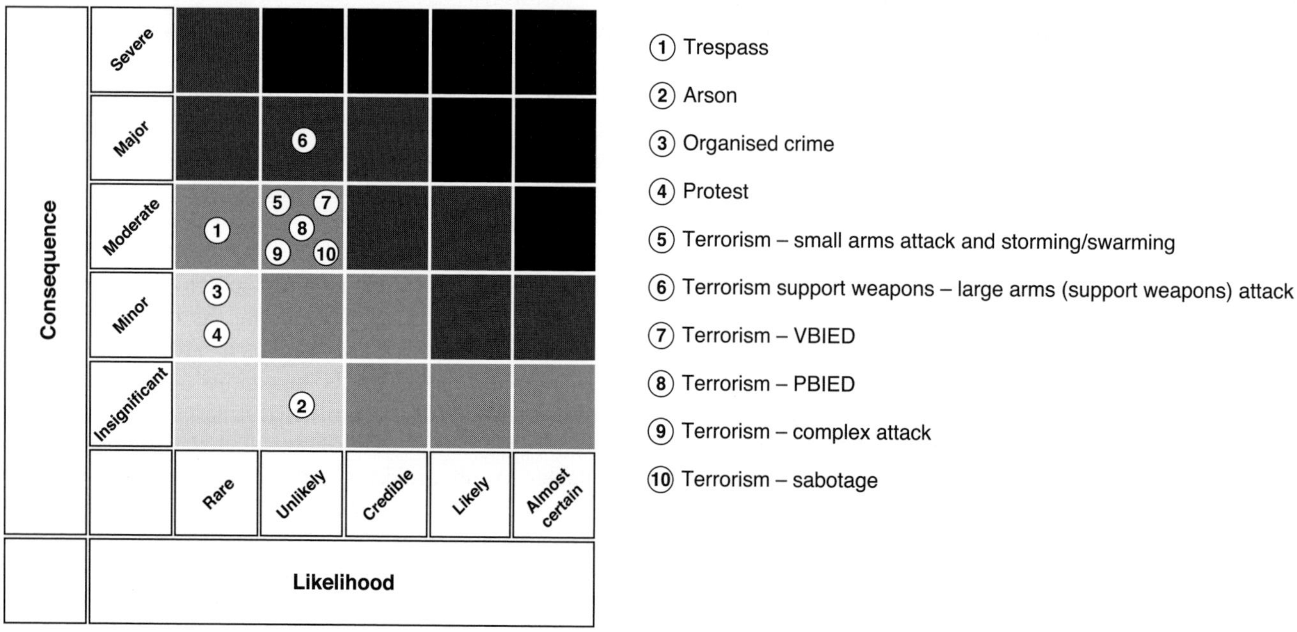

security-minded building information modelling, digital built environments and smart asset management' (BSI, 2015), sets out a holistic approach to security, which allows an assessment of the overall security threat to a built asset so as to inform appropriate and proportionate security risk management. The PAS addresses the steps required to create and cultivate an appropriate security mind-set and secure culture across many partners and encourages an appropriate, proportionate need-to-know approach to the sharing and publication of information about digital built assets.

Once the initial risk assessment process has been completed and the palette of mitigation options agreed, the process of design can then be undertaken. The options considered should be based on a number of core principles, namely:

- resilience and connectivity
- layout and landscaping
- access control
- surveillance and awareness
- integration
- flexibility.

It is important to note that at this stage an early assessment of the cost consequences of the proposed mitigation options should be undertaken. This is to ensure the cost versus benefit is evaluated at this, and other, key stages throughout the design. This evaluation is essential as, in some cases where mitigation of security risks is paramount, the capital investment in security systems can represent in excess of 1% of the total construction budget for a stadium. This consequently can necessitate a high ongoing maintenance and replacement budget over time, as technology typically becomes outdated or obsolete after 5–10 years. Therefore, a suitable allocation of budget for maintaining these systems should be considered from the outset of a development.

Resilience and connectivity

This principle covers a wide range of planning and design aspects, perhaps most notably the appropriate selection of materials and designs to provide physical security against vandalism, accidental damage and deliberate sabotage. Resilience and connectivity should be combined so that building supplies, including power, data and other utilities, are planned and designed with failure in mind. As a minimum it would be expected that a stadium would have diverse power and communication routing to reduce the number of single points of failure. This should also be supplemented with a standby power system that can support the full load of the life safety and essential services, which would also include all security technical systems.

Wireless communication systems should not be relied upon exclusively, as it is relatively straightforward to jam and impair wireless communication signals. It is therefore essential that this is factored into designs, to maintain connectivity and ensure that communications are resilient. Similarly, portable communications systems should be able to operate in all areas of a stadium, including basement areas, to avoid communication 'blackouts'.

Openings within the building (e.g. air intakes at low level) can create vulnerabilities to malicious attacks such as deliberate contamination of the air supply with chemical, biological or radiological agents. Generally such openings should be secured and placed out of easy reach at a high level, or designed with physical systems that prevent circumvention. Where necessary, appropriate physical security measures should be included; it is recommended that these are independently specified and certified by the Loss Prevention Certification Board (LPCB) or equivalent.

Glazing systems often present vulnerability in buildings. Anti-bandit, ballistic- and blast-resistant glazing have different characteristics in order to provide the required level of protection. Studies have shown that up to 80% of injuries in a blast are caused by flying glass fragments. It must be emphasised that there is no one type of glazing that fits all scenarios (Wedding, 2010).

The stadium structure, such as columns, beams and slabs, should be sufficiently resilient to prevent disproportionate and progressive collapse in the event of a blast. However, it is much better to provide physical methods to ensure that possible blast threats are kept away from the building (see layout and landscaping, below). This is known as 'stand-off', and exploits the relationship between blast impact and distance, which decays exponentially from the source. In some instances the explosion itself may not be significant, but secondary blast effects such as ground shock and fragmentation should be considered for their impact on services and structural and glazing composites (see Chapter 7).

Adjacent developments and municipal/wide-area systems should be securely connected and integrated with the stadium communication systems. This is required to enhance the range of awareness and allow authorised personnel outside the stadium to monitor and coordinate dynamics factors such as traffic, transit systems and pedestrian flows in the surrounding streets and other affected key public interchanges.

Accessible and inaccessible voids should be designed out as far as possible; where this is not possible they should be secured appropriately and be independently checked and certified before sealing. Similarly, horizontal surfaces such as window sills at high level should be sloped steeply to prevent opportunities to deposit packages or devices.

Figure 6.3 Resilient power infrastructure should be embedded in the stadium design and plans

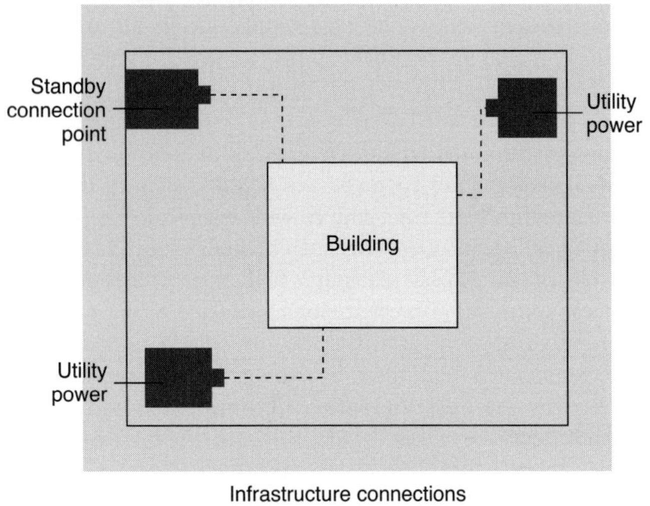

Case study – Protection against sabotage

In 1997 a betting syndicate tried to sabotage Premier League football matches in the UK (BBC News, 1999). At the time, bookmakers paid out if a match was abandoned, so the syndicate stood to make huge financial rewards if it was successful. The sabotage was accomplished by tampering with the circuit-breakers that control the floodlights.

This type of incident can be avoided by ensuring that there are no single points of failure for essential equipment (Figure 6.3), and that all access areas to stadium infrastructure are secured and monitored appropriately. It is also useful to ensure that key stadium systems support dual-factor authentication so that key systems cannot be accessed or changed without the input of more than one senior person.

Layout and landscaping

Stadium security can benefit greatly from a well-planned layout and surrounding landscape. The shape and orientation of stadiums can have security implications and should be coordinated with security and blast specialists from the outset.

Suitably designed (untraversable) landscaping can be used to provide low-cost or even 'free' security to help keep unauthorised vehicles and pedestrians away from sensitive areas, to increase and enforce stand-off, and to demark the area in terms of responsibility.

Enforced stand-off is an important design tool that can help protect buildings such as stadiums from the large blast forces generated from a VBIED. The stand-off distance refers to the distance between public-vehicle accessible space and the stadium building. There are no hard and fast rules on stand-off distance, typically a minimum of 30 m, but as every metre counts, seek to maximise whenever possible.

The development layout should encourage a sense of place and self-organisation through intuitive routes through the stadium and surrounding areas. This will keep the number of signs required to a minimum (inappropriate signage can create confusion, clutter and, if not coordinated, surveillance blind spots and places to conceal explosives) and will also help to prevent users from straying into no-access areas and ensure that evacuation in the event of an emergency is much more efficient.

Other design measures that need to be considered are the incorporation of crime prevention through environmental design (CPTED) and secured-by-design methodologies, to ensure that a baseline level of physical security is achieved throughout the stadium.

Case study – Emirates Stadium 'letters' – hostile vehicle mitigation (HVM) and architecture

One known and significant threat to a stadium or crowded space (such as the podium of a stadium) is the delivery of an explosive device by means of a large vehicle carrying an explosive payload. An effective method of mitigating this threat is the use of successfully tested and correctly spaced (max 1.2 m air gap) hostile vehicle mitigation (HVM) physical measures in order to prevent a vehicle penetrating the vehicle security barriers. In the case of Arsenal's Emirates Stadium (Figure 6.4), these measures were designed to look like 'letters', rather than the conventional 'bollard' design, to reflect the architectural needs of the site.

The use of these HVM measures prevents unauthorised vehicles entering the pedestrian bridges and the potentially crowded spaces around the podium areas; in addition, this maintains a stand-off from a vehicle being used to encroach or penetrate and detonate an explosive device adjacent to or within the stadium.

Typically the baseline standards for achieving a known level of performance for these HVM measures are IWA 14-1:2013 and IWA 14-2:2013 (the pre-cursor to these are the UK publications PAS 68, PAS 69 (BSI, 2013a, b) and CWA 16221 (BSI, 2010)). The design and incorporation of HVM measures should also be developed in conjunction with the following guidance 'Integrated Security Guide', TAL 1/11, TAL 2/13 (CPNI, 2011, 2013, 2014), and *Blast Effects on Buildings* (Cormie *et al.*, 2009).

Figure 6.4 The letters outside Arsenal's Emirates Stadium that are innovative HVM measures

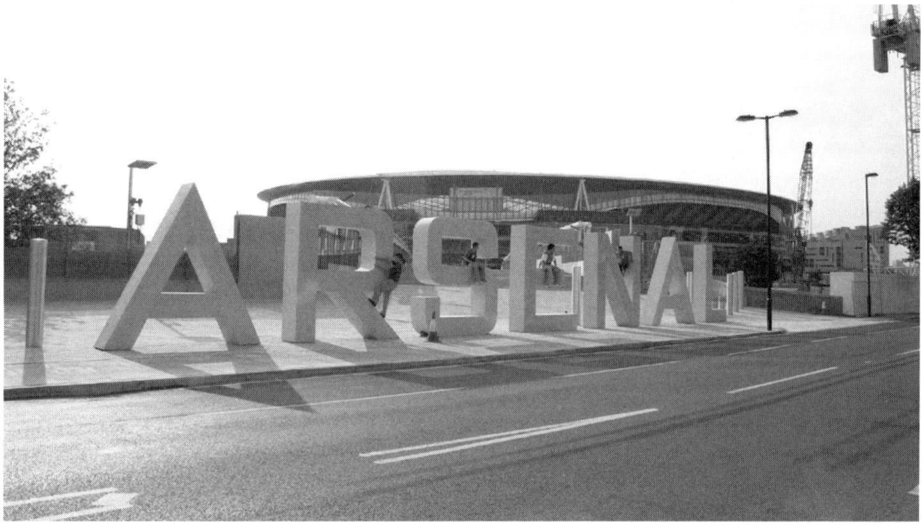

Access control

Creating security zones throughout the stadium is an essential part of developing the security strategy. Zoning helps to determine which areas of the stadium require access control. Restricted security zones such as server rooms, command and control facilities, and data centres (due to their critical role in ensuring an event can be staged) should have the highest levels of access control provided.

Controlling access into back-of-house areas is essential to ensure that visitors and members of the public cannot access private or staff-only areas. Certain spaces should also be restricted only to senior staff members who require access (e.g. stadium management and IT support staff). It is, therefore, of vital importance that the access-control strategy and design is developed in conjunction with the stadium operational team.

There should be no reason for modern stadiums to rely on conventional key systems, which are easily copied and can create significant operational expenditure in the event of a loss or compromise. Consideration should be given to the use of a number of authentication layers of the access devices that are appropriate for the application (i.e. biometric access-control systems may be used to minimise the risk of stolen access cards for the most sensitive areas of the stadium, while less-sensitive areas may be equipped with conventional radio-frequency identification (RFID) or proximity-card-based systems).

Locking systems, especially for high-security locations should be latch or bolt based. Maglocks should be avoided, as the locks themselves can reduce the size of the aperture, often have much lower holding forces than latch-based locks, and can suffer from the effects of residual magnetism, which can cause safety issues if not designed properly.

Tailgating is the practice where two users enter a site under the credentials of a single user by passing closely or 'piggybacking' with another user. Access-control systems should, where necessary, include measures to help prevent tailgating, and consideration should be given to enhanced measures to cater for high-risk temporary events (Figure 6.5).

Dedicated entrances should be provided for separate user groups as far as is possible. This helps to make stadium management more efficient and allows security personnel to spot unauthorised entry more easily (Figure 6.6).

It is important to consider all modes of stadium operation, including the key mode when 'no' events are taking place. The site and building will need to be zoned so that areas can be securely 'locked down' while maintaining the 'day-to-day' facilities management operations. Carefully designed, this mode can prevent unnecessary provision of manned guarding for the locked down areas, and a combination of electronic access control and zoned intruder detection should be considered.

Surveillance and awareness

High levels of natural surveillance should be planned both within and outside the stadium. Generally, façade recesses or alcoves should be avoided as their inclusion can encourage entrapment and lurking to occur, together with locations in which to conceal explosive devices. Removing such locations reduces the opportunity for criminals to operate or indeed allowing a device to be deposited whilst concealed.

Figure 6.5 Temporary biometric and turnstile access-control solutions can be deployed for high-risk situations

Furthermore, during the design phase a void certification plan should be developed to identify concealed voids, and an inspection regime adopted during the construction and operational phases to ensure a high degree of confidence that these voids remain free from concealed devices.

Good sightlines that provide unrestricted views of the event or activity are usually a feature of modern-day stadiums, as these will help to deliver high levels of natural surveillance. This should be extended across the whole development, both within and outside the stadium.

Figure 6.6 Access control can benefit from user-group separation as shown in this diagram of stadium access-control points

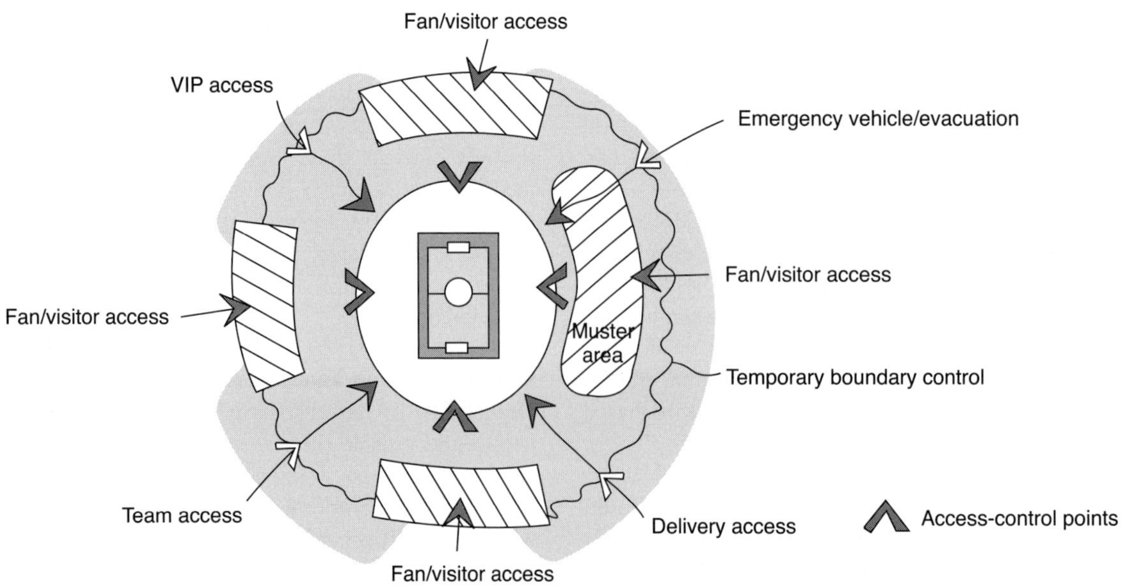

Natural surveillance should be supplemented by high-quality active surveillance (i.e. CCTV) supported by good artificial lighting levels and colour rendering to cover night-time operations. This should be considered in line with the UK Home Office's Centre for Applied Science and Technology (CAST, formerly Home Office Scientific Development Branch) operational requirements (UK) (Cohen *et al.*, 2009), which gives high levels of coverage to all public areas, especially the entrances, crowded areas overlooking the activity areas/field of play, concourses and entrances to all private and back-of-house spaces. This will create a high degree of situational awareness, which is important for dealing with incidents and extreme events.

Where deployed, CCTV should be designed to have sufficient quality, screen size, recording frame rate and duration to meet the operational requirement (OR) and support the stadium and stakeholder operations. CCTV is an essential tool for law-enforcement agencies, and has been used regularly to secure convictions. CCTV is also a useful training tool for stadium personnel, as near misses, incidents and events can be recorded and reviewed.

Pitfalls

The structural design of many high-capacity stadiums allows for some structural flexibility, and thus movement, to take place under full load. It is important to consider that any CCTV cameras mounted on these structures will move accordingly, and that any images viewed at a distance (i.e. at full zoom) may be significantly impaired due to the effects on the image-processing system. This can occur to the point where any images become unusable, particularly during a high-capacity or major event.

Intruder-detection systems (IDSs) also need to be considered throughout the stadium, in the context that the correct placing of alarm sensors within sensitive or key areas provides an effective mechanism for detecting an abnormal condition (e.g. an insecure door or movement within a locked-down area).

Command and control facilities should also be designed to be ergonomic (the starting point should be ISO 26800:2011) and resilient, and it is usually desirable to physically separate day-to-day building command and control operations from event-day operations, as these are essentially required for different operational stances. The locations of the supporting hardware for these systems (i.e. server, recoding racks, etc.) should also be physically adjacent but separated from these operational spaces; this is to cater for the servicing of systems and to ensure that the correct environmental and security conditions are provided for each facility. In addition, the system hardware should be designed to be resilient and allow dual functionality (i.e. diverse primary and secondary power supplies, and dual-array recording systems that can seamlessly operate from either control facility embedded into the design).

Security provision at large public events does not end at the delineation point of the stadium footprint, and typically a planning condition of the development is to provide technical interfaces (usually CCTV) in a local authority or police control room. These interfaces allow flexibility both to export and to import live CCTV images in order to assist with security and public-safety issues beyond the stadium footprint.

It is important that all stadium security systems are appropriately commissioned and tested against the design specifications and security strategy. Of particular importance is the CCTV system, which should be tested 'end to end' to ensure that image quality, resolution, frame rate and other characteristics all meet performance requirements (Figures 6.7 and 6.8). Local and international standards should be referenced. However, a useful starting point (although now an old document) is the Police Scientific Development Branch publication *Performance Testing of CCTV Perimeter Surveillance Systems* (Aldridge and Gilbert, 1996), which outlines useful test principles.

Integration

A fully coordinated approach should be taken throughout the planning, design and construction phases of the stadium development, as this will help to ensure that aspects of the design do not conflict (e.g. lighting works to support the CCTV system and does not impair it; alarm system detectors are not blocked by unnecessary building obstructions). Regular reviews should take place to ensure that safety and emergency measures do not create security vulnerabilities. Often this can be as straightforward as rerouting emergency escape routes, which is obviously much easier to undertake at the design stage than post-construction.

Stakeholders should be regularly consulted to ensure that they support the plans and the design. Often they will have useful input, and the development process should include protocols to manage stakeholder requirements and comments.

Modern stadiums now have integrated security, building management, vertical transportation and life safety systems as a minimum; this should not be the gold standard of design but the baseline. It is essential that when a security breach is detected an effective response is dispatched and the matter closed in a swift and controlled way. Typically, software-based security management systems are used to support this important operational command and control function, and these systems are designed to integrate all the key technical security systems (i.e. CCTV, IDS, intercoms, electronic access control systems, etc.). These systems can be used both as an effective 'reactive' tool to provide a link between 'detection' and 'reaction' of an incident, and as a 'proactive' tool in order to effectively allow for post-incident review and analysis of events. The information provided by these management 'tools' can lead to more effective training

Figure 6.7 Test targets such as the popular Rotakin are useful commissioning tools for CCTV systems. In this image the detail on the test target is barely visible due to the lack of uniformity in the lighting design

of security staff with regard to their awareness and readiness of the stadium operations.

It is important to integrate all aspects of the design with operations (Figure 6.9), as this should help to ensure that specified systems perform as the operational team expect and, perhaps more importantly, that money is not spent on systems that will not be used. Experienced operational leaders should be in place from the earliest stages to ensure that the physical design is coordinated with the operational plan, including the security policies and procedures.

Searching and screening

Whether for safety or security reasons, search and screening regimes are increasingly being deployed at stadiums and arenas, and any requirement for such measures should be based on a robust risk assessment spanning, for example, public safety, crime, protests and terrorism. The risks identified should then inform the overall risk assessment and security strategy for the venue.

Whilst attention is normally focused on screening spectators (and their personal possessions), consideration should also be given to what measures are required "back-of-house". People, their possessions, vehicles, mail, deliveries and supply chain entering the site, all need to be considered.

A variety of screening technologies and techniques are available, with their use tailored to meet the needs of the particular venue. For example, manual search and visual inspection

Figure 6.8 This image shows that environmental conditions such as the reflection from the puddle in dull lighting conditions can affect CCTV performance

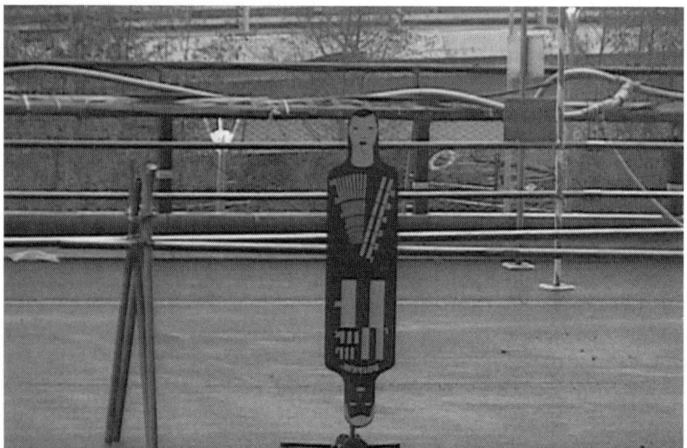

Figure 6.9 The security management system integrates the other security systems into a single front end

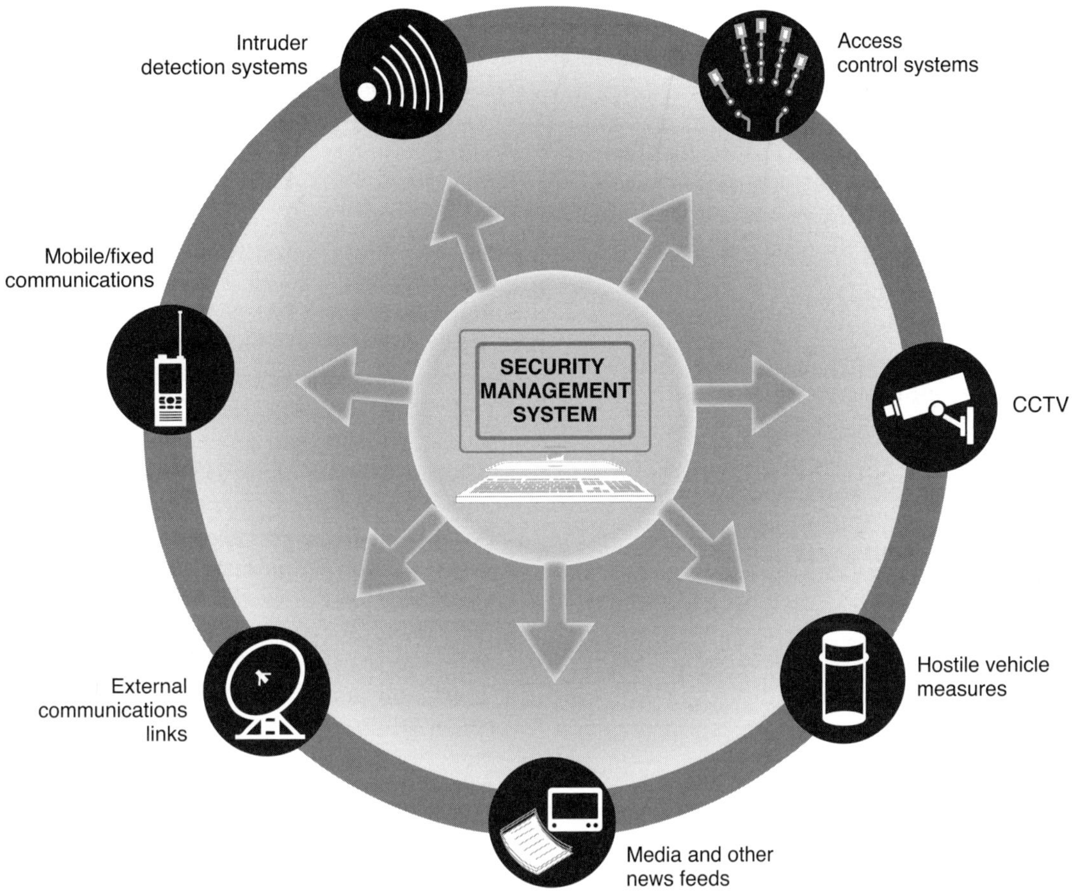

techniques can range from a quick (a few seconds per person) check that no large threat items are present in bags, coats, etc., to slower (around a minute per person), methodical searches to check for the presence of smaller items such as penknives. Similarly, established technologies such as walk-through metal detectors and X-ray machines commonly seen in airports can be used to achieve efficient and effective screening for key threats at major event venues.

Detection dogs are also a potentially valuable tool, not least as they can be deployed anywhere on the approach to the venue where there is sufficient space, and can act as a highly visible deterrent. Care must be taken to ensure that any dogs deployed have been trained to detect a relevant selection of threats (usually explosives or pyrotechnics/flares/fireworks) and that the capability they offer is fully understood.

By considering searching and screening from the outset, this allows the measures to be integrated alongside other security measures for the venue, together with complementing the overall spectator ingress process. Further guidance on the design and delivery of checkpoint-based screening measures can be referenced from PAS127:2014 (BSI, 2014).

Flexibility

It is inevitable that change will occur over the lifetime of any stadium and arena. It is not possible to foresee every scenario, so as much flexibility as possible should be incorporated. Plan for change and unexpected events, including:

- *Incidents* – these could include a major event such as a terrorist attack within or close to the venue.
- *Different modes of operation* – modern-day stadiums have to be flexible to maximise their utilisation. During the week a stadium may be used as a conference facility for local business, while at weekends the mode of operation will be very different for a sports event.
- *Changes in threat* – modern-day stadiums face very different threats to stadiums designed 50 years ago. The different terrorist threats alone can be extremely

challenging to plan for, and include a whole range, from suicide bombers to chemical, biological, radiological and nuclear threats.

- *Special events* – the security required for day-to-day events such as football matches, music concerts, etc. can often be very different from the security needed for a major event such as a World Cup or the Olympic Games. Flexibility should be planned into stadium developments to ensure that, if hosting international tournaments is likely, security requirements are embedded in the plans and design.

Typically, one of the most effective ways of maintaining flexibility and scalability after the construction of a stadium is to ensure that an effective security management regime is embedded in the governance model of the business. This level of representation will enable this key aspect of the business not only to support the day-to-day running of the stadium, but also to allow flexibility within the security capability, enabling quick adaptation to new threats and vulnerabilities that may arise from local, national or international events.

Human factors

The security environment around stadiums and arenas involves significant human interaction and resources in order to operate and maintain the security aspects of the development, i.e. equipment/policy/guarding. Therefore careful consideration of the key aspects of the human factors is vital during the design phases of the project, i.e. control room layout, guard-force facilities (see CPNI, 2014a, b).

REFERENCES

Aldridge J and Gilbert C (1996) *Performance Testing of CCTV Perimeter Surveillance Systems*. Police Scientific Development Branch, PSDB 14/95. See http://www.ipcctv-solutions.com/whitepapers/Home%20Office%20-%20Performance%20Testing%20of%20CCTV%20Perimeter%20Surveillance%20Systems%20-%20Rotakin.pdf (accessed 05/01/2015).

BBC News (1999) *Man Guilty of Floodlight Plot*. See http://news.bbc.co.uk/2/hi/uk_news/426092.stm (accessed 05/01/2015).

BSI (British Standards Institute) (2010) CWA 16221:2010. Vehicle security barriers. Performance requirements, test methods and guidance on application.

BSI (2013a) PAS 68:2013. Impact test specifications for vehicle security barrier systems. BSI, London, UK.

BSI (2013b) PAS 69:2013. Guidance for the selection, installation and use of vehicle security barrier systems. BSI, London, UK.

BSI (2014) PAS127:2014. Checkpoint security screening of people and their belongings – guide.

BSI (2015) PAS 1192-5:2015. Specification for security-minded building information modelling, digital built environments and smart asset management.

Cohen N, Gattuso J and MacLennan-Brown K (2009) *CCTV Operational Requirements Manual*. Home Office, Scientific Development Branch, St Albans, UK. See https://www.gov.uk/government/uploads/system/uploads/attachment_data/file/378443/28_09_CCTV_OR_Manual2835.pdf (accessed 05/01/2015).

Cormie D, Mays G and Smith P (eds) (2009) *Blast Effects on Buildings* (2nd edition). Thomas Telford Limited, London.

CPNI (2011) TAL 1/11 Vehicle security barriers within the streetscape. https://www.gov.uk/government/uploads/system/uploads/attachment_data/file/4389/1-11.pdf.

CPNI (2013) TAL 2/13 Bollards and pedestrian movement. https://www.gov.uk/government/uploads/system/uploads/attachment_data/file/245355/tal-2-13-bollards-pedestrian_movement.pdf.

CPNI (2014) *Integrated Security Guide*, 2nd Edition. http://www.cpni.gov.uk/documents/publications/2014/integrated%20security%20-%202nd%20edition.pdf?epslanguage=en-gb.

CPNI (2014a) Personnel Security and guard force motivation. http://www.cpni.gov.uk/advice/Personnel-security1/Guard-force-motivation/.

CPNI (2014b) CCTV control rooms and human factors. http://www.cpni.gov.uk/documents/publications/2014/2014001-human_factors_cctv_control_rooms.pdf?epslanguage=en-gb.

Deloitte (2012) *Deloitte Football Money League*. See http://www.deloitte.com/assets/Dcom-Sweden/Local%20Assets/Documents/FootballMoneyLeague%202012-uk-sbg-dfml-2012-final120209.pdf (accessed 05/01/2015).

ISO (International Organization for Standardization) (2011) ISO 26800:2011 Ergonomics – General approach, principles and concepts. ISO, Geneva, Switzerland.

ISO (2013a) ISO IWA 14-1:2013. ISO International Organization for Standardization) (2013) ISO IWA 14-1:2013, Vehicle security barriers – Part 1: Performance requirement, vehicle impact test method and performance rating. ISO, Geneva, Switzerland.

ISO (2013b) ISO IWA 14-2:2013. Vehicle security barriers – Part 2: Application. ISO, Geneva, Switzerland.

Van Riper T (2010) *North America's Most Lucrative Stadiums*. See http://www.forbes.com/2010/01/05/lakers-cowboys-yankees-business-sports-stadiums.html (accessed 05/01/2015).

Wedding WC (2010) *Experimental Study of Blast Resistant Glazing System Response to Explosive Loading*. Master's thesis, University of Kentucky, Lexington, KY, USA. See http://uknowledge.uky.edu/cgi/viewcontent.cgi?article=1028&context=gradschool_theses (accessed 05/01/2015).

Stadium and Arena Design
ISBN 978-0-7277-5790-6

http://dx.doi.org/10.1680/saad.57906.083

Chapter 7
Blast engineering in sports stadiums

Jurek Tolloczko

National and international sports events are opportunities for all of us to see the best of competitive entertainment. Stadium and arena events are becoming larger and more spectacular. They attract global interest, drawing huge audiences at the event and, very often, millions more in the home through the power of global media.

Revolutionary advances have been seen in the scale and quality of venues. These venues and events bring together many international sportsmen and sportswomen, thousands of spectators, VIPs, international businesses and the media. It is always the planned intention that the event will be enjoyable, well organised and memorable. It is always anticipated that the event will be remembered for the correct reasons. Sadly, this is not always the case.

In today's reality, a violent act or incident at, or near to, an event could transform an atmosphere of enjoyment into a scene of desperation witnessed by millions as it unfolds. Regrettably, some might view this as an end goal.

Prevention is always better than cure, and a combined physical, structural and active security strategy – that is to say, the complete package – should be considered. Access control and stringent monitoring and vigilance, before and during an event, make up the best approach to preventing incidents from occurring.

However, the climate in which we live and work has changed. How do we guarantee that an individual (or group of individuals) that has no regard for his or her own life, or the lives of others, can be stopped from carrying out the unthinkable?

Should an incident occur, prior planning and built-in physical security will protect people, minimise damage, prevent structural collapse, ensure integrity for evacuation and prevent breach of command/control facilities.

Explosion is an extreme event. However, if we consider it well in advance, and particularly during the early stages of design, we can implement structural solutions that are neither intrusive nor result in a stadium or arena that looks more like a fortress than an architectural icon. If we consider explosion as a design event (however potentially extreme), then structural solutions can be achieved cost-effectively. Design against explosion has to be performed as part of a holistic approach to the design of the complete facility. This, by common sense, will involve a genuine relationship between the client (owner and operator), architect, structural engineer, contractor and specialist suppliers, with input from the appropriate security agencies.

The main message is: consider explosion and fire from the very first day that consideration is given to developing a new facility or an upgrade of an existing facility. This will result in a better and more cost-effective solution than if blast is considered as an afterthought.

It was previously stated that explosions are extreme events, but what is meant by 'explosions, accidental or deliberate'? In the event of an accidental explosion, caused by, say, an inadvertently discharged firecracker, a cigarette or an electrical fault, localised damage can occur. If not contained or controlled, such an incident can escalate into a fire. Obvious subsequent damage can occur.

Man-made devices, delivered in a package, plastic bag, sports bag, container, vehicle, etc., cause deliberate explosions. Identifying the probable size and location of such explosions is of key importance, as is identifying subsequent consequences and protective solutions.

We have all seen, through the power of the media, the consequences of an explosive device. The risks from such a deliberate incident can be multiple:

- *Primary projectiles*: a vehicle-delivered device adds to the risk of flying debris of various size, mass and energy.
- *Secondary projectiles*: glass, cladding, doors, architectural-type features, structural components and street or venue furniture are all potential fragments.

- *Blast*: can result in injury to humans as a result of over-pressure or blast wind.
- *Local structural collapse*: localised failure of structural members can make inoperable areas of a building that might be key to evacuation or command and control.
- *Total collapse*: occurs when a localised incident escalates, through the mechanism of progressive structural collapse, to the total destruction of the building or facility. Examples of such events are obvious but extreme.

The resulting loss of life, damage to assets and public confidence are factors that regrettably now have to be addressed. All reasonable measures to mitigate such consequences must be seriously considered.

Designing stadiums, auxiliary buildings and facilities to be fully blast resistant is not a viable option. Designing and building 'fortress stadiums' is technically possible but not realistic. However, knowledge and systems do exist by which new and existing facilities can be enhanced to significantly reduce the effects of explosions. So why have such solutions not been considered in the past?

First, in many countries terrorist attack has either not been considered or the local population has become accustomed to such incidents. Solutions have been developed and implemented that are proportional to the perceived threat. In the UK, terrorist activity has been a reality for many years. However, September 11 changed the global perspective on terrorism with regard to the scale, the frequency and the type of attack that may be expected to occur.

Second, the required information to develop appropriate solutions is not readily available to the traditional practising architect and engineer. Knowledge and expertise in this area is vested in a finite number of individuals and companies that, for obvious reasons, do not openly publicise their services.

In designing a facility to resist explosive blast a logical philosophy can be adopted:

- Deflect an attack by demonstrating through layout, planning, intelligence, and proactive security and defence that the chances of success for a terrorist are small.
- Disguise the valuable parts of potential targets so that, if security is breached, the emphasis of the attack is focused on the wrong area.
- Disperse potential targets so that an attack would not cover an area large enough to cause significant loss of life or destruction – consider especially areas of access to central facilities.
- Stop the attack reaching areas of potential targets by making use of the strategic interaction of physical barriers and active security.
- Blunt the attack, should it reach a target, by strengthening key areas and structure.

A threat assessment is normally the first step in the process of providing effective security or protection for any event or facility. The process of assessing the threat can be undertaken as a risk analysis, which necessitates that all elements constituting a possible threat have been identified and evaluated, together with the degree of probability that a specific risk will occur and its potential consequences.

Many attack scenarios can be considered:

- devices located in access tunnels and culverts
- mortar-type devices
- package-type devices
- vehicle devices
- incendiary devices.

The resulting damage will depend on the device type, size and location.

Help in identifying possible threats can be obtained from local, national and international agencies.

A threat assessment 'risk analysis' should not itself propose solutions or countermeasures but should be a method of assessing and agreeing what might be the type of attack. This assessment will also be venue and event related, and might change over time, particularly for larger events that involve a significant planning period.

Protection against explosive threat should be developed and included in the facility's security/protection plan. Protection against attack can never be guaranteed. The cost of protection has to be assessed against the resulting benefit. This is a difficult calculation because the consequences of loss vary – some loss might be incremental but certain losses may be catastrophic.

It is not possible, and perhaps not appropriate, in this chapter to cover the type of explosion loads and resulting consequences. However, some structural design issues should be raised.

Traditional facilities and buildings, for in-service use, are designed to well-defined static loading conditions (and, in some cases, dynamic loads), where strength and deformation criteria are comprehensively defined in codes of practice and standards. Loads resulting from explosions are of much greater magnitude and are further complicated by their time-varying nature. To enable us to analyse structural response due to explosion loads we have to consider many other factors. This adds to the complexity. So we have to use more sophisticated methods of analysis and design.

In determining structural loads, the criteria of explosive charge size, type of explosive, location, interaction and reflection from the local built environment have all to be considered and quantified. Zoning and banding of potential threats help focus on key scenarios rather than trying to design against every conceivable threat.

When the loads have been developed, the definition of protection needs to be quantified. Structural acceptance criteria are considered from a strength and deformation limits point of view. In some cases, acceptance of a degree of permanent structural deformation may not only be tolerated but may also be an essential feature of the design. This criterion may, therefore, be applied in a markedly different way for blast loading compared with conventional design.

Blast design methodology will, therefore, include:

- the time-varying nature of the load
- enhanced material characteristics resulting from strain-rate effects
- post-elastic material and component behaviour
- large displacements and rotation – avoiding progressive collapse
- suitable application of safety factors.

This approach can be applied to panel-, beam- and column-type structures, on either a component or system basis. However, there is one obvious point that is frequently missed to the detriment of a solution. Physical barriers or walls can, and are, provided to resist explosions. This is done on a daily basis. What is found is that the provision of such walls is rarely a problem. The problem faced is that such walls are often asked for too late in the main structural design process. This creates unnecessary challenges and cost when the load transfer to the main structure and/or foundations has already been designed and built in complete isolation.

Practical solutions to protect against explosions do exist. To optimise their inclusion, and costs of inclusion, it is essential to think ahead.

Stadium and Arena Design
ISBN 978-0-7277-5790-6

http://dx.doi.org/10.1680/saad.57906.087

Chapter 8
Fire engineering in sports stadiums

John Dowling British Constructional Steelwork Association
Justin Garman and Jason Pritchard Mott MacDonald Limited
Florian Block Buro Happold FEDRA

Introduction

This chapter has been prepared with the intention of bringing to the construction industry examples of the best practice in structural fire engineering in sports stadiums in the UK.

The concept of designing for building safety in fire is not new. One can look back to events such as the Great Fire of London, and the rebuilding of the city afterwards, when provision was made for space separation and fire-resistant walls between houses. In modern times, fire precautions are installed in buildings in the UK in response to the requirements of the Building Regulations.

The regulations are presented in a manner that is functional rather than prescriptive. That is, the regulations outline what is required but leave the building designer free to meet the requirements in the manner in which he or she prefers and which can be proven to meet the requirements of the regulations. For example, Requirement B3(1) of the Building Regulations for England states that: 'The building shall be designed and constructed so that, in the event of a fire, its stability will be maintained for a reasonable period.' The equivalent requirement (2.3) in the Scottish Building Regulations states that: 'Every building must be designed and constructed in such a way that in the event of an outbreak of fire within the building, the load-bearing capacity of the building will continue to function until all occupants have escaped, or been assisted to escape, from the building and any fire containment measures have been initiated.'

The manner in which these, and the other fire-related regulations, are met is at the discretion of the building designer but, to assist in this, the governments of the various regions of the UK publish documents that provide guidance on the means by which compliance may be achieved. The most widely used of these is England's *Approved Document B (Fire Safety)* (Department of Communities and Local Government, 2006).

Approved Document B contains detailed provisions for fire safety in buildings. These are intended to provide guidance for some of the more common building situations. Assistance on what are considered appropriate periods of fire resistance for different building occupancies are given in Table A2 of *Approved Document B*. However, these provisions are not mandatory and the document states that:

> There is no obligation to adopt any particular solution contained in the Approved Document if you prefer to meet the relevant requirement in some other way.

It goes on to suggest what it means by this phrase by stating:

> Fire safety engineering can provide an alternative approach to fire safety. It may be the only viable way to a satisfactory standard of fire safety in some large and complex buildings and in buildings containing different uses.

This has been a welcome development, because increasing innovation in design, construction and usage of modern buildings has created a situation where it is sometimes difficult to satisfy the functional requirements of the Building Regulations by any other means. Fire safety engineering offers a way around this problem. It can be seen as an integrated package of measures designed to achieve the maximum benefit from the available methods of preventing, controlling or limiting the consequences of fire. The Institution of Structural Engineers says of structural fire engineering, a subset of fire safety engineering in general: 'By adopting a performance based approach to structural fire engineering ... more economic designs can be achieved and more innovative and complex buildings can be constructed.'

This chapter describes two case studies of the value that fire safety engineering can deliver in modern sports stadiums.

Fundamentals of fire safety engineering

Fire safety engineering as a separate and distinct branch of science has been in existence for only a relatively short time. It had its origins in Sweden but was taken up and developed within the UK, which can now claim many of the world's leading practitioners and researchers.

Fire safety engineering is aimed at adopting a rational, scientific approach to address specific problems in the provision of fire precautions in buildings. A fire safety engineering assessment of a building can be carried out in isolation or as part of a wider fire engineering analysis that takes into account many other aspects of building fire safety requirements. It can lead to considerable cost savings by matching the performance to the risk, but this should not necessarily be thought of as the main purpose of the exercise. That can be considered as putting fire precautions where they are most necessary and avoiding them where they are not necessary.

Guidance on the procedures for carrying out fire safety engineering analyses in buildings is contained in BS 7974:2001 (BSI, 2001). The introduction to this standard describes its purpose as being: 'to provide a framework for developing a rational methodology for the design of buildings using a fire safety engineering approach, based on the application of scientific and engineering principles to the protection of people, property and the environment from fire.' BS 7974:2001 is accompanied by a series of published documents giving detailed guidance on the principles of fire engineering, fire development, spread of smoke, structural response, fire detection, fire service intervention, evacuation and risk assessment.

Fire safety engineering is applied across a wide range of building types. Typical of these is the design of sports stadiums. Modern developments incur considerable investment, and clients are seeking alternative means of attracting revenue on capital outlay. This means that most new sports stadiums can no longer be described as simple steel, concrete and blockwork structures for the sole purpose of watching sport. Instead, they are mixed occupancy, often containing shops, restaurants and conference facilities. This can create difficulties in developing fire safety policies consistent with the objectives of the Building Regulations by the use of government published guidance alone.

Fire safety engineering requires the cooperation of the entire design team if its full potential is to be realised, and it is strongly advised that fire safety be part of the remit of the team from the start of the design process. Structural engineers, fire engineers, architects, clients and representatives of the local authority need to communicate throughout the design process. It is important to understand that the value which the fire engineers can deliver is directly proportional to the input they have in the design. The fire engineer must be given the opportunity to work closely with the architect and engineer to understand the features of the structure, and to be able to communicate changes to detailing that are sometimes required to allow the development of advanced capability in fire.

Fire safety engineering is sometimes considered predominantly in terms of maintaining structural stability. There are usually three steps to be addressed in a full assessment to address this:

- Predicting the likely fire scenario, heating rate and maximum temperature of the atmosphere inside the fire compartment. This involves assessing the fire load (i.e. the quality and type of combustible material), the ventilation available and the thermal characteristics of the compartment linings.
- Predicting the temperature of the structural members. This is a function of the temperatures and duration of the fire, the location and size of the structural element and any protection applied.
- Predicting the stability of the structure. This is a function of the temperature that the element achieves, the load and also any effects due to composite action, restraint and continuity from the remainder of the structure.

A structural fire engineering assessment is most economically effective when it can be shown that the structure, or parts of the structure, have sufficient inherent fire resistance to the temperatures generated in a fire to either reduce the level of fire precautions or to avoid the need to apply any fire protection.

The following examples demonstrate how a fire engineering assessment can deliver client value while maintaining adequate levels of life safety in sports stadiums.

Fire engineering design of Wembley Stadium (Mott MacDonald in conjunction with Aurecon)

Architects: World Stadium Team comprising Foster + Partners and HOK Sport

Engineer: Mott Stadium Consortium (Mott MacDonald, Aurecon and Jacobs)

With a capacity of 90 000, Wembley is the UK's biggest sports stadium and is the home of English football (Figure 8.1). The stands are covered by the world's largest single-span roof structure, supported by a 315 m wide arch, visible from over 20 km away. The stadium was designed to allow unobstructed views of the pitch from every seat, and the stands themselves are steeply raked to bring spectators as close as possible to the action. Thanks to its iconic architecture and intense atmosphere, it is arguably one of the most spectacular sports stadiums in the world.

Figure 8.1 The new Wembley from above

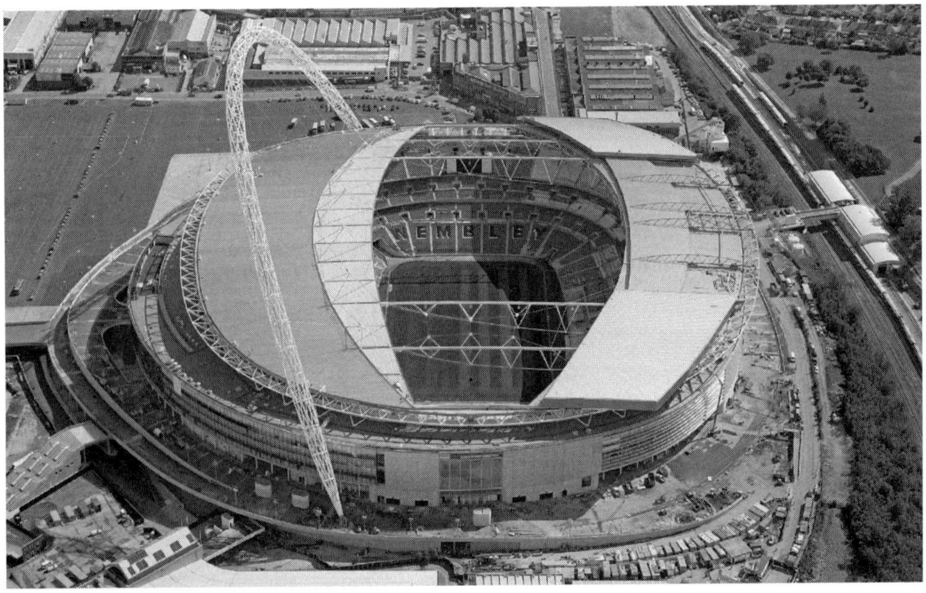

The highest tiers of the stadium are easily accessed by means of escalators and there is an atrium that wraps around the building, capable of accommodating up to 40 000 spectators. This atrium forms part of the main circulation routes and is one of a number of areas where a fire engineered approach was utilised. Due to the unique size and configuration of the space, fire safety was based on first-principles calculations. However, it is more than just a sports stadium. It is also capable of hosting world-class athletic and music events and, in addition, contains a number of banquet halls and associated amenities.

North core atrium

The atrium interconnects all levels, including the lower basement, and has an effective height of 40 m. It serves as part of the main circulation path within the stadium. It is not completely fire or smoke separated from the adjacent congregation area; therefore a means of escape had to be considered both from the atrium and from the associated areas. These escape routes are protected from the effects of heat and smoke by a smoke-exhaust system, sprinklers and controlled fire load.

The strategy used was that, where there are open interconnections to the atrium, these areas are effectively controlled fire load spaces. Fire and smoke are, therefore, not likely to spread into the atrium. Nevertheless, the fire engineering design guarantees that, in the event of a fire, only a minimal smoke layer can form. This is achieved by the use of an exhaust system which ensures that smoke is collected locally. This facilitates occupant egress within the atrium and limits the spread of smoke out of this area.

Zoned smoke control is provided in strategic locations throughout the area. The requirements for this and the size of the systems required to maintain tenable conditions were determined by fire modelling. First-principles calculations were used to determine appropriate fire sizes based either on predicted fuel load or on predicted fire size and sprinkler activation.

Service tunnel fire engineered solution

The geometry of the building is based on a radial grid plan and its elevation is parabolic (Figure 8.2). This made it difficult to obtain the vertical connectivity that is required for fire-fighting shafts, and the basement service tunnel did not line up or connect with the shafts.

The team's innovative solution was to create an environment where the fire tenders would enter the service tunnel and fight fires within the stadium structure itself. To enable this, the service tunnel was divided into six primary zones. Smoke extraction is provided from the incident zone and pressurisation from the two adjacent zones to provide conditions comparable to a fire-fighting lobby.

An analysis of conditions during a fire within the basement was completed, discussed and evaluated with the relevant stakeholders. Of particular importance was the temperature and visibility for fire-fighters. The predicted conditions were under the tenability threshold for fire-fighter life safety as agreed with the London Fire and Emergency Planning Authority (LFEPA). Fire and smoke modelling (computational fluid dynamics) was used to demonstrate that infinite tenability could be obtained for the agreed fire scenarios.

Figure 8.2 The new Wembley at night. (This figure is reproduced in the colour-plate section as Plate X)

Fire protection

The prescriptive guidance of *Approved Document B* is generally governed by building height. If followed strictly, this would have required elements of the structure at the new Wembley Stadium to have 2 hours of fire protection. The alternative approach adopted by Mott MacDonald was to use fire modelling to calculate the temperature and duration of the fires that were likely to occur in the buildings, and then use this information to calculate the temperatures that the steel was likely to reach should a fire actually occur in the building. These temperatures were then compared against failure temperatures using a limiting-temperature approach based on BS 5950-8:2003 (BSI, 2003). This created a situation where it was possible to specify the fire protection required on the basis of the likely risk rather than the global 2 hour requirement of *Approved Document B*. In some areas, this resulted in a situation where it was possible to omit fire protection completely, as the inherent fire resistance of the steel sections was sufficient to meet any risk created by the fire. (A more detailed description of this process is given later in this chapter.) The protection strategy took cognisance of the fact that columns are key elements in the stadium's structure and that any column failure would potentially cause structural collapse.

Special attention was paid in the structural fire protection strategy to the potential for local buckling of beams. This can occur because of the resistance to the thermal expansion of heated beams by the very rigid structure. To take this into account, beam-to-column connection details were verified by the structural engineer for added shear arising from potential local buckling in the lower flanges and part of the web.

Fire engineering design of Arsenal FC's Emirates Stadium (Buro Happold FEDRA)

Architects: HOK Sport Venue Event Architecture

Engineer: Buro Happold

Located at Ashburton Grove in North London, this stadium has a capacity of just over 60 000 (Figure 8.3). The primary use of the stadium is hosting Arsenal FC first team home Association Football games, as well as other domestic and international cup ties. The stadium facilities are also commonly used for non-match-day events that include private conferences, receptions and corporate sponsored events.

The design comprises five storey levels, a single service level (level S1) and four levels of spectator accommodation (levels 0 to 4). The primary access is at level 0 and is provided by a raised podium concourse that runs around the full perimeter of the stadium externally.

There are lower and upper concourses for general admissions located at level 0 and level 3. The concourses consist primarily of circulation space but also accommodate a number of common spectator facilities such as toilets and concession units. Additional levels of viewing accommodation comprising corporate and spectator box viewing facilities are located on levels 1 and 4, which also contain a number of restaurant facilities and bar/lounge areas to the rear of the viewing accommodation.

The service level (S1) is a semi-underground space that circumvents the playing area and comprises different uses that support

Figure 8.3 The new Emirates at dusk

the day-to-day operations of the stadium, including vehicle parking, commercial food production, player changing facilities and media/press facilities.

Fire safety design

The fire strategy developed for the stadium utilised the principles of fire engineering to provide a flexible solution, and enabled the design to be considered under its own merits while ensuring that the functional requirements of the Building Regulations and other relevant fire safety legislation were achieved.

The extent of fire safety measures contained within the fire engineering design enabled specific levels of risk present within the various parts of the stadium to be identified and appropriate fire protection measures to be specified.

Means of escape

The means of escape from within the spectator areas were designed with reliance on the following:

- variations in evacuation time to address different levels of fire risk
- enclosure of key fire risk areas
- limitations of fire load
- automatic fire detection to address specific risks
- fire-fighting operations (including first-aid fire-fighting provision).

A key aim was to utilise primary circulation routes as means of escape. This negated any need to provide secondary dedicated escape routes, on the principle that people are often unwilling to utilise escape routes with which they are unfamiliar and that a significant proportion of occupants will utilise their route of entry for escape in an emergency.

The means-of-escape strategy was based on achieving escape to a place of relative safety (an area free from the products of fire, i.e. heat and smoke) in a time of less than 8 minutes and ensuring the safety of escape routes for periods in excess of this maximum time. Each area was addressed individually to determine the most suitable escape strategy.

The accommodation to the rear of the spectator viewing areas at all levels contains a multitude of uses, including concession units, catering, restaurant, bar and lounge areas, all of which introduce potential fire load into the escape routes serving the spectator viewing areas.

The means-of-escape strategy at levels 0 and 3 involved designating the concourse areas as places of relative safety. The primary fire risk at these levels is represented by the concession units. In order to reduce the risk posed by these units, a package of fire safety measures was proposed, including:

- each concession unit is enclosed in a fire-resisting construction
- the serving counters are protected by controlled automatic fire shutters
- automatic fire detection
- active fire-suppression systems are provided for hot cooking facilities
- all ventilation/extract ductwork systems to the units are fire rated (integrity and insulation).

This permits a maximum 8-minute evacuation time to be employed for these levels, as they could be categorised as low fire risk.

The means of escape strategy at levels 1 and 4 utilises the primary circulation and hospitality spaces to the rear of the spectator viewing areas for means of escape. The spectator viewing area vomitories discharge directly into lounge and dining areas. While recognising that the lounge and dining areas could not be designated as a place of relative safety, due to the contents and future use of the space, the fire strategy quantified that there was no need for the traditional design method of providing fire-separated corridors through these areas from the vomitories.

In order to address the risk posed by spectators escaping through these lounge and dining areas, a series of realistic fire tests was undertaken by Buro Happold FEDRA. The fire test data validated that the likelihood of fire spread within a typical restaurant space, in the event of an ignition source being introduced, was unlikely, and consequently the risk to escaping occupants was minimal. In order to provide sufficient degree of robustness to the design, the fire strategy adopts a 2½-minute evacuation time at these levels.

Fire spread

Level S1

Level S1 forms a basement that circumvents the playing area. This level is separated from the spectator accommodation by a 60-minute fire resisting compartmentation.

The vehicle parking facilities are designed to accommodate cars, coaches and large outside broadcast unit vehicles. This meant that utilising the recommendations for basement car parks contained in contemporary fire safety guidance, such as *Approved Document B*, to the whole of this area was inappropriate (Figure 8.4).

While the fire risk of small vehicles such as cars in car parks is well defined, and it is recognised through fire research and testing that they generally represent a controlled fire load, the same cannot be said for larger commercial type vehicles, such as coaches and articulated lorries. Consequently, the fire safety design addressed these issues by adopting a different set of fire-protection methods for the different areas of risk represented at this level. The commercial vehicle parking areas, coach and articulated outside broadcast vehicles are provided with automatic sprinkler protection. A hybrid smoke clearance system is also provided by use of natural ventilators open to external air located in areas around the perimeter of level S1. The natural ventilation system is supplemented by a number of strategically located ceiling-mounted impulse fans. The impulse fans are primarily provided for environmental purposes to reduce the risk of stagnant air pockets forming in the oval shaped enclosure of level S1. However, the presence of these fans can also be taken advantage of in fire conditions, to assist in the removal and clearance of smoke and hot gases. The use and operational requirements of this hybrid system were tested and validated using computational fluid dynamics.

This design methodology and the analysis undertaken facilitated commercial vehicle parking areas to be open to the main volume of the basement area, and enabled automatic sprinkler protection to be applied only where necessary.

Upper tiers

The upper tiers of spectator accommodation are formed from precast concrete units supported by steel raking members. These structural members, in accordance with contemporary fire safety guidance, are required to exhibit a 90-minute fire resistance period.

It was recognised that the steel structural members could only be heated due to a fire at the lower levels. Consequently, advantage was taken of structural fire engineering assessments to validate that the application of applied fire protection to the steel raking members was not necessary in order to achieve the required degree of fire resistance. This design methodology takes advantage of the inherent fire-resisting properties of the actual steel raking members.

A structural fire-engineering assessment was conducted on a number of the upper tier raking members. The assessment calculated the design temperatures (the temperature likely to occur in a realistic fire scenario) of steel resulting from a fire at a lower level. The resultant temperatures were then compared

Figure 8.4 The new Emirates from the East Stand

with the limiting temperature (failure temperature) of the steel raking members. More details on this process are available in BS 5950-8:2003.

The assessments undertaken showed that the design temperature was always lower than the limiting temperature. This analysis validated that the upper tier raking members did not require applied fire protection. A number of safety factors were included within the design, such as:

- The raking members were sized to resist the dynamic loading at ambient temperature. This requirement is not necessary in a fire condition, and consequently the steel has reserves of strength in the fire limit state.
- The analysis considered the upper tiers in a fully loaded condition. The majority of the imposed load on the upper tier rakers is from the spectators. In the event of a serious fire, the occupants will have been evacuated, and consequently under such circumstances the upper tier rakers are likely to support little or no imposed load.
- The limiting temperature is derived from test data utilising a BS 476-20:1987 (BSI, 1987) standard fire curve. It is unlikely that in a fire the external steel elements can achieve similar temperatures to that of the internal steel elements.

The worst-case fire for a building is usually a short-lived fire with the maximum atmospheric temperature occurring shortly after the fire starts. In this case, the steel temperature lags behind the atmospheric temperature over this period, and a 'steady-state' temperature is not reached. The assessment assumed no temperature lag. This is conservative, especially for the upper tier rakers, where the section sizes are relatively large and will exhibit temperature lag.

REFERENCES

BSI (British Standards Institution) (1987) BS 476-20:1987. Fire tests on building materials and structures. Method for determination of the fire resistance of elements of construction (general principles). BSI, London, UK.

BSI (2001) BS 7974:2001. Application of fire safety engineering principles to the design of buildings. Code of practice. BSI, London, UK.

BSI (2003) BS 5950-8: 2003. Structural use of steelwork in buildings. Code of practice for fire resistant design. BSI, London, UK.

Department of Communities and Local Government (2006) *Approved Document B (Fire Safety)*. See http://www.planningportal.gov.uk/buildingregulations/approveddocuments/partb (accessed 12/02/2015).

Stadium and Arena Design
ISBN 978-0-7277-5790-6

http://dx.doi.org/10.1680/saad.57906.095

ice
Institution of Civil Engineers
publishing

Chapter 9
Stadium signage

John Pascoe Electrocomponents
Mark Steinkamp Daktronics

Stadium signage began with hand movements, calligraphy and bill posters but has, over recent decades, become an advanced technology of crucial importance to stadium planners, designers and operators. Not so long ago, big outdoor displays relied on tube technology and quick-connect fittings to simplify removal and replacement of components over limited array/module lifespans. The invention of light-emitting diodes (LEDs) changed all that and opened up exciting new possibilities, illustrated by the examples in this chapter. Displays have become much bigger and much better. Recent innovations include the incorporation of Internet Protocol Television (IPTV) to drive high-definition video and all kinds of other content to displays of multiple technologies throughout facilities and catchment areas.

First strikes

Once stadium signage related principally to direction (entrance, exit), safety (mind the step, keep off the grass) and bodily functions (refreshments, toilets). Appreciation of the marketing opportunities provided by a captive audience of like-minded sports fans brought about rapid improvements in the quality and diversity of stadium signage. Up until the early years of the 20th century, such signage was created by sign writers or applied by billboard posters. The results could be pretty impressive as, for example, at Exposition Park, Pittsburgh (Figure 9.1), photographed during the Pittsburgh versus New York baseball match on Saturday 5 August 1905.

It can be argued that George A. Baird of Chicago initiated the evolution of the high-tech stadium when he invented baseball's first electronic scoreboard in 1908. This seems primitive now in that – while electric lights illuminated the balls, strikes and outs – the line score still had to be changed manually. But Baird's invention centralised and revolutionised the scoring function because the new scoreboard was attached by wire to a keyboard behind the umpire. This meant the scoreboard operator could look out from the scoreboard (Figure 9.2) and no longer had to work from the outfield fence, using binoculars to try to interpret the umpire's decisions. Figure 9.2 shows the scoreboard at the end of the first game of the 1946 season at Briggs Stadium, Detroit (53 000 capacity), which was a 1937–1938 reconstruction of Navin Field (built 1911–1912), which was itself a reconstruction of Bennett Park.

Health and safety signage

In the UK, the Green Guide (SGSA, 1997) stated that clear, comprehensive signposting should be provided in order to show the routes to different parts of a sports ground (Figure 9.3). It stressed the importance of the design, layout and signposting of entrances, exits, entry routes and exit routes to ensure these can cope safely with the number of spectators using them. Attention was drawn to the need for signs, and especially fire or other safety signs, to be sited so that they could be easily seen and readily distinguished by people with impaired vision or impaired colour perception. Directional signs are provided to encourage crowds in any particular section of the ground to move in one direction when leaving the ground. Wherever practicable, such signage provides information on the destination of the exit route (e.g. northside car park, high street) so as to instil confidence in the people using them, especially if they are evacuating the premises.

In the 1980s, the International Organization for Standardization (ISO) tested symbols for signage with a multilingual audience, and developed the symbol of a running person incorporating a door, indicating both the direction of egress and which way the door will open at the point of exit. This symbol was the first to be based on comprehensive communications research.

While compliance with British standards is not a legal requirement, following the principles and practices of British standards goes a long way towards meeting the requirements of the UK's Health and Safety (Safety Signs and Signals) Regulations 1996 No. 341. This legislation was formed from and satisfies the

Figure 9.1 Pittsburgh versus New York, Saturday, 5 August 1905. (© R W Johnston)

European Community Safety Signs Directive EC Directive EEC/92/58. The UK regulations offer design criteria, which can be adapted within sensible limits, and the EC directive offers guidelines as opposed to a code.

The Health and Safety (Safety Signs and Signals) Regulations were introduced to create standardisation of signs so that a given symbol will instantly convey a given message using a combination of geometrical shape, colour and pictorial symbol. These regulations did not specify the circumstances in which signs were to be displayed. Decisions of this kind were the responsibility of those in control of the premises – normally the club or stadium operator working in conjunction with professional advisors and the certifying authority. Neither the safety sign regulations nor British standards specified requirements for the cleaning and maintenance of safety signs. But any sign must be clean and undamaged in order to convey its message effectively. Activities such as cleaning, maintenance and replacement should, therefore, be carried out on a regular basis. Care should also be taken that directional and safety signs are not obscured by advertising during a stadium event.

A new British standard, BS 5499-4:2000 (BSI, 2000), for safety signs was published in September 2000. It includes fire safety signs and a code of practice for escape-route signing. The maximum viewing distances given in the code were devised using a Japanese visual acuity study of normal-sighted people based on Landolt rings. All signs conforming to BS 5499-1:2002 (BSI, 2002) and BS 5499-4:2000, used in conjunction with a formal risk assessment, satisfy all requirements under building, fire precaution and health and safety at work regulations and legislation.

Stadium signage is now so important that it should be considered at the stadium concept design stage. When this happens, the results can be dramatic. At the concept stage of the City of Manchester Stadium, host of the 2002 Commonwealth Games, three particular principles were established. The first was for 'a masted roof structure integrated with spiral circulation towers

Figure 9.2 Briggs Stadium, Detroit, 28 July 1946. Scoreboard operator can be seen below '0', adjacent to '2'. (© Detroit News (Styrlander))

Figure 9.3 Hampden Park, Glasgow. (© Corus (John Clarke))

Figure 9.4 Etihad Stadium, Manchester (also known as City of Manchester Stadium). (© Arup)

providing signal, orientation and safe, stepless access'. In the completed stadium (Figure 9.4) the structural masts, of maximum height 64 m (210 ft), elegantly signal the stadium location and, by rising through the main circular access ramps, orientate spectators for safe access to and egress from the seating areas. This powerful safety communication by the stadium structure was achieved at no extra cost because the designers of the stadium thought about signage at the outset of the project.

Messaging and marketing

Signage within and around stadiums is not only necessary to meet the health and safety criteria but also has immense brand-building powers. Architectural sign systems for stadium developments have been available for many years. In recent years, opportunities have arisen to coordinate such systems with a vast new range of perimeter, concourse and inner bowl signage in the forms of channel-cut signs, LED displays, rotational systems, scoreboard panels, wall murals, billboards, bench signage and penalty box signage.

Examples of applications in the 'virtual entertainment' sector of the stadium signage industry include a wide diversity of electronic scoreboards and video displays designed for sports grounds and stadiums throughout the world.

Chase Field, Phoenix, AZ, USA

Chase Field in downtown Phoenix is the home of the Arizona Diamondbacks. The stadium opened in 1998 and was the first in the USA to be built with a retractable roof. The original scoreboard has since been replaced with a multi-million dollar state-of-the-art high-definition scoreboard (Figure 9.5) measuring 14 m high × 41 m wide (46 ft × 136 ft). In addition to showing match replays and animations, the scoreboard displays crowd images beamed onto it between innings by high-definition cameras located around the stadium. Fans can, for a credit card donation to the Arizona Diamonds Foundation, purchase scoreboard messages online. In such cases the Diamondbacks email an image of the message to the purchaser 5–10 weeks after its display.

Lucas Oil Stadium, Indianapolis, IN, USA

This multipurpose sports stadium (Figure 9.6) opened in 2008 as the new home of the National Football League's (NFL's) Indianapolis Colts, enabling the Colts' previous home – the RCA Dome – to be demolished to make way for an extension to the Indiana Convention Center. The naming rights were bought by Lucas Oil for $121 million over 20 years. The new stadium features a retractable roof and window wall, making it possible for the Colts to play both indoors and outdoors on their FieldTurf surface. Other features include two high-definition

Figure 9.5 Chase Field, Phoenix. (© barnhill88) (This figure is reproduced in the colour-plate section as Plate XI)

Figure 9.6 Lucas Oil Stadium, Indianapolis. (© Matt Kryger/ Indianapolis Star)

scoreboards each 16 m high × 30 m wide (53 ft × 97 ft), located in the northwest and southeast corners of the stadium. The stadium capacity (63 000 seats) can be expanded to accommodate 70 000 spectators for big events such as the Super Bowl. An estimated 111 million Americans watched the Super Bowl in 2011, which makes for a lot of eyes on the scoreboards.

Faurot Field, Columbia, OH, USA

Much stadium signage and/or communication enhancement is carried out within stadium refurbishment works. The horseshoe-shaped, 68 349 seat Memorial Stadium at Faurot Field, Columbia – home of the University of Missouri Tigers football team – was built in the 1920s. In 2009, the university installed a new $5 million 9.1 m high × 24.4 m wide (30 ft × 80 ft) north end-zone scoreboard (Figure 9.7) with updated video capabilities and a new sound system to replace the existing 10-year-old equipment. The 'Mizzou' fans were delighted with a HARMAN JBL VLA end-zone loudspeaker solution wrapped around the Tigers' new scoreboard, designed by Wrightson, Johnson, Haddon, & Williams of Dallas, TX. The JBL VLA speakers' 60° horizontal reach allowed the audio engineers to attain very uniform sound throughout the sunken stadium. This speaker line is the first designed to look and sound like a line array, but actually allow for complete side fill with only one rig. The scoreboard was fitted with 28 VLA601H-WRC loudspeakers in four arrays, augmented by three PD743-WRC loudspeakers. Through an existing fibre-optic infrastructure, the system was connected using HARMAN's HiQnet System Architect and BSS London processing. (Harman HiQnet is the world's first connectivity and control protocol that integrates all product categories in the signal chain for professional audio systems of all types, size and application. Harman HiQnet eliminates the need for the user to manage multiple disparate operating systems or be responsible for programming individual signal processors, speaker controllers, wireless microphone systems and mixing consoles.)

Commonwealth Stadium, Lexington, KY, USA

The University of Kentucky has added an integrated, high-definition football video and sound system (Figure 9.8) at the 67 942 seat Commonwealth Stadium, home of the Wildcats. The multi-million dollar system made its debut at the Wildcats' 2011 home opener on 10 September against Central Michigan. The most visible components of the system are the two high-definition HD-X video displays installed behind each end zone. Measuring approximately 11.3 m high × 24.4 m wide (37 ft × 80 ft), each display provides live and recorded video in high definition, with picture-in-picture capability and multiple zones to show scores, statistics and sponsor information. In terms of

Figure 9.7 Faurot Field, Columbia. (© Jared White)

Figure 9.8 Commonwealth Stadium, Lexington. (© wsonradio)

consumer TV sets, it would take more than 550 42-in. flat-panel screens to fill the space of each new end-zone LED video board. In addition to the end-zone displays, fans enjoy additional full-colour content, including additional statistics, scoring and timing information, crowd-pleasing animation and other motion graphics on 14 digital ribbon displays located around the seating bowl. The digital ribbon boards surround the interior of the seating bowl with more than 549 m (1800 ft) of digital technology. The longest two of these displays measures more than 148 m (485 ft) and spans the side lines of the stadium. A custom sound system was also provided to enhance the Wildcats' game days. The Sportsound system delivers pounding bass energy, smooth mid- and high-range frequencies, and high-impact entertainment. Components include a main speaker cluster behind the end zone, under-balcony speakers and concourse speakers, so that each fan can hear and enjoy all the sounds of the game. The video and audio equipment is operated with a high-definition front end control system.

Sports Authority Field at Mile High, Denver, CO, USA

In their book *Stadium Engineering*, Culley and Pascoe (2005) described how a contract for an LED screen at Sports Authority Field (then Invesco Field) was won by preparing a bid based on an alternative steelwork design, substituting hollow structural sections (HSSs) for the heavier wide-flange members in the original design. The higher strength-to-weight ratio of the HSSs enabled the same size scoreboard to be designed with 50% less weight, a saving of 150 tons. Much has changed at the home of the Denver Broncos since then. The latest video displays to be installed are 12.2 m high × 67 m wide (40 ft × 220 ft), with 952 lines of resolution, and feature a high-definition pixel layout, multiple levels of protection from the elements and a very robust cabinet design with an improved ventilation system. The new displays can show one large image, to highlight live video and instant replays, or can be divided into separate windows to show a variety of graphics, animations, statistics, scoring information and sponsor advertisements.

Bucky Bronco, on top of the south scoreboard (Figure 9.9), was cast in 1975 from the model that Roy Rogers used to make a statue of his horse Trigger. Rogers' sole condition for lending the model was that the horse should have its own name. Made of fibreglass and steel, Bucky stands 8.2 m (27 ft) tall and weighs 726 kg (1600 lb). Stadium scoreboards are subjected to high wind loads, and those at Mile High were designed to withstand

Figure 9.9 Sports Authority Field at Mile High, Denver. (© Daktronics) (This figure is reproduced in the colour-plate section as Plate XII)

Figure 9.10 Marlins Park, Miami. (© Roberto Coquis)

wind forces up to 152 km/h (95 mph). The slenderness of the boards and the high strength and rigidity of the HSSs optimise the wind-engineering solution.

Marlins Park, Miami, FL, USA

Figure 9.10 shows the main scoreboard in the stadium with, in the foreground, the first pitch of the Major League game between the Marlins and the St Louis Cardinals on 4 April 2012. Figure 9.11 shows flat panel video displays in the restaurant area communicating a very different type of content. Hunt/Moss, a joint venture, was responsible for introducing systems that incorporate the stadium's 30 m (99 ft) wide, high-definition video display, more than 274 m (900 ft) of digital ribbon displays, and two additional displays positioned in the left field wall. Daktronics Show Control System drives the entire system from a single control room, pushing content to all the displays around the field and in the IPTV system.

Figure 9.11 Marlins Park restaurant area. (© Daktronics)

Figure 9.12 The Big Screen, XX Commonwealth Games, Glasgow 2014. (This figure is reproduced in the colour-plate section as Plate XIII)

The Big Screen, XX Commonwealth Games, Glasgow 2014, UK

Signage is at the heart of all modern sports and stadium developments. The XX Commonwealth Games held in Glasgow, Scotland, from 23 July to 3 August 2014, featured Europe's largest-ever LED screen, known as the 'Big Screen'. The 11 m high × 100 m wide (36 ft × 328 ft) screen (Figure 9.12), weighing 38 tonnes, ran the entire length of Celtic Park Stadium's south stand for the Games' opening ceremony on 23 July. Sports Technology was responsible for the supply, installation and operational support of video display screens and public address and audio systems linking across the Games' competition venues.

Kyle Field, Texas A&M University, College Station, TX, USA

Texas A&M aspired to redevelop Kyle Field into the finest collegiate football stadium in the USA. The redevelopment (to 102 500 capacity) in 2014 included the design, manufacture and installation of a 14.3 m high × 50 m wide (47 ft × 163 ft) LED video display (Figure 9.13). This is the largest in USA college football and the first to feature 1080 resolution, with a 13HD pixel layout conferring superb clarity and wide-angle viewing. As at Glasgow, teams in the USA are installing and connecting billboards, marquees and the 'street to the seat' experience. Texas A&M is installing new street-furniture displays outside its stadium in 2015. These displays provide information to the crowd walking up to the game, specials and general advertising.

EverBank Field, Jacksonville, FL, USA

Two of the world's largest LED screens were installed in summer 2014 at EverBank Field (Figure 9.14), home of the Jacksonville Jaguars. Each display measures 18.3 m high × 110 m wide (60 ft × 362 ft), giving a total of 2018 m^2 (21 720 ft^2) per display. Combined, these displays consist of 35.5 million LEDs used to light up the stadium with live video,

Figure 9.13 Kyle Field, Texas A&M University, College Station. (© Daktronics)

Figure 9.14 EverBank Field, Jacksonville. (© Cyd Hoskinson/WJCT)

Table 9.1 Appropriate pixel pitch for display application (based on traffic speed and desired display content)

Speed	Pixel pitch					
	7.6 mm (0.3 in.)	12 mm (0.47 in.)	20 mm (0.78 in.)	34 mm (1.33 in.)	46 mm (1.8 in.)	89 mm (3.5 in.)
5 mph 8 km/h	T&G	TO	TO			
15 mph 24 km/h	T&G	T&G	TO	TO		
25 mph 40 km/h		T&G	T&G	TO	TO	
35 mph 56 km/h		T&G	T&G	T&G	TO	
45 mph 72 km/h			T&G	T&G	T&G	TO
55 mph 89 km/h				T&G	T&G	TO
65 mph 105 km/h					T&G	T&G
75 mph 121 km/h						T&G

T&G, text and graphics recommendation; TO, text only recommendation
© Daktronics – Brookings – SD – USA

instant replays, animations and graphics, statistics, sponsor messages and other game information. Each display features a 13HD pixel layout, providing excellent image clarity and wide-angle visibility.

Distance and speed reading

In this chapter we have written about signage for a captive audience in a stadium seating bowl (i.e. the movement is by the message and not by the viewer). For a moving audience – people approaching or passing a stadium or sports facility – there are additional criteria relating to signage to consider. To understand a message, a viewer must have sufficient time to read the message content. This necessitates consideration of maximum and minimum viewing distances, mode of travel of the viewer (on foot or bicycle, or in a car) and the speed at which the viewer is travelling. Tables 9.1 to 9.3 are used by clients for signage products and, in particular, for those specifying Galaxy text and graphics displays.

Looking forward

The industry is starting to reach a balance between the brightness and impact of displays and premium resolution and legibility. Surface mount technology is now bright enough for outdoor applications and, specifically in terms of this chapter, for stadiums and arenas. Surface mount device (SMD) LEDs deliver excellent legibility of text both at distance and in close-up viewing (e.g. concourse areas and shaded auxiliary displays) (Figure 9.15). Traditional through-hole technology still provides brightness appropriate for displays exposed to sunlight and offers great visual impact (see Figure 9.14).

In general, along with bigger display size, the industry is looking at higher resolution for maximum impact of digital real estate (Table 9.4). Higher resolution line-spacing options now include 10 mm and 8 mm.

Table 9.2 Closest approximate distance when light emitted from pixels begins to blend into a continuous picture

Pixel pitch	Minimum viewing distance
12 mm (0.47 in.)	6 m (20 ft)
20 mm (0.78 in.)	14 m (45 ft)
34 mm (1.33 in.)	23 m (75 ft)
46 mm (1.8 in.)	32 m (105 ft)
89 mm (3.5 in.)	61 m (200 ft)

© Daktronics – Brookings – SD – USA

Table 9.3 Text viewing ranges (approximate)

Maximum viewing time (seconds)								Character size:		Maximum viewing distance:	
5 mph 8 km/h	15 mph 24 km/h	25 mph 40 km/h	35 mph 56 km/h	45 mph 72 km/h	55 mph 89 km/h	65 mph 105 km/h	75 mph 121 km/h	in.	mm	ft	m
13.7	4.6	2.7	1.9	1.5	1.2	1.1	0.9	2	51	100	30
41.1	**13.7**	**8.2**	5.8	4.6	3.7	3.2	2.7	6	152	300	91
61.6	**20.5**	**12.3**	**8.8**	6.8	5.6	4.7	4.1	9	229	450	137
89.0	29.7	**17.8**	**12.7**	**9.9**	8.1	6.8	5.9	13	330	650	198
123.3	41.1	24.6	**17.5**	**13.7**	**11.2**	9.5	8.2	18	457	900	274
164.4	54.8	32.8	23.4	**18.2**	**14.9**	**12.6**	**10.9**	24	610	1200	366
246.6	82.2	49.2	35.1	27.3	**22.3**	**18.9**	**16.4**	36	914	1800	549
328.8	109.6	65.6	46.8	36.4	29.8	**25.2**	**21.8**	48	1219	2400	732

Values in bold are acceptable exposure times
© Daktronics – Brookings – SD – USA

Figure 9.15 Camp Randall Stadium, University of Wisconsin, Madison. (© Daktronics)

Table 9.4 Display viewing distances

Model	Actual pixel pitch:	Minimum viewing distance:		Smooth image viewing distance:	
	mm	ft	m	ft	m
LED video displays – indoor					
DVN-4	4.66	10.51	3.2	52.56	16.02
DVN-6	6.60	14.89	4.54	74.44	22.69
DVN-10	9.90	22.33	6.81	111.66	34.03
DVN-15	15.85	35.75	10.9	178.77	54.49
DVN-20	19.81	44.69	13.62	223.4	68.09
LED video displays – outdoor					
DVX-8 SMD	8.14	18.36	5.60	82.63	25.18
DVX-10 SMD	10.16	20.63	6.19	103.13	31.44
DVX-13 HD	13.06	29.46	8.98	147.73	44.9
DVX-15 HD	15.24	34.38	10.48	171.89	52.39
DVX-15 SMD	15.24	30.94	9.43	154.7	47.15
DVX-16 MT	16.63	37.51	11.43	187.56	57.17
DVX-20 MT	20.32	45.84	13.97	229.18	69.86

Pixel pitch	Indoor					Outdoor						
	4 mm	6 mm	10 mm	15 mm	20 mm	8 SMD	10 SMD	13 HD	15 HD	15 SMD	16 mm	20 mm
True pixel pitch	4.66 mm	6.60 mm	9.90 mm	15.85	19.81	8.14	10.16	13.06 mm	15.24 mm	15.24 mm	16.63 mm	20.32 mm
Minimum viewing distance	10.51 ft 3.2 m	14.89 ft 4.54 m	22.33 ft 6.81 m	35.75 ft 10.9 m	44.69 ft 13.62 m	18.36 ft 5.60 m	20.63 ft 6.19 m	29.46 ft 8.98 m	34.38 ft 10.48 m	30.94 ft 9.43 m	37.51 ft 11.43 m	45.84 ft 13.97 m
Smooth image distance	52.56 ft 16.02 m	74.44 ft 22.69 m	111.66 ft 34.03 m	178.77 ft 54.49 m	223.4 ft 68.09 m	82.83 ft 25.18 m	103.13 ft 31.44 m	147.3 ft 44.9 m	171.89 ft 52.39 m	154.7 ft 47.15 m	187.56 ft 57.17 m	229.18 ft 69.86 m

Minimum viewing distance: the minimum viewing distance can be defined as the closest distance at which the colours from an individual pixel begins to blend with the surrounding pixels. The bottom table lists the minimum viewing distances for some common sizes
Smooth image distance: the point at which the video display is at its purest distance in relation to where the viewer is standing
© Daktronics – Brookings – SD – USA

Acknowledgements

The authors are indebted to Julia Eberhart and Justin Ochsner of Daktronics for their help with content and review of this chapter.

REFERENCES

Culley P and Pascoe J (eds) (2005) *Stadium Engineering*. Thomas Telford, London, UK.

BSI (2000) BS 5499-4:2000. Safety signs, including fire safety signs. Code of practice for escape route signing. BSI, London, UK.

BSI (2002) BS 5499-1:2002. Graphical symbols and signs. Safety signs, including fire safety signs. Specification for geometric shapes, colours and layout Specification for fire safety signs. BSI, London, UK.

SGSA (Sports Grounds Safety Authority) (1997) *Guide to Safety at Sports Grounds*, 4th edn. See http://www.safetyatsportsgrounds.org.uk/publications/green-guide (accessed 12/02/2015).

FURTHER READING

Benson M (1989) *Ballparks of North America*. McFarland, Jefferson, NC, USA.

BSI (British Standards Institution) (1980) BS 5378-1:1980. Safety signs and colours. Specification for colour and design. BSI, London, UK.

BSI (1986) BS 5499-2:1986. Fire safety signs, notices and graphic symbols. Specification for self-luminous fire safety signs. BSI, London, UK.

BSI (1990) BS 5499-3:1990. Fire safety signs, notices and graphic symbols. Specification for internally-illuminated fire safety signs. BSI, London UK.

Daktronics (2014) *Daktronics Hits Unprecedented Number of Professional Football Installs*. See http://www.daktronics.com/news/Pages/Daktronics-Hits-Unprecedented-Number-of-Professional-Football-Installs.aspx (accessed 12/02/2015).

Daktronics (2014) Jacksonville Jaguars to Unveil World's Largest HD Video Displays in Sports from Daktronics on July 26. See httpwww.daktronics.com/news/Pages/Jacksonville-Jaguars-to-Unveil-Worlds-Largest-HD-Video-Displays-from-Daktronics-on-July-26.aspx (accessed 12/02/2015).

Daktronics (2015) See http://www.daktronics.com (accessed 12/02/2015).

Daktronics (2015) *Daktronics Selected to Provide New LED Technology for Sports Authority Field at Mile High and Denver Broncos*. See http://www.daktronics.com/Company/NewsReleases/Pages/DaktronicsSelectedbyDenverBroncos.aspx (accessed 12/02/2015).

European Council (1992) EC Directive EEC/92/58. European Safety Signs Directive. European Community, Brussels, Belgium.

Fire Safety Advice Centre (2015) *Health and Safety (Safety Signs and Signals Regulations) 1996*. See http://www.firesafe.org.uk/health-and-safety-safety-signs-and-signals-regulations-1996 (accessed 12/02/2015).

HSE (Health and Safety Executive) (1996) *The Health and Safety Regulations (Safety Signs and Signals)*. Statutory Instrument No. 341. HSE Books, London, UK.

HSE (1996) *Safety Signs and Signals. The Health and Safety Regulations (Safety Signs and Signals). Guidance on Regulations*. HSE Books, London, UK.

HSE (1999) *The Event Safety Guide. A Guide to Health, Safety and Welfare at Music and Similar Events*, 2nd edn. HSE Books, London, UK.

ISO (International Organization for Standardization) (1987) ISO 6309. Fire protection – safety signs. ISO, Geneva, Switzerland.

ISO (1994) ISO 9002. Quality management standard. ISO, Geneva, Switzerland.

McGee K (2011) *Daktronics Refurbishes Stadium with HARMAN JBL System*. See http://livedesignonline.com/blog/daktronics-refurbishes-stadium-harman-jbl-system (accessed 12/02/2015).

Peckham G (2004) New developments in US and international safety sign technology. *ASSE Professional Development Conference and Exposition, June 7–10 2004, Las Vegas, Nevada*. American Society of Safety Engineers, Park Ridge, IL, USA.

Sports Technology (2015) See http://www.sportstech.uk.com/index.html (accessed 12/02/2015).

Stadium and Arena Design
ISBN 978-0-7277-5790-6

http://dx.doi.org/10.1680/saad.57906.109

Chapter 10
Stadium acoustics

Jim Griffiths

Acoustic disciplines and their effects

The acoustic, noise and sound requirements of a stadium are interrelated and should not be considered in isolation. The successful integration of the behaviour of sound in acoustic environments is achieved through:

- providing effective sound reinforcement throughout the building
- allowing each activity in a given space to operate without interference
- providing the appropriate natural acoustics (atmosphere) to enhance the experience of the event
- minimising the risk of environmental noise impact to local communities.

The main acoustic disciplines that need to be considered and their effects on the environment and sound quality are shown in Figure 10.1.

The relative importance of these issues with regard to other engineering factors is a function of the client brief. It is critical to establish the proposed operational uses of the stadium, which are linked to the business plan, not only on opening but also in the future. This will enable priorities to be ranked, which will be used model the final design.

The design and impact associated with these disciplines need to be coordinated with the design team and, in particular, the architect, structural engineer, mechanical and electrical engineer, and the quantity surveyor. The interface of the acoustic consultant with other engineering and architectural disciplines is critical, as acoustic engineering can have a significant effect on the overall design, and hence the cost, of the project. These primary acoustic issues are discussed further below.

Bowl acoustics

The natural acoustics of the bowl have a significant effect on three main issues related to the stadium. These are:

- the ambience/atmosphere to enhance spectator enjoyment of sporting events
- assistance in providing the necessary speech intelligibility of the stadium sound system
- the effect on sound quality during music events.

The bowl acoustics are governed primarily by the size, shape and building fabric of the stadium, and can be variable in nature when considering that modern-day facilities may incorporate such features as closing roofs or movable video screens which can alter the acoustic environment (Table 10.1).

As in most fields, there is a conflict in acoustic design requirements between spectator enjoyment for sporting events and speech intelligibility of the sound system and music quality for concerts. In general, for sporting events the acoustics need to be as reverberant ('live') as possible to provide atmosphere. This leads the design to many hard, reflecting surfaces and a roof shape that provides constructive reflections from opposite stands. Conversely, for good intelligibility of speech over the sound systems, the acoustic environment needs to tend towards anechoic (as 'dead' as possible) such that there is a high ratio of the direct sound to the reverberant sound. This would move the design towards acoustically absorptive surfaces. This issue is further complicated if the stadium is to be used for music concerts. In this scenario, where large touring sound systems are used, important long-delayed reflections need to be reduced, as these unwanted echoes can be a significant source of annoyance to patrons. There are examples of many hundreds of people leaving or requiring re-seating due to this acoustic defect.

Clearly there is a balance between the dead, live and long-delayed reflection requirements. Acoustic modelling is required so that the design measures can be investigated such that the acoustics can be optimised for most stadium activities (Figure 10.2).

The sound behaviour in a stadium is dictated by complex wave theory which, coupled with a complex stadium space, cannot be

Figure 10.1 Acoustic disciplines and their effects

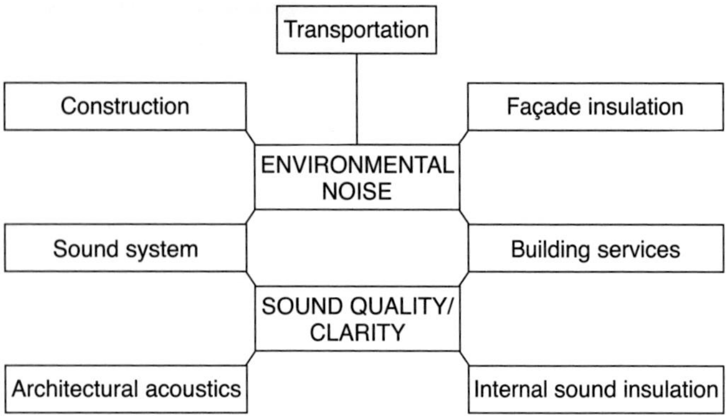

readily modelled directly by wave theory. Most models use ray-tracing techniques, which approximate wave behaviour by the use of multiple ray reflections, acting on the principle of the angle of sound incidence being equal to the angle of sound reflection. Other factors that are taken into account during modelling are the sound source characteristics, attenuation due to distance, air absorption, interactions (reflections) with physical boundaries and absorptive surfaces, diffraction edges and transmission through structures. An example of a ray-tracing 3D image and associated echogram is illustrated in Figure 10.3. The paths of various reflections are shown in a stadium from the roof, side walls, etc., and acoustic defects can be investigated by inspection of the echogram (Figure 10.4), such that late reflections with high sound energy can be related to the surface in the 3D image.

In practice, generic treatments have generally been found to be effective, although it must be stressed that only a detailed modelling study will identify the problems, and hence the most effective treatments, for each specific case and operational use.

Figure 10.2 Raynoise – geometrical acoustics modelling

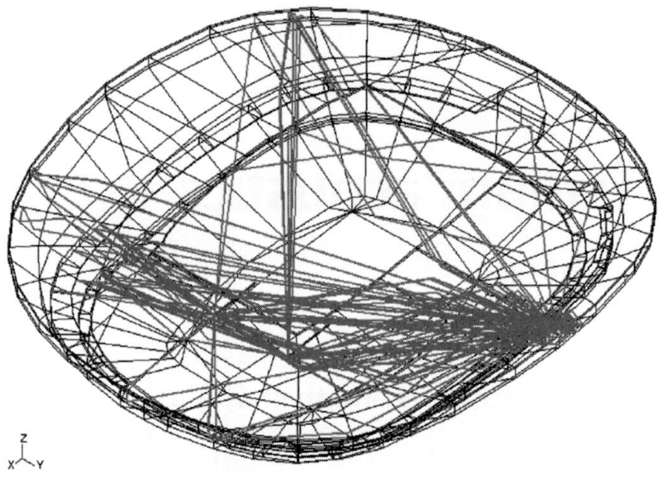

The atmosphere within a stadium is becoming an important feature of the stadium bowl and is part of the 'fan experience', which is at the heart of a sporting venue. The atmosphere is not just related to acoustics and sound but is a combination of many variables, starting with the teams. These factors are shown in the atmosphere wheel in Figure 10.5.

Façade sound insulation

To determine the acoustic requirements of the bowl sound insulation it is critical to establish two primary variables: the source noise levels and the acceptable receptor noise levels. Both are variable by virtue of the potential range of proposed activities

Table 10.1 Generic acoustic treatment for the bowl

Area	Acoustic issues	Potential treatment
Roof	Can provide good reflection for sporting events to retain atmosphere but detrimental to PA systems	Provide some acoustic absorption to rear areas of the roof. This will assist speech intelligibility of the stadium PA system and reduce unwanted reflections by way of the rear-wall roof geometry. Will retain some reflections for atmosphere
Balcony fronts	Vertical surfaces can produce long-delayed reflections during concerts	Treat with acoustic absorption or diffusers. Leave open with hand rails. When advertising is required, screen-printed cloth can be provided with acoustic absorption infill
Vertical surfaces	Vertical risers and rear walls can produce long-delayed reflections during concerts	Treat with acoustic absorption appropriate to the surface. Create panel absorbers and 'bass traps' as part of the design
Executive box fronts	Glazing on boxes and function suites can produce long-delayed reflections	Angle glazing 5° left, 5° right, 5° up and 5° down, and repeat to provide a more diffuse sound field and reduce focusing

Figure 10.3 3D model of a stadium with ray-tracing and associated echogram

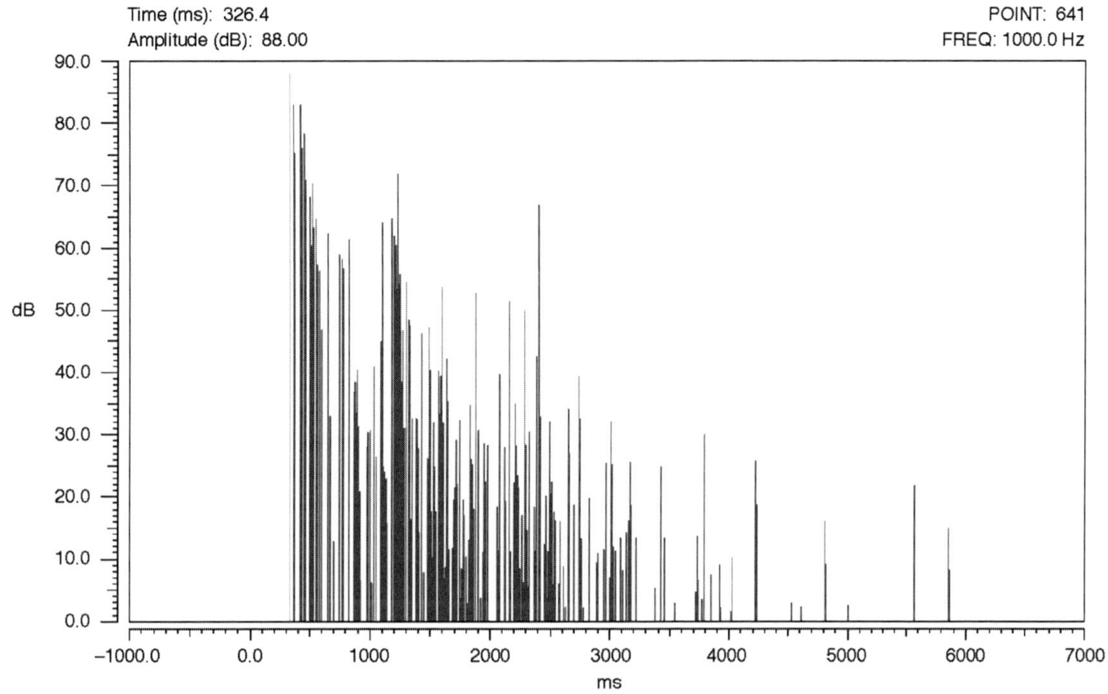

in the stadium, which in turn may affect the noise-rating level applicable to community receptors.

Source noise

A wide range of events is likely to be considered to be held at modern-day stadiums in order to create a commercially viable multi-use sports and entertainment facility. Such events may well include a diverse range of activities, such as football, rugby, American football, hockey, show jumping, athletics, boxing, greyhound racing and speedway racing. Other leisure and cultural activities are likely to include pop or rock concerts, choral, opera and orchestral events as well as non-musical

Figure 10.4 Echogram

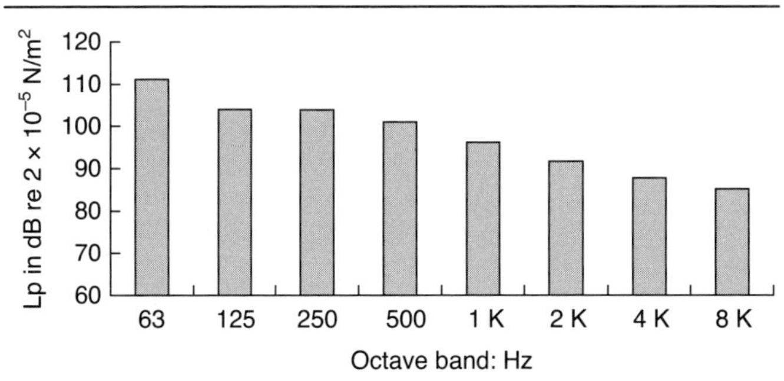

Figure 10.5 Atmosphere wheel. (Vanguardia Consulting, 2014)

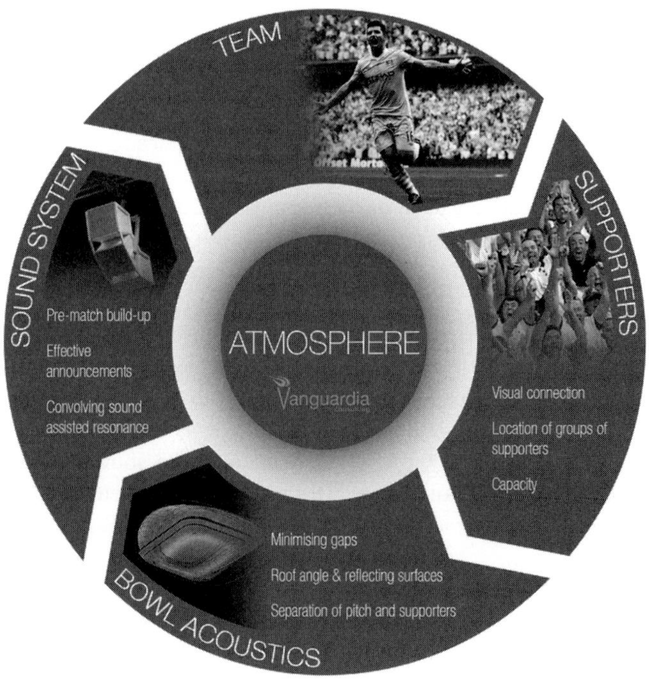

Table 10.2 Concert environmental guidelines

Concert days per year, per venue	Venue category	Guidelines
1 to 3	Urban stadiums and arenas	MNL should not exceed 75 dB(A) over a 15-minute period
1 to 3	Other urban and rural venues	MNL should not exceed 65 dB(A) over a 15-minute period
4 to 12	All venues	MNL should not exceed the LA90 by more than 15 dB(A) over a 15-minute period

MNL is the L_{Aeq} of the music measured at 1 m from the façade

activities such as conferences, theatres and rallies. The activity of relaying 'live' matches or concerts ('beam-backs') by way of the stadium daylight screens and sound system from other venues is also becoming popular and is a further potential activity, and hence a source of noise.

In general, the noise from the majority of events is generated by the audience, which often produces high peak sound levels but for short durations, and often for events during the daytime. Pop concerts, however, generate by far the highest continuous sound levels, with an unbalanced frequency spectrum biased towards the low frequency (Griffiths, 1989). A typical concert spectrum is shown in Figure 10.3. Furthermore, these events are usually held into the late evening, which further increases the community awareness of this source of noise, along with the low-frequency content and high noise emission.

If it is proposed to hold pop concerts at a stadium on a long-term basis, and as noise is one of the four primary factors used to determine the granting of an entertainment licence for concert events (Home Office, 2003), concerts should be used to establish the worst-case source noise level and hence the highest sound insulation performance of the building façade.

Noise data from concerts have been collected at over 400 events in the UK and overseas and a noise study of concert noise levels commissioned by the Health and Safety Executive (HSE) was published in 1991 (HSE, 1991). More recently, further noise levels were reported by Vanguardia Consulting and RPS for Defra (2010a). From these data and experience, it has generally been acceptable to control outdoor concert noise levels to an equivalent a continuous noise level over 15 minutes (L_{Aeq}; 15 min) of 100 dB(A) at 40 m from the sound system while maintaining an effective form of entertainment for the audience, although levels can vary from 96 to 107 dB(A).

UK environmental standards

Acceptable receptor noise levels for concerts have been developed over the past 30 years. The first notable guidance was published by the Greater London Council (GLC) in 1976, and this has undergone several revisions (GLC, 1985). In this publication the environmental noise criterion was based on the increase in L_{Aeq} over the ambient L_{Aeq} level with respect to complaints data. These data were published in 1985 (Griffiths, 1985) and are reproduced in Table 10.5.

Further refinements to the guidelines were developed from research (Griffiths and Dove, 1992) in the early 1990s which reviewed variables such as baseline noise levels, venue types and concert numbers, as well as the response to the rhythmic low-frequency bass sound of modern music (Griffiths *et al.*, 1988). The revised guidelines were published in the Noise Council Code of Practice (Noise Council, 1995) and are summarised in Table 10.2. The Noise Council Code guidance is under review following research completed by Edinburgh Napier University and Ipsos MORI for Defra (2011).

Predicting sound insulation

Given the source noise level, the source pattern characteristics, the additional reverberant sound energy and the propagation of sound due to spherical spreading and excess attenuation, the composite bowl sound insulation can be determined to meet the appropriate environmental criteria. For open stadiums, barrier effects must also be included in the evaluation process, as sound is diffracted at the edge of roofs, etc.

Meteorological data must be considered in the analysis. This was identified by Griffiths and Kamath (1987) who reviewed the effect of meteorology as one of 11 variables for concerts held at Wembley Stadium.

As for the bowl acoustics, ray tracing is often adopted to predict the overall impact of noise on the community. The three main areas that contribute to the overall sound insulation, and hence noise propagation to the community, are:

- the main wall/underside of the seating tier, which, for the most part, forms the vertical interface between the stadium and external environment
- the roof or partial roof

- penetration through either of the above for the purpose of, for example, personnel access, vehicular access, ventilation and other servicing requirements.

Sound systems

In addition to the traditional performance requirements of sound systems (life-safety announcements and commentary), sound systems for modern venues are often required to perform the following functions:

- relay important messages to all public areas of the venue
- provide commentary related to the event
- be suitable for the reproduction of a full range music to enhance the fan experience (this is typical at NFL events) and to convey messages from sponsors and advertisers to interface with screens and other visual equipment
- interface with concert sound systems to provide direct sound to the more distant parts of the stadium
- provide the full audio requirements for the venue attractions (tours, museums, etc.)
- provide a communications system for the hard of hearing.

The systems will need to serve all public areas, including but not limited to, the bowl, the pitch and track, the concourses, executive boxes, function rooms, lounges, restaurants, operational areas, turnstiles, exit gates, external walkways and car parks within the curtilage of the site.

The sound systems also provide emergency evacuation communications for civil commotion and bomb alerts, and must be interlinked to the fire-detection system as a voice alarm. They must also be designed with a surveillance system to report any faults. These and other aspects are required in order to meet British Standard Codes of Practice (BSI, 1997b, 2011a, 2013) enforced under the powers of the regulatory authorities.

The sound system design should be incorporated within the architectural acoustic modelling, so that the optimum quantity and types of loudspeakers can be established for each given acoustic space to meet the necessary performance criteria. In terms of meeting relevant standards, one of the primary issues is to meet a speech intelligibility index (SII) of not less than 0.5 (BSI, 2011b). This index can now be readily measured, and hence the quality of the sound system is not just a subjective assessment.

The designs must also have due regard to the environmental impact, in particular external loudspeakers. Where possible, the sound from the system should be designed not to exceed the existing background noise in the community.

The sound systems installed at the old Wembley Stadium (1990) and at the Millennium Stadium are examples of systems developed to perform many of the aforementioned functions now required for modern venues. The Wembley system (Jones and Griffiths, 1991), in particular, was the first of its kind to be fully integrated with concert touring systems, and has been used by every artist performing at the stadium. The system, using digital signal processors for equalisation, signal delay and routing, improved the external environment. During concerts, the use of this system reduced external music noise levels by up to 3 dB(A) as compared with the stadium system when it was not in operation. The sound quality in the stands was also generally improved, as the sound was focused into the seating areas by local speakers designed for the acoustic space rather than being served by a multi-purpose concert system located some distance from the audience, where both unwanted reflections and high-frequency absorption are common problems.

The new Wembley Stadium sound system was based on the same operational principles, with a fully integrated sound system for life safety, music reproduction and concert interface.

Internal acoustics and sound insulation

As with most developments, the acoustics of internal rooms and the sound insulation between them requires careful consideration. This is particularly the case in entertainment venues, where there are normally a number of areas requiring specific low-noise acoustic environments while other areas generate high noise levels (e.g. plant rooms). From the outset, therefore, careful space planning with respect to acoustic separations between high- and low-noise environments needs to be addressed.

Areas requiring specific acoustics treatment are TV and radio broadcast suites, presentation and function suites, and conference areas. TV and radio suites are particularly important as, in modern venues, the suites are often built in the grandstand so that commentators have a clear line of sight of the playing field or performance area. Therefore, to conduct quiet interviews while being adjacent to high noise levels from the audience requires a high level of sound insulation, normally in excess of 60 dB (Rw). A 'room within a room' design may be required, especially when 'live' broadcasts are transmitted to the audience by way of the sound system, where feedback problems can arise.

Room data sheets giving the full technical specification for each room are normally prepared. With respect to acoustics, the performance for most areas is specified by surface finishes with typical absorption coefficients, sound insulation values for each given partition and noise rating (NR) values for the noise environment dependent on the activity in the room. Often, generic wall constructions are identified to meet various levels of sound insulation, examples of which are identified in Table 10.3.

Clearly, other factors, such as suspended ceilings, suitable door construction, glazing requirements, flanking transmissions and

Table 10.3 Examples of sound insulation for wall constructions

Sound criteria	Minimum Rw	Typical wall construction
Low level of insulation acceptable for general areas	40 dB	Stud wall partition, 12.5 mm wallboard each side of 70 mm metal studs, 25 mm mineral wool in cavity
Medium level of insulation normally acceptable for presentation and conference areas	50 dB	Stud wall partition, two layers of 12.5 mm wallboard each side of 70 mm metal studs, 50 mm mineral wool in cavity
High level of insulation for broadcast areas and strictly private rooms	60 dB	Ideally, 'room within a room' construction or high specification staggered stud construction

A/V mounts for plant, need to be specified as appropriate, taking into account the activity within the room and the activities in adjacent areas.

Building services

Noise generated by heating, ventilation, air-conditioning and associated generator systems can affect both the internal areas and the external environment. The transmission of noise from these sources should, therefore, be controlled before it is radiated internally and externally. This is normally achieved by the careful selection of acoustic attenuators, to minimise the sound level at both the inlet and outlet of the ventilation systems. Vibration isolation is also required for the mechanical plant to avoid structure-borne noise.

The most appropriate criteria for assessing the environmental impact of this type of noise source and, indeed, the standard preferred by most local authorities is BS 4142:1997 (BSI, 1997a). As this noise rating relates to the existing background noise level, which is likely to vary from one community area to another around the site, the acceptable noise standard will also vary. During the detailed design of Wembley Stadium, this was addressed by recording the background noise in each area. The stadium was then divided into eight zones applicable to each nearby community area, and target noise levels were predicted 1 m from the façade of the stadium in each zone, taking into account the distance attenuation and intervening barriers from the community sites to the venue. Target noise levels ranged from 65 to 78 dB(A) for the various locations around the perimeter of the stadium. These were specified to assist in the location of noisy plant in the least-sensitive areas, to specify the performance of the attenuators in each zone for given plant sound power levels, and to demonstrate to the local planning authority that the plant was being designed in relation to environmental noise conditions.

For the internal environment, NR criteria are normally provided for each given room function, and are included in the technical information given in the room data sheets. This enables the appropriate size of attenuators to be calculated for each area. Typical NR values for stadium areas are identified in Table 10.4.

Environmental noise impact

As shown in the spider diagram in Figure 10.1, consideration of environmental noise is important throughout the design, planning and construction process, as it can have a significant effect on the operational use when the stadium is finally opened. Invariably, an environmental impact statement and associated environmental impact assessment are required by the planning authority before the stadium can be built. The issues that are normally covered in such an assessment for stadiums include:

- operational noise (e.g. sporting, concerts)
- stadium sound systems
- plant and machinery
- noise from patrons when going to and from the stadium and while in the stadium
- transportation noise
- construction noise.

The majority of these issues have already been discussed in this chapter; however, transportation and construction warrant a brief discussion.

The most common complaints associated with transportation are generally related to the inconvenience the additional (road) traffic caused, in terms of congestion, parking, etc., rather than noise aspects directly. Transport consultants are, therefore,

Table 10.4 Typical NR values for stadium spaces

Area	Range of NR values
Stadium bowl	45–50
Gymnasia	40–50
Conference areas	25–35
Broadcast suites	20–30

Plate I El Djem, Tunisia. (© Octopus at Slovenian Wikipedia) (This figure is discussed on p. 5)

Plate II Mound Stand redevelopment, Lord's Cricket Ground, London, 1987. (© Arup) (This figure is discussed on p. 17)

Plate III Skydome, Toronto, 2007. (© Marcus Obal) (This figure is discussed on p. 18)

Plate IV Shading elements reducing solar radiation at Sydney Olympic Stadium (ANZ Stadium), Australia (© Christine Goodman, 2014) (This figure is discussed on p. 33)

Plate V View of Oita Stadium from the north. (© KKAA) (This figure is discussed on p. 42)

Plate VI Toyota Stadium: the closed roof and the movable LED display screen. (© KKAA) (This figure is discussed on p. 46)

Plate VII Interior view of Oita Stadium with closed roof. (© KKAA) (This figure is discussed on p. 49)

Plate VIII A crowd density heat map taken from the London 2012 Olympic Stadium micro-simulation crowd model. (This figure is discussed on p. 68)

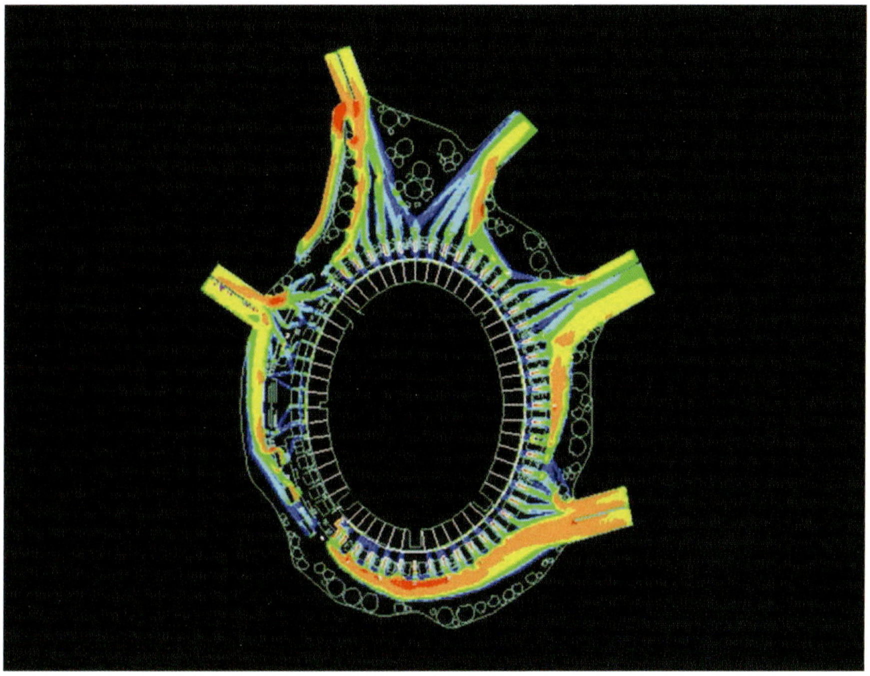

Plate IX Three-dimensional visualisation taken from a micro-simulation crowd model of the security screening area proposed for Arena Cuiaba, a venue for the 2014 FIFA World Cup in Brazil. (This figure is discussed on p. 70)

Plate X The New Wembley at night. (This figure is discussed on p. 90)

Plate XI Chase Field, Phoenix. (© barnhill88) (This figure is discussed on p. 98)

Plate XII Sports Authority Field at Mile High, Denver. (© Daktronics) (This figure is discussed on p. 100)

Plate XIII The Big Screen, XX Commonwealth Games, Glasgow 2014. (This figure is discussed on p. 102)

Plate XIV Construction of the seating tiers at Melbourne Cricket Ground (MCG). (This figure is discussed on p. 138)

Plate XV Precast Y-columns at the Peter Mokaba Stadium. (This figure is discussed on p. 140)

Plate XVI Architectural precast concrete can be designed to blend in with existing structures and add a personal element or accommodate almost any design. (Courtesy of Smith Midland Corporation) (This figure is discussed on p. 148)

Plate XVII New English National Stadium, Wembley. (© Simon Atkinson) (This figure is discussed on p. 155)

Plate XVIII Miller Park, Milwaukee, WI: roof closed. (© Tim Griffiths 2001) (This figure is discussed on p. 159)

Plate XIX Miller Park, Milwaukee, WI: roof open. (© Tim Griffiths 2001) (This figure is discussed on p. 159)

Plate XX Scottish Rugby Union, Murrayfield, Edinburgh. (John Clarke, © Corus) (This figure is discussed on p. 163)

Plate XXI Manchester United FC redevelopment: under-terrace. (John Clarke, © Corus) (This figure is discussed on p. 170)

Plate XXII Ken Rosewall Arena, seating and access. (© Christine Goodman, 2014) (This figure is discussed on p. 171)

Plate XXIII Saitama Super Arena, Japan. (© Ellerbe Becket) (This figure is discussed on p. 172)

Plate XXIV Hazza Bin Zayed Stadium bowl interior and the sweeping parasol roof. (© Dennis Gilbert, courtesy of Pattern Design Ltd) (This figure is discussed on p. 204)

Plate XXV Hazza Bin Zayed Stadium and public realm at night. (© Dennis Gilbert, courtesy of Pattern Design Ltd) (This figure is discussed on p. 209)

Plate XXVI Wimbledon Centre Court celebration, May 2009 (Courtesy of AELTC) (This figure is discussed on p. 212)

Plate XXVII Wimbledon Centre Court 3D image of existing structure. (This figure is discussed on p. 213)

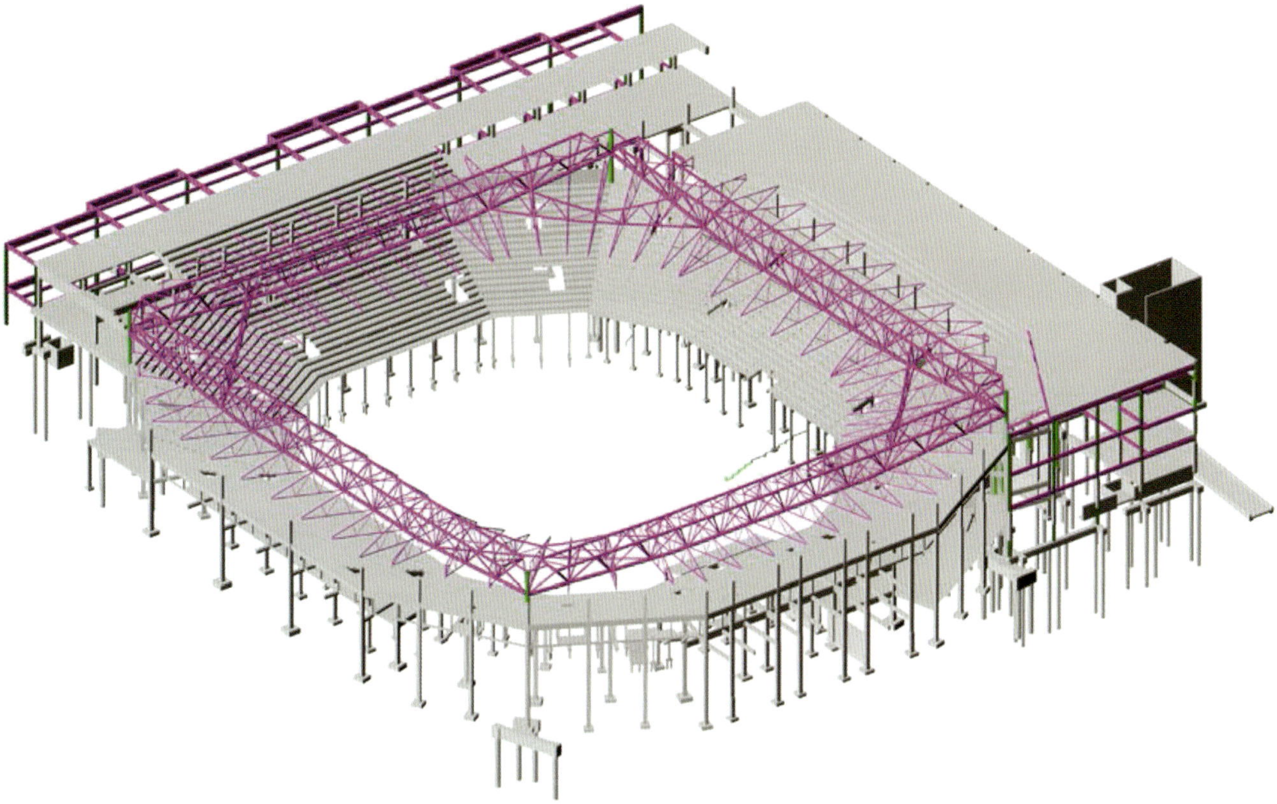

Plate XXVIII Aerial view of Wimbledon during play (Courtesy of AELTC) (This figure is discussed on p. 218)

Plate XXIX Stadium Medal collection, John Pascoe (© Simon Atkinson, 2014) (This figure is discussed on p. 235)

Table 10.5 Construction noise

Period	Noise limits		
	Hours	Time period hours: T	Noise limit L_{Aeq}: T
Monday to Friday (except public holidays)	07:00–08:00	1	65
Monday to Friday (except public holidays)	18:00–19:00	1	70
Monday to Saturday	08:00–18:00	10	75

No noise from the site to be audible at the curtilage of residential property at any other times, without prior approval

required to develop an adequate scheme to ensure that the infrastructure supports the development.

Although, as mentioned, it is generally not the noise directly from the transport that causes problems, if these issues are not addressed properly it tends to sensitise the local residents and makes them far less tolerant to any noise, be it emanating from the events or indeed due to transport. Noise is, in fact, often used as a proxy, when the actual cause of complaint is the inconvenience of traffic. It is important, therefore, that, in addition to transportation noise directly, the other implications of traffic control are dealt with properly.

Nevertheless, transportation noise on event days can be a significant problem in its own right, and should be assessed and dealt with accordingly. When assessing the impact of noise from transport it is necessary to establish the breakdown of attendance patterns into car, bus, rail, foot, etc., which is usually provided by specialist highway planners. Using this information and knowledge of the baseline situation (i.e. traffic levels on similar days without events), the impact in terms of increases in or absolute levels of noise can be determined.

The actual noise levels from road and rail traffic are calculated using the methodologies contained in the *Calculation of Road Traffic Noise* (DoT, 1988), and the *Calculation of Railway Noise* (DoT, 1995), respectively. The approach to achieving acceptable noise levels for all sources can be found in the *Noise Policy Statement for England* (Defra, 2010b) and the *National Planning Policy Framework* (DCLG, 2012), or a particular local authority may have adopted its own standards. The frequency and number of events held at the stadium should be borne in mind when selecting an appropriate standard, recognising that events are normally held relatively infrequently at major stadiums.

Construction

The term 'construction' encompasses both construction and demolition activities. This is normally the case for stadiums where new developments are often to be located on the same site as the existing facility. The development of the stadium is a major construction activity in an area, often being positioned close to sensitive premises and lasting in excess of 2 years.

Clearly, both noise and vibration are inherent in all types of demolition and construction activities, and they can never be completely eliminated. The levels of noise and vibration generated will depend primarily on the type of plant and equipment used, the period of operation and the time of the day they are operated.

To provide a representative assessment of the likely noise aspect of construction operations, it is important to have detailed discussions with the main shell-and-core contractor to establish the preferred construction methods and activity period for each phase of the programme of works. This information can then be used to predict receptor noise levels using BS 5228-1:2014 (BSI, 2014a). Typical plant used for construction and demolition activities for stadiums include various piling rigs, hydraulic crushing grabs, lancing equipment, pneumatic breakers, excavators, tower cranes, generators, air compressors, assorted power tools and lorry movements.

Typical noise and vibration limits for stadium development are outlined in Tables 10.5 and 10.6. These values are based on the commonly adopted daytime limit of $75L_{Aeq}$ 08:00–18:00 hours, with a reduced level for the hour either side of the day period to allow the contractor to make some noise during mobilisation on

Table 10.6 Vibration limits

Criteria	Period	Limit
Disturbance		
Residential	Daytime	0.40 $m/s^{1.75}$ VDV
Other buildings	Daytime	0.80 $m/s^{1.75}$ VDV
Building damage	All periods	12.5 m/s PPV

PPV, peak particle velocity; VDV, vibration dose value

site and the gradual closure of the site in the early evening. The levels normally apply to existing habitable properties and, for new proposed residential developments, sometimes part of the application (master plan for the area), guidance is sought from BS 8233:2014 (BSI, 2014b), which provides advice on the design of buildings to meet good internal noise environments for living rooms, bedrooms, etc. The vibration levels are based on BS 6472:1992 (BSI, 1992) for minimising disturbance and BS 7385 (BSI, 1993) for reducing the risk of structural damage.

Summary

With patrons' increasing expectations of good standards, which include those of the sound and acoustics, coupled with the general emphasis on environmental enhancement and sustainability, the need to properly address all aspects of sound, noise and acoustics for stadiums is becoming of greater importance in order to complement the traditional engineering and architectural disciplines required for this type of development.

REFERENCES

BSI (British Standards Institution) (1992) BS 6472:1992. Guide to evaluation of human exposure to vibration in buildings (1 Hz to 80 Hz). BSI, London, UK.

BSI (1993) BS 7385-2.1993. Evaluation and measurement for vibration in buildings. Guide to damage levels from groundborne vibration. BSI, London, UK.

BSI (1997a) BS 4142:1997. Rating industrial noise affecting mixed residential and industrial areas. BSI, London, UK.

BSI (1997b) BS 6259:1997 Code of practice for the design, planning, installation, testing and maintenance of sound systems. BSI, London, UK.

BSI (2011a) BS 7827:2011. Code of practice for designing, specifying, maintaining and operating emergency sound systems at sports venues. BSI, London, UK.

BSI (2011b) BS EN 60268-16:2011. Sound system equipment. Part 16: Objective rating of speech intelligibility by speech transmission index. BSI, London, UK.

BSI (2013) BS 5829-8:2013. Fire detection and fire alarm systems for buildings. Part 8: Code of practice for the design, installation, commissioning and maintenance of voice alarm systems. BSI, London, UK.

BSI (2014a) BS 5228-1:2009 + A1:2014. Code of practice for noise and vibration control on construction and open sites. Noise. BSI, London, UK.

BSI (2014b) BS 8233:2014. Guidance on sound insulation and noise reduction for buildings. BSI, London, UK.

DCLG (Department for Communities and Local Government) (2012) *National Planning Policy Framework*. DCLG, London, UK.

Defra (Department for the Environment, Food and Rural Affairs) (2010a) *Noise Monitoring at Pop Concerts, NANR 297*. Defra, London, UK.

Defra (2010b) *Noise Policy Statement for England (NPSE)*. Defra, London, UK.

Defra (2011) *Research into Attitudes to Environmental Noise from Concerts, NANR 292*. Defra, London, UK.

DoT (Department of Transport) (1988) *Calculation of Road Traffic Noise*. HMSO, London, UK.

DoT (1995) *Calculation of Railway Noise*. HMSO, London, UK.

GLC (Greater London Council) (1985) *Code of Practice for Pop Concerts*. GLC, London, UK.

Griffiths JET (1985) Noise control techniques and guidelines for open air pop concerts. *Proceedings of the Institute of Acoustics* **17(3)**, n.p.

Griffiths JET (1989) Sound control at British open air pop concerts. *Inter-noise '89 Proceedings*. Noise Control Foundation, Poughkeepsie, NY, USA.

Griffiths JET and Dove A (1992) Environmental noise guidelines proposed for the new Health and Safety Executive guide for pop concerts. *Proceedings of the Institute of Acoustics* **14(5)**, n.p.

Griffiths JET and Kamath SS (1987) Revised environmental noise guidelines for pop concerts. *Reproduced Sound* **3**.

Griffiths JET, Patel S and Kamath SS (1988) Low frequency sound from open air pop concerts. *Proceedings of the Institute of Acoustics* **10(7)**, n.p.

Home Office (2003) Licensing Act 2003. See http://www.legislation.gov.uk/ukpga/2003/17/contents (accessed 12/02/2015).

HSE (Health and Safety Executive) (1991) *A Survey of Sound Levels at Pop Concerts*, Report No. 35/1991. HSE, London, UK.

Jones SP and Griffiths JET (1991) Wembley Stadium – a large scale sound reinforcement system. *91st Convention, New York*. Audio Engineering Society, New York, NY, USA.

Noise Council (1995) *Code of Practice on Environmental Noise Control at Concerts*. Noise Council, London, UK.

Part III

Elements of construction

Stadium and Arena Design
ISBN 978-0-7277-5790-6

http://dx.doi.org/10.1680/saad.57906.119

Chapter 11
Computerised design of stadiums

Martin Simpson Associate Director, Arup, London, UK

Introduction

In the intervening years since the previous edition of this book, the extent to which computational techniques are used in the design and construction of modern stadiums has increased significantly. It is not true to say that modern stadiums cannot be designed or constructed without the use of such computation techniques, but it is true to say that modern stadiums would not be completed within the project timescales without the use of computerised techniques. However, a computer is only a tool to be used by competent designers and engineers to enhance the process of design and construction. This chapter looks at some of the significant computational techniques currently available for use in the design and construction of stadiums.

Parametric design

One of the most powerful computational techniques to make an impact on the design of stadiums is parametric design. Instead of representing a design as a series of fixed quantities, parametric design represents the design as series of relationships. These relationships can relate to non-geometrical characteristics, such as cost or time, but a good stadium-specific example also involves the geometry of a seating bowl formed from precast concrete terrace units.

The going and rising are primary dimensions defined by the architect, and are related to the requirements for sightlines. The specific relationship between the geometry of the seating bowl and sightlines is covered in more detail in Chapter 14. There are many more dimensions to fully define the geometry of the precast unit. Some of these dimensions will only be finalised when the supplier of the terrace units is appointed.

The precast units can be visualised as either a:

- series of unique objects with individual dimensions – a stadium seating bowl could contain many thousands of such objects
- families of components, where the dimensions are represented by parameters – the values of the parameters are dependent on the location of the precast unit.

Therefore, all the terrace units can be represented with relatively few parametric components.

The use of parametric components is an important technique, but it starts to deliver real benefits when aligned with *associative design*. *Associative or constraint modelling* occurs when the relationships and interactions between several components are specifically defined. In the example of a stadium seating bowl, each precast terrace unit can be represented as a parametric component, however the sightlines of the bowl can only be calculated when all the parametric terrace units are linked together. In fact it is the requirement for the sightline that will ultimately define the values of the unique dimensions for each precast terrace unit.

The use of parametric components with associate modelling is extremely powerful in allowing multiple scenarios to be tested rapidly. In practical terms there are usually several conflicting requirements by different stakeholders over the same components. For example, the parabolic curve that defines the optimum bowl geometry is conflicted with minimising the number of unique precast units, which in turn has a cost impact. Using parametric and associative modelling techniques it is possible to test the effect of standardising units while meeting requirements for sightlines. Figure 11.1 shows a parametric workflow detailing how sightlines in section are linked to objects in a building information modelling (BIM) model. This exercise and the use of parametric design assisted the design team in creating an efficient seating bowl that brings everyone in the stadium as close to the action as possible.

The large size of stadiums means that even subtle changes in the geometry of the profile can have significant impact on the overall size of the stadium, which in turn can affect the area of roof, floor slabs and external cladding. On large stadiums, an increase in the radius of the stadium by 1 m can add roughly 1000 m^2 to the roof area.

In stadium design the geometry of all components flows from the geometry of the bowl. A lot of progress has been made in the

Figure 11.1 Parametric workflow in the design of a seating bowl (a) sightline control; (b) workflow in McNeel Grasshopper; (c) sightline values in McNeel Rhinoceros; (d) precast components in Autodesk Revit

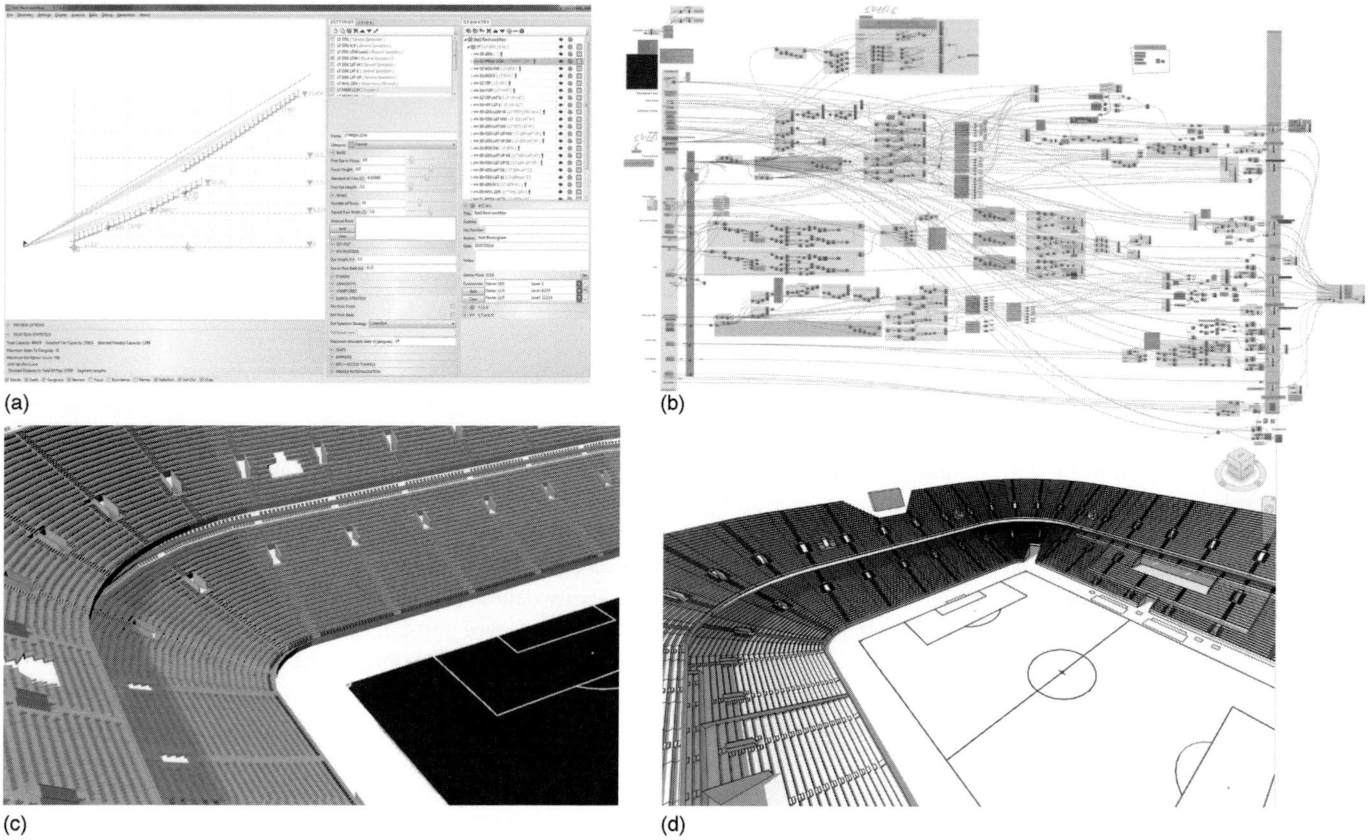

parametric design of seating bowls, and most specialist sports architects will now use some form of parametric bowl design software. Future challenges will lie in extending parametric design beyond the bowl, into the frame and beyond.

Simulation techniques

We live in a physical world, and in order to make decisions during the feasibility and design phases of projects we need to be able to simulate how the project will behave in reality. This process is called *virtual prototyping*.

- It is *virtual* in that it does not physically exist; it exists only within the realms of computerised simulations.
- It is a *prototype* because it is the first from which all others will be derived.

Virtual prototypes allow designers and engineers to test multiple scenarios, essentially correcting mistakes and errors before committing to the actual physical project. Some critics argue that virtual prototyping in the architecture, engineering and construction (AEC) industry is not appropriate because, unlike the automotive or manufacturing industries, we do not fabricate the same product many times, and therefore we do not gain advantages of mass production. The opposing argument is that, because many construction projects are unique there is not the scope for evolution or refinement during a production cycle; therefore, it is essential to get it right first time. The completion date for many stadiums and sports venues is fixed in relation to a major sporting event such as the Olympics or the start of a sporting season. The projects are time critical, and so it is imperative to reduce the risk of errors that may cause delays. Finally, the cost of correcting errors post-construction is an order of magnitude higher than correcting them before committing to fabrication (i.e. correcting them during the design process).

There are two distinct types of simulation:

- *Geometric simulations* represent the tangible objects that clients and spectators will eventually interact with in the physical world. These simulations can be used to ensure that physical requirements of brief and specifications are adhered to. Examples may be the minimum width of concourses, head height of escape stairs or net lettable space of offices. Geometric simulations are also essential to ensure that physical components will fit together correctly when the project is built. Finally, the geometric information contained within these simulations is essential input data for technical simulations.

- *Technical simulations* are required to ensure that the completed project will behave as required. There are many specialist architects, designers, engineers and consultants who contribute to a successful stadium project. Each specialist will have a series of processes and calculations they will need to complete in order to discharge their professional responsibilities. Many of these processes have been automated in the form of computerised design tools. Examples of some of these discipline specific simulations are outlined below.

Structural simulations

Structural design is concerned with ensuring that the stadium will resist and safely transmit all loads and forces to the foundations and, ultimately, the ground. On typical buildings, structural engineering is primarily focused on strength, stiffness and stability and such finite-element structural analysis tools are well documented (Figure 11.2). However, on stadiums and sport venues the influence of loads that vary with time or dynamic loads can have more significance:

- *Crowd–structure interaction*. Stadiums are particularly unique in the potential for the interaction of large crowds of spectators. Spectators can interact with the structure through jumping and/or bouncing which, if the structure is designed incorrectly, could cause resonance, resulting in

Figure 11.2 Simulations to support structural engineering: (a) finite-element analysis on the Singapore Sports Hub; (b) dynamic analysis for crowd–structure interaction; (c) Donbass Arena in a wind tunnel; (d) shrapnel damage on a spiral strand cable

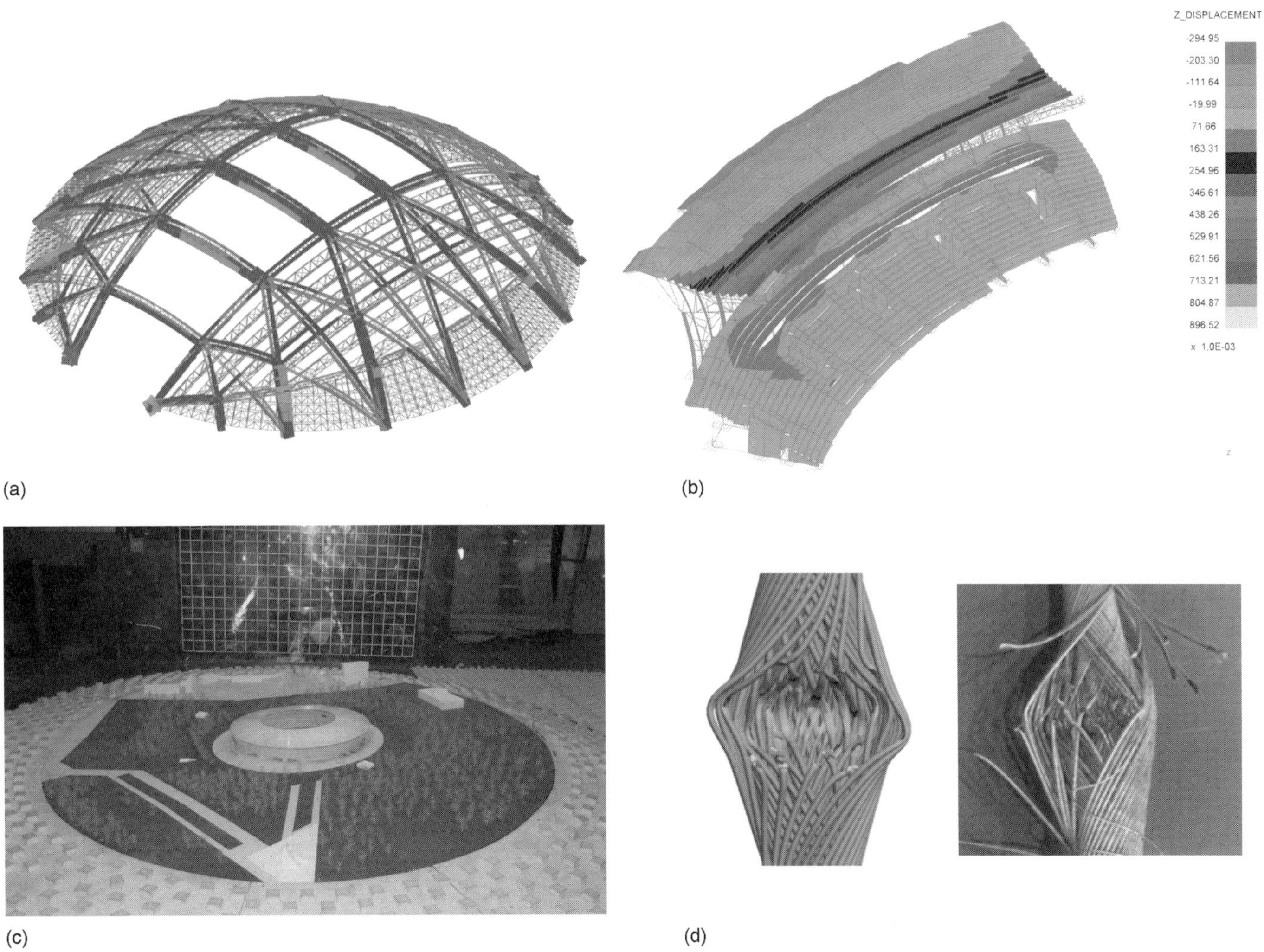

discomfort and, in extreme cases, failure. Chapter 12 discusses in depth the requirements of the dynamic design of stadiums and sport venues. There is now a much greater level of understanding of crowd–structure interaction and sophistication in the simulation of both the dynamic load (i.e. the spectators) and the interaction of the structure with this load. Modern performance-based approaches using simulation techniques provide far better value for money for the client than purely specifying a natural frequency limit (ISE, 2008).

- *Wind–structure interaction.* Stadiums are often required to form unique iconic landmarks at either the club, city or even country level. This is often manifested in the design of the roof and external façade. Stadium roofs pose a unique series of challenges. They must span large distances, in a cost-effective manner, with no internal supports and usually with a large central hole. Understanding loading is crucial to efficient structural design. Wind loading tends to govern the design of lightweight roof structures but its dynamic nature is the least understood.

There are two components of structural response to wind: the static component, which deals with the pressure of the wind on a surface; and the dynamic component, which deals with the structural response to that pressure. The fact that stadium roofs are often lightweight and relatively flexible compared with traditional buildings means that the usual design code methods for wind dynamic response are not appropriate. They can be simultaneously conservative for some elements while being non-conservative for others.

The testing of scale models in a 'wind tunnel' remains the most appropriate manner of obtaining both the static portion of the wind load and information relevant to the dynamic responses, even with advances in computational fluid dynamic (CFD) simulation. Methods of deriving the peak loads by applying a dynamic factor to the static loads are known to be unsafe at times. There is an emerging understanding that loads in individual elements within the roof system can be much more susceptible to dynamic magnification than previously thought. Multiplying the static wind pressures in every element by the same dynamic factor may significantly underestimate the forces in some elements and overestimate them in others. By using the fluctuating pressure information from the wind tunnel to calculate the dynamic responses, it is possible to gain a more complete picture, including the different dynamic responses affecting each element directly.

- *Seismic simulations.* In many areas of the world seismic requirements often govern the structural design. Stadium structures are atypical in that they are relatively low-rise and stiff compared with other building types. The stiffness of a stadium is a benefit when dealing with resistance to crowd-induced dynamic loads, but a detriment when dealing with seismic loads (a stiff building attracts more seismic force). Many of the simplified code-based approaches contained in international codes of practise for low seismic hazards have been developed on the basis of empirical evidence for rectilinear medium- to high-rise buildings. However, these approaches are not appropriate to stadiums, which do not behave in the same ways as these buildings. Even for relatively low ground accelerations a response spectrum type of analysis should be carried out. For more extreme seismic hazards a time–history analysis should be carried out, using linear or non-linear material properties depending on whether elastic or plastic structural behaviour is being investigated.
- *Blast simulations.* As significantly large crowds can gather in stadiums and other sport venues, the sensitivity of structures to blast, both accidental and deliberate, needs thorough investigation. Certain risks can be mitigated through management procedures, but usually a level of structural simulation is required to inform and justify security provisions such as stand-off distances.

Internal environmental simulations

Major international sporting events such as the FIFA World Cup or Olympic Games are held across all continents and in a wide range of climates. It is necessary for venues hosting sporting events at the highest level to ensure the appropriate environment for spectators and competitors.

In order to ensure, and in some cases justify, the decision to host such sporting events requires considerable technical simulation relating to the internal environment of the stadium. Bowl cooling is gaining popularity as a method of ensuring that a country's natural climate is not a barrier to sport.

The following are the key drivers for bowl-cooling design:

- supply air-temperature variation based on time of day/year
- control of delivery for venue flexibility and energy saving
- effective ventilation strategy for comfort, energy and delivery
- plant space limits are determined by the stadium envelope, sightlines and the need to maximise seats while minimising volume of stadium
- minimise capital expenditure and operational expenditure.

Recent experience has shown that the most efficient method of providing spectator comfort in hot humid climates is to provide a 'cool zone' or microclimate within the spectator zone, rather than to cool the entire volume of the stadium. This is done by providing cooled air directly to the spectators through vents within the seating bowl attached to a cool air supply through ducts or plenums behind the seating bowl.

A number of simulation tools are used to assist in the design of bowl-cooling strategies, including dynamic thermal modelling (DTM), CFD modelling and daylight simulation.

DTM is used to predict operational temperatures, while CFD is used to provide a better understanding of the air movement around the spectators and associated temperatures. CFD is far more computationally expensive than the DTM, as computers have become more powerful and, especially with parallel processing, it is possible to analyse whole stadiums instead of 'strips', and even to take into account phenomena, such as wind interaction, which can have the effect of scouring cool air from the stadium (Figure 11.3).

Environmental simulation tools currently take significant time to set up and run and for the results to be interpreted. Design programs need to reflect the current constraints on the technology, especially if the results are fundamental to the successful design of the venue. It is also essential, as with all simulation techniques, to understand that the results from the simulation are valid and calibrated against reality.

Figure 11.3 (a–c) CFD analysis demonstrating the conditioning of the spectators and pitch for a stadium in the Middle East; (d) illustrating wind scour effect

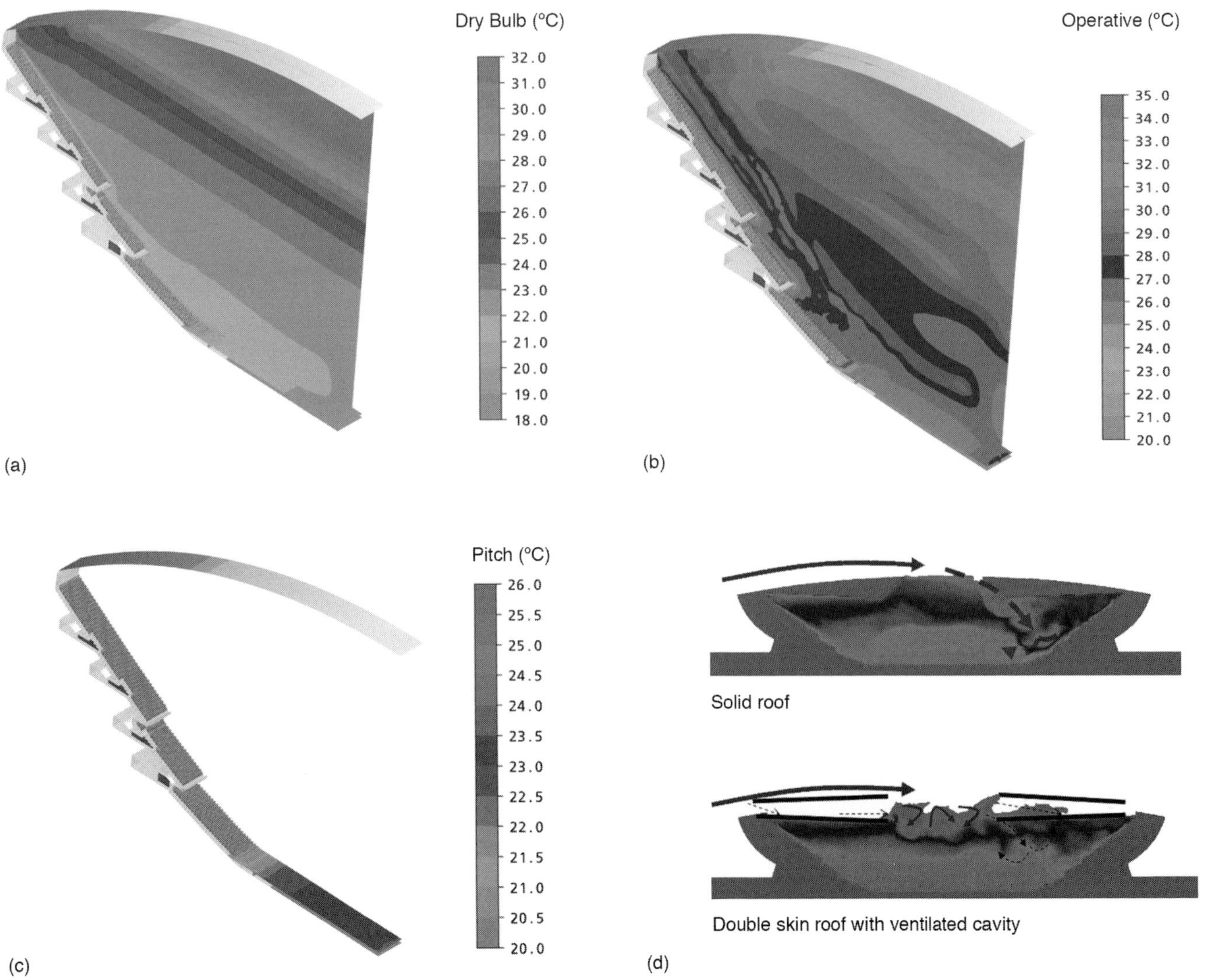

Acoustic simulations

Internal and external acoustic simulations are important in stadium and sport venue design.

In many cases new venues are sited in locations where there was no previous stadium. In other cases the business case behind enhancements and extensions of existing venues may require holding concerts or other events. Noise breakout from the venue (i.e. from crowd and sound systems) can be a significant disturbance and a nuisance to the surrounding areas. Planning authorities will often require a demonstration of how noise from the venue will affect the surrounding area.

Noise mapping is a simulation technique used to predict noise and advise on measures needed to control noise emissions from the stadium. The design team can then derive solutions that meet the requirements stated by the planning authorities.

In addition to the effect on the external environment, the internal acoustic design of stadiums and sport venues is of critical importance. At a fundamental level, stadium acoustics must be designed to provide clear voice messaging to spectators in the event of an emergency, to ensure controlled and efficient evacuation.

In contrast, a reverberant acoustic atmosphere can be beneficial in creating a good sporting environment. The roof geometry and construction material play a major role in how much sound from the crowd is reflected.

Sound system performance modelling is an important simulation technique used to provide a good environment for sport and to ensure that the sound system meets the requirements for sport entertainment and safety (Figure 11.4).

Lighting simulations

Modern sporting events are broadcast to global audiences. Recent advances in broadcasting include technologies such as high definition, super slow motion and 3D TV. Stadium lighting, therefore, needs to support the broadcasting requirements, in particular shadow-free, flicker-free and glare-free environments for the players, officials and media. The development of super-slow-motion cameras, in particular, means that it is now necessary to comply with specific values of flicker factor in order to satisfy international television broadcast requirements.

Modern stadiums require many floodlight units to provide appropriate vertical and horizontal luminance on the field of play, and additional stadium bowl lighting is provided as both a visual backdrop for the TV images and for the safety and orientation of the spectators.

Specialist lighting simulation is used to assess the position and orientation of floodlights and the stadium bowl lighting, in order to provide the necessary confidence in the proposed solution (Figure 11.5). Carried out early in the design the simulation ensures that gantries are correctly located and accessible to ensure compliance with both broadcasting requirements and, in many cases, the requirements of specific sport governing body.

Other applications for lighting simulations within the sport venue design are:

- Shadow – to ensure that the roof geometry and material specifications do not provide hard shadows, which can be detrimental to both the broadcasting requirements and competitors/judges.
- Architectural/feature lighting – to ensure that the stadium is suitably iconic at night.

Simulations to support fire engineering

It is necessary to simulate and plan for various emergencies to ensure the safety of people in and around the stadium. In addition to an overall fire safety management system, safety is a primary factor in the design of stairs, gangways and seating.

National safety laws governing evacuation of typical building typologies have generally not been developed with stadiums and other sport venues in mind. Theatres are usually the closest building type, but these rarely have the number of spectators that can be found in stadiums. In many cases, approving authorities wish to see the result of evacuation simulations, together with the simulation of how smoke builds up and is controlled to justify their decision on the design and, ultimately, prove the venue is safe to operate (Figure 11.6).

Additional complications arise when looking at the movement of spectators beyond the confines of the building in order to dissipate crowds and the effects on the wider transportation system.

Optimisation

One of the significant advantages that computerised techniques provide is gaining results within a reasonable timescale. Varying the input data using parametric design techniques together with technical simulations can provide access to a wide range of scenarios. Some of these scenarios will perform better than others. Finding the parameters that produce the best result is called 'optimisation'. There are different forms of optimisation. Finite-terminating and iterative-convergent methods provide a definite answer to a particular question. However, many engineering problems are so large in scope or interdependent on so many variables that only heuristic techniques can yield answers. Heuristic methods try to improve a solution iteratively with regard to a measure of success, but these methods can only provide a better solution than those tested. There is no

Figure 11.4 Acoustic simulation: (a, b) noise breakout from the new Anfield Stadium; (c) noise-wave propagation in the new Anfield Stadium; (d) sound system design using Ray Bounce

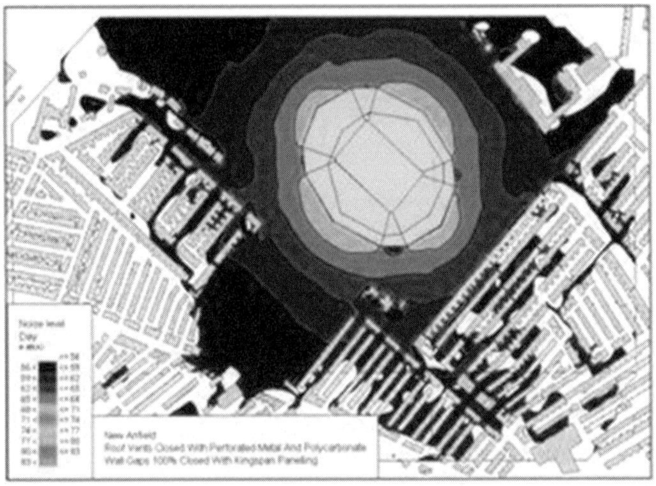

(a)

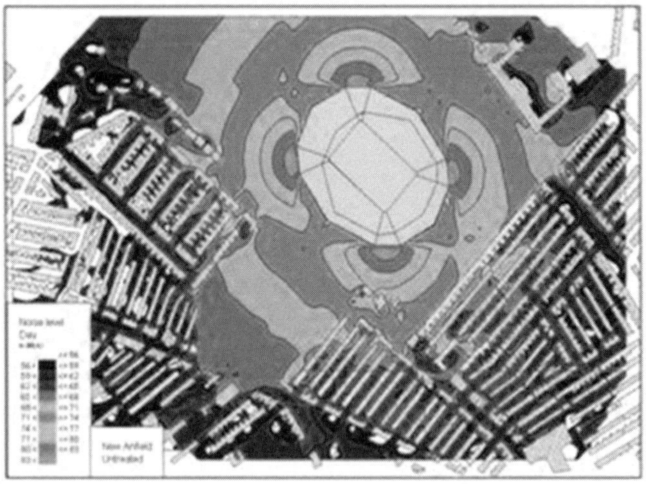

(b)

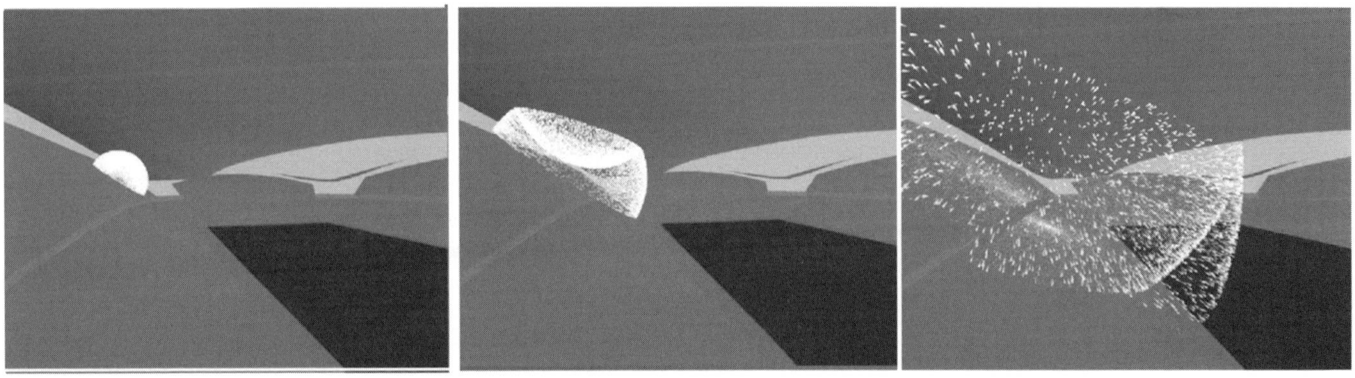

(c)

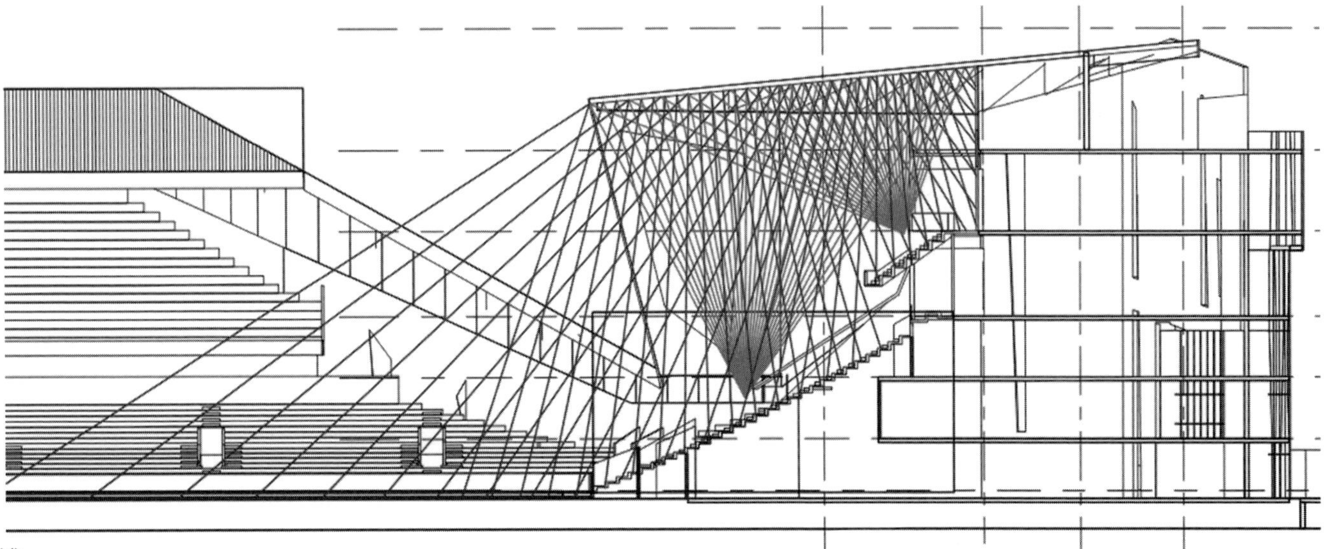

(d)

Figure 11.5 Simulations to support lighting: (a, b) shadowing due to the path of the sun; (c) field of play lighting; (d) glare analysis; (e, f) flicker reduction for slow motion TV

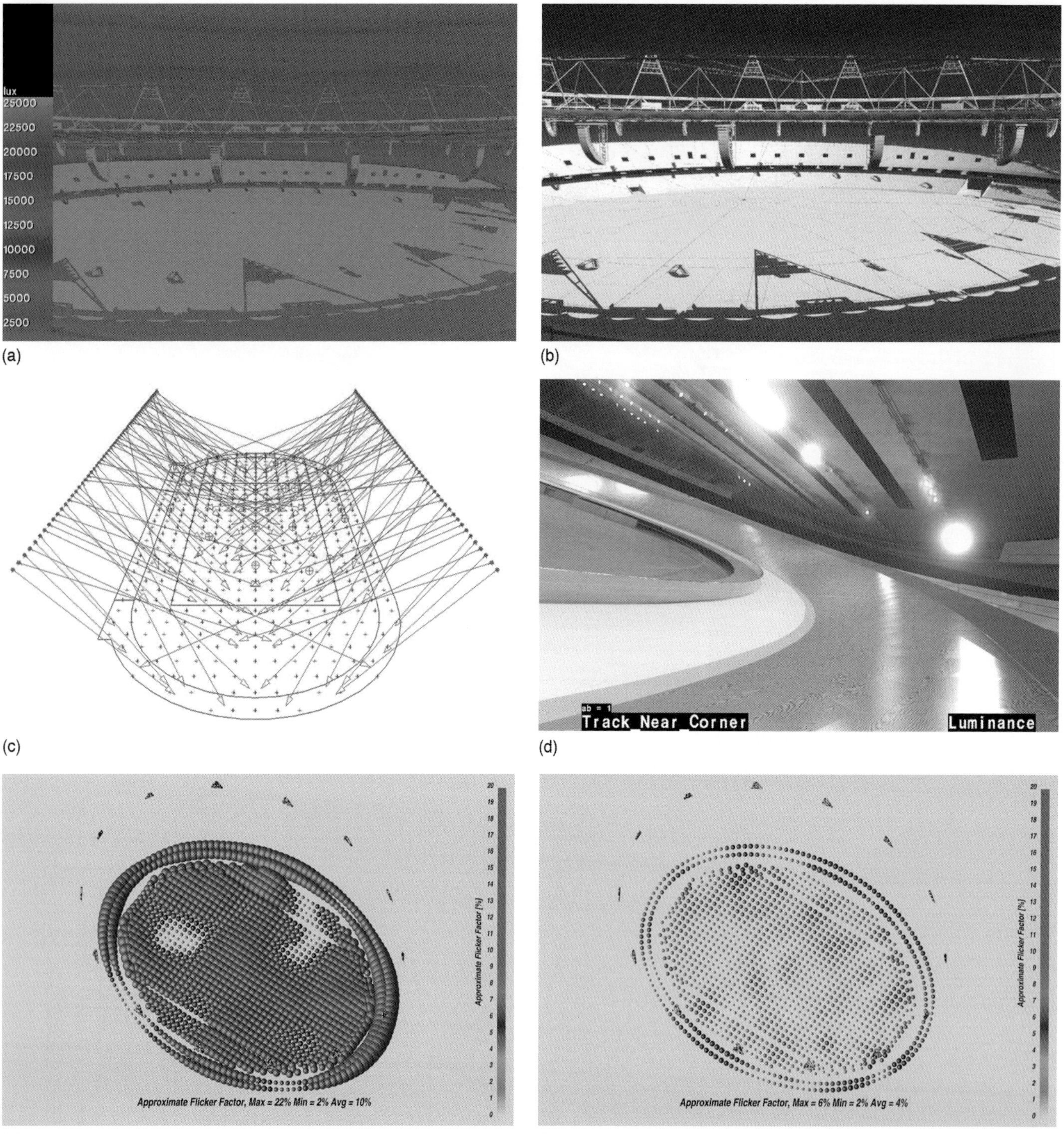

Figure 11.6 Simulation to support fire design: (a) smoke build up; (b) evacuation modelling

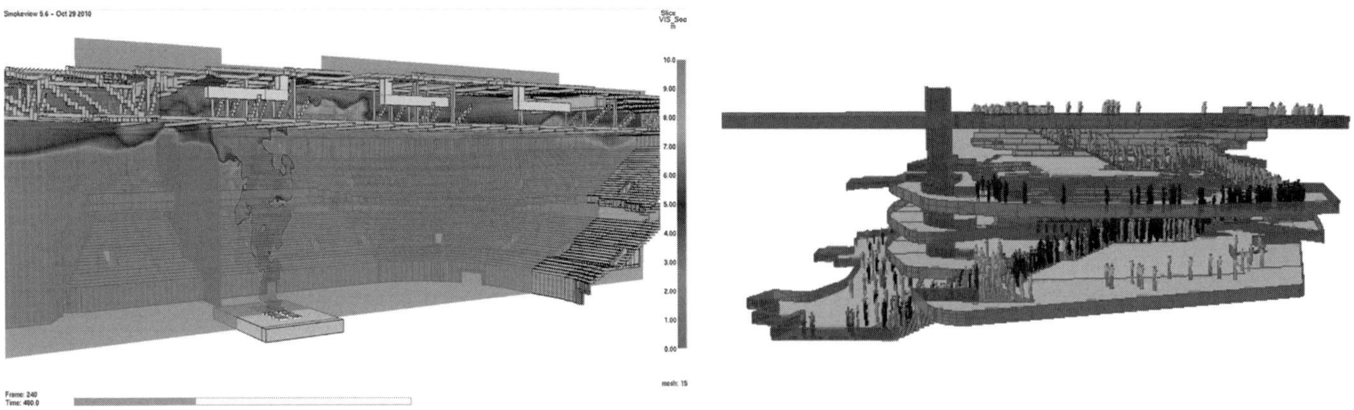

(a) (b)

guarantee that a better solution does not exist with different parameters.

There are many different types of heuristic optimisation techniques. For example:

- genetic algorithms
- hill climbing
- particle swarm
- ant colony.

A straightforward example of optimisation with regard to stadium design would be minimising the structural material and, therefore, weight. The shape of the structure could be adjusted (topological optimisation) and/or the member sizes could be adjusted (elemental optimisation) (Flager *et al.*, 2009).

Importantly, minimum weight does not mean minimum cost. Typically, in Europe the material cost for a stadium roof may account for only a third of the cost, while fabrication and erection would each account for another third. In other parts of the world this ratio may be different. Depending on the procurement strategy, data relating to fabrication and erection are usually unavailable within the design cycle unless specialist fabricators are onboard to inform the design team.

Minimising material is a relatively isolated single-discipline example of optimisation (Figure 11.7). In many cases design disciplines have competing requirements over the same parameter. Computation techniques and the increasing speed of simulation offer significant potential for multidisciplinary simulation and optimisation. As computers continue to increase in power, it may be possible to gain real-time feedback from design decisions and how they affect behaviour and interaction between design disciplines.

Communication

One of the most fundamental improvements in stadium design arising from computerised techniques is the communication of ideas and information. Great design is only possible if ideas can be communicated between all project stakeholders.

Visualisation

The quality of computer visualisations continues to improve to almost photo-realism. These techniques have exceptional value in terms of marketing and communication with the public. The embellishment of visualisations is tempting but, wherever possible, the geometric prototype should be used as the basis for the visualisation, as it is the closest to reality in term of sizes, scale and behaviour. It is not inconceivable that clients who are delivered a reality that does not match visualisations may seek redress.

As technology continues to evolve, the still render or the predefined fly through become less relevant. Technology from the computer games industry provides unparalleled opportunities for clients and stakeholders to interact with prebuilt concepts in immersive environments. Designers are experimenting with overlaying the results from several simulation technologies into a single immersive environment.

Engagement of clients, users, fans and designers increases the scope for contribution and, therefore, enhancement of designs during the design phase.

Rapid prototype technology

Clients, designers and other stakeholders can interact with the virtual prototypes in many ways. However, tactile scale models continue to be an important vehicle for the communication of ideas. 3D printers are becoming increasingly popular due to better affordability and quality. Using such printers

Figure 11.7 Examples of optimisation techniques: (a) minimising weight of steel for a stadium in the Middle East; (b) using rectangular instead of trapezoidal panels on a sloping façade for Donbass Arena

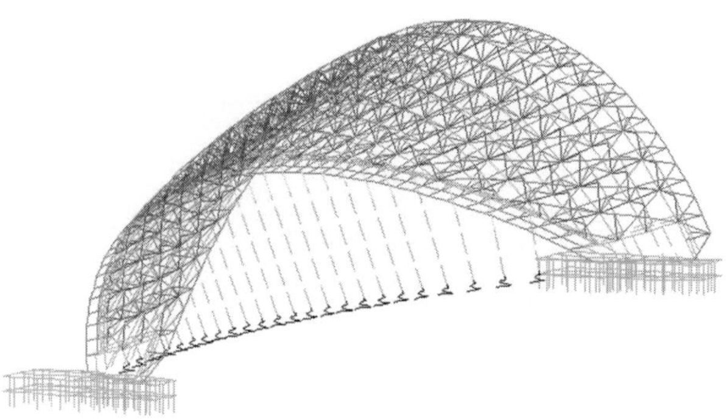

Design Method	Design Cycle Time	Total Design Cycles	Total Time
Conventional	4 hrs	39	9360 min
Serial Opt (1 CPU)	6 min	8042	48250 min
Parallel Opt (128 CPUs)	3 sec	8042	402 min

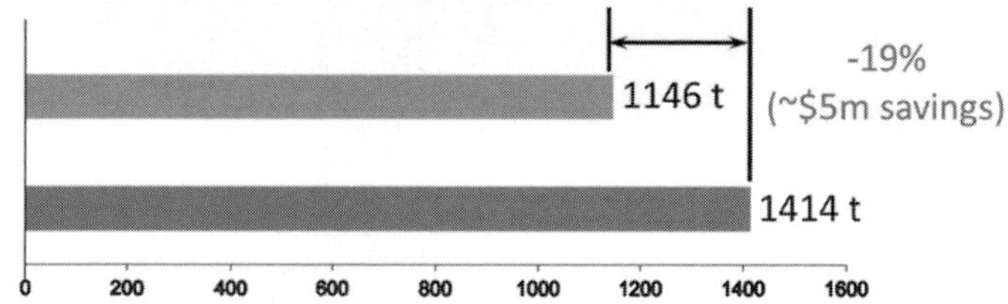

(a)

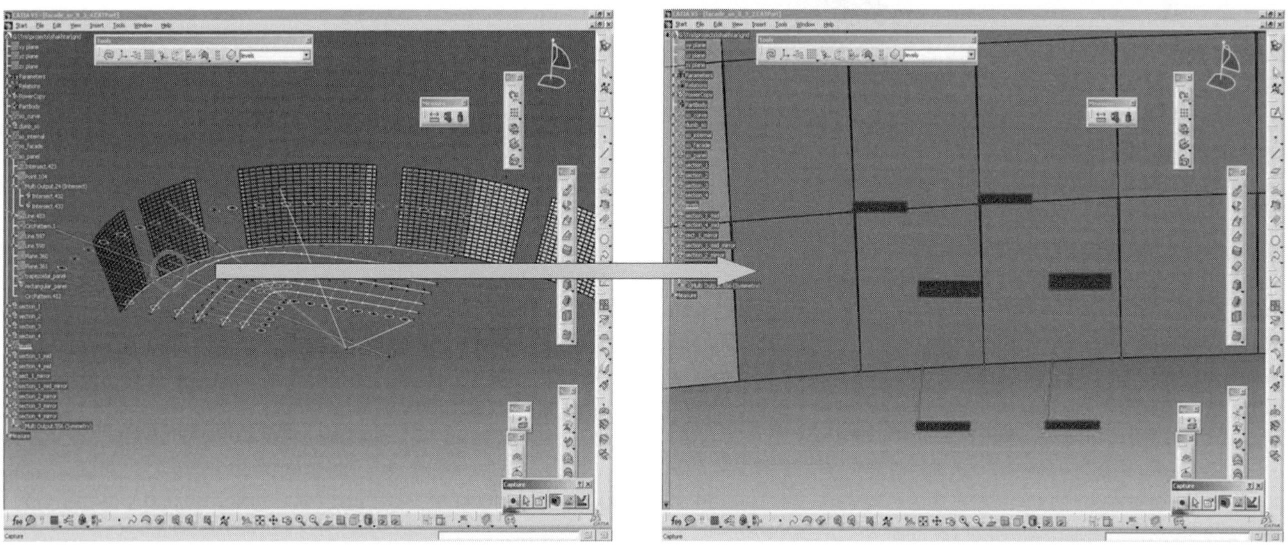

(b)

facilitates the interaction with tactile scale models without the expense or time delays of commissioning traditional scale models.

Accurate scale models are also important in other simulations. Relevant to stadium design is the use of the rapid-prototype within the wind tunnel. Using 3D printing technology it is possible to quickly test different configurations and the effect of design changes. It is also now possible to incorporate the tubes for the pressure testing sensors within the scale models during the printing process.

As the technology continues to evolve and become more accessible, rapid prototyping will become more common. It is also highly likely that the technology is not only used to produce a prototype but also the actual final product, allowing mass customisable objects and leading to a paradigm shift comparable in magnitude to the first industrial revolution. Examples are already starting to occur such as Arup's work on 3D printed structural nodes in metal for tensegrity structures and an experimental pavilion. The latest results show the design freedom the technique allows for as well as the reductions in size and weight that can be achieved (Figure 11.8).

Information communication technology (ICT)

As the computerised techniques for stadium design increase in scope, capability and complexity, the need to share and manage data between project stakeholders also increases. This is perhaps the least recognised area of computation design technique, but is fundamental to successful projects. Holistic design or 'total architecture' (Arup, 1970) can only occur if people make the right decision at the right time using the right information. This, in turn, can only occur with the correct data management and collaboration tools.

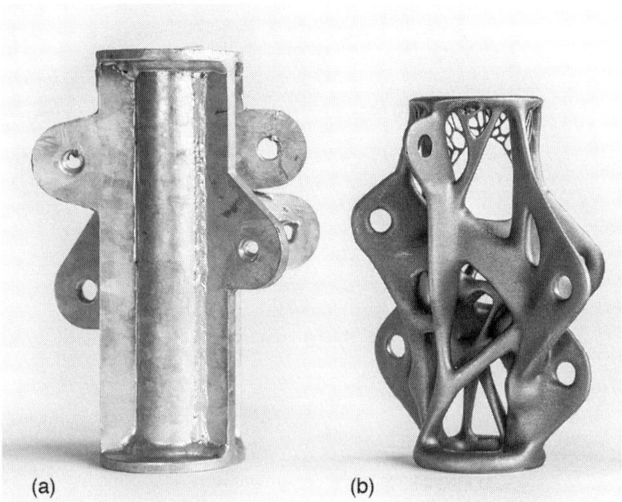

Figure 11.8 A metal node showing (a) traditional manufacturing; (b) metal 3D printing

National and international collaboration is a fundamental requirement for the success of large stadiums and sport venues. The emergence of distributed collaboration systems using cloud-based technology will undoubtedly play a significant role in the future of these types of project, but they are not without risks. The challenges that need to be addressed are security of data, protection of intellectual property and longevity of data. If such systems are used to store operational data as well as design or construction data, then there are additional challenges in terms of both the volume of data that will be generated over the life-time of a venue and also how this quantity of data can be distilled to allow meaningful decisions to be taken.

Construction

Computational techniques continue to be an essential tool in the design of stadiums and other sport venues. Good design should always consider the practical implications of fabrication and construction. In addition to the design tools previously mentioned, it is also worth considering the computational techniques that are being used in the construction of stadiums and venues.

4D simulation

The fourth dimension is often attributed to time (Figure 11.9). With sport venues such as stadiums, time is usually critical as there is often an unmoveable deadline such as an opening ceremony or the first game of a season. Managing the process of construction is extremely complex but the ability to simulate the construction sequence is a powerful tool.

Sport venues can present other challenges to construction that are not immediately obvious. In many cases, new stadiums are actually extensions or enhancements to previous grounds. In these cases the construction sequence needs to be carefully planned and managed to ensure that the stadium can remain operational during the modifications. In other cases the pitch may be natural grass that needs to be pre-seeded. The requirements for growing pitches may not be immediately obvious to construction professionals but can be fundamental to the needs and requirements of the sport.

One of the significant challenges with stadiums is the construction of the roof. Roofs can be extremely complex systems and have significant implications on the construction sequence. In some cases, the construction sequence can become a critical-load case which, as in the case of the FC Twente stadium roof (Structural Safety, 2012), can be fatal if not adequately considered.

Offsite manufacture

In many areas of the AEC industry productivity is decreasing. However, the steel and precast concrete industries are showing significant increases in productivity through the adoption of improved manufacturing techniques (Eastman and Sacks, 2008).

As a building typology, stadiums are very suited to a manufacturing approach. They are assembled from a relatively well-understood series of components that have clearly defined relationships. Parametric design for precast terrace units and other components could be extended to include the manufacturing process.

Temporary venues that need to be demounted and reconstructed in other countries require, by default, an 'assembly' approach to construction. Therefore, there is significant opportunity to drive innovation in a factory-based approach to construction for temporary venues for the world's major sporting events.

Ultimately, total productivity will increase when construction projects are assembled on site rather than being built on site. Whether the AEC industry can ever fully adopt the concurrent design processes seen in other manufacturing industries remains untested. A fundamental change in philosophy is actually required, and at this moment there are considerable barriers to adoption. Perhaps the most obvious is that the concurrence between manufacturing and design can only occur if the manufacturers and the designers are actually working concurrently.

Operation

Computational techniques have made a significant contribution to design and are becoming more common in construction. An area that has yet to see significant adoption of computerised techniques is the operation phase of sports venues. By contrast, this is the phase during which the majority of whole-life cost is accrued and the phase that the spectators/fans are most interested in. There are emerging computational techniques that

Figure 11.9 4D simulation of Singapore Sports Hub (© Dragages Singapore/Arup)

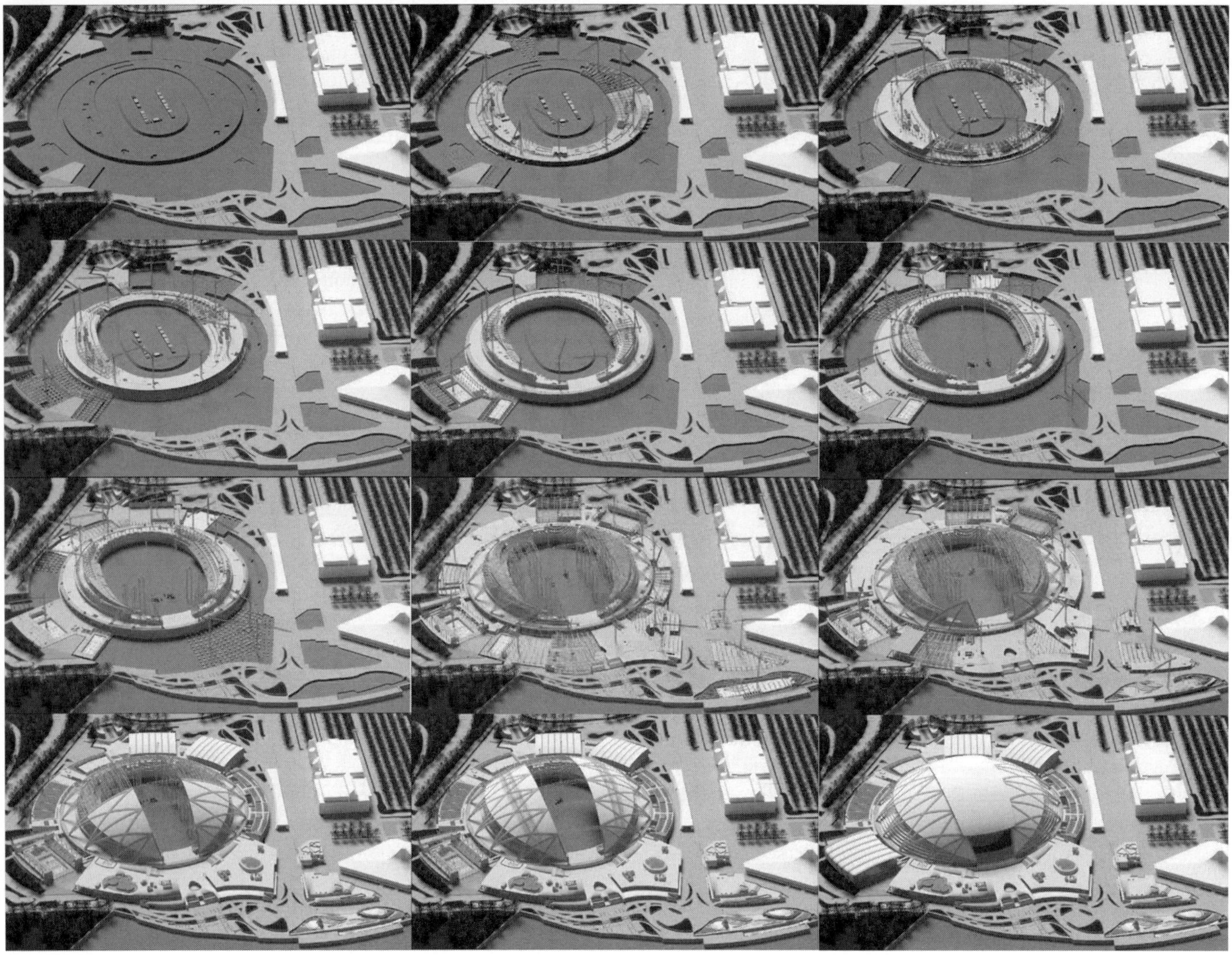

are applicable to operation. The study of venues during operation is also required to drive future design and thereby close the life cycle of the project.

Spectator simulation

Design tools originally developed to simulate evacuation of spectators can also be used to model normal operation of sporting venues. Queuing is an unfortunate product of the nature of some sport events, with a relatively short interval to visit concessions and/or toilets. A client may feel that when spectators are queuing for toilets they are not buying goods from concessions. By using simulation techniques it is possible to model spectator behaviour and test layouts to see how they perform during operation, thereby enhancing both spectator experience and client cash flow (Figure 11.10).

Asset management

Computerised techniques already benefit clients through better buildings and lower project risk. Emerging technologies also provide additional scope to assist clients in the operation of sport venues:

- *Operational efficiencies* can be achieved by linking sensor data from equipment and components to centralised building management systems that can collate and compare how assets are performing and highlight undesirable behaviour or conflicts.
- *Scheduled maintenance* can be improved by collating operational and maintenance data in a central database to record the frequency of services/replacements to equipment and components. Using this information to

Figure 11.10 Simulation of spectators during normal operational events

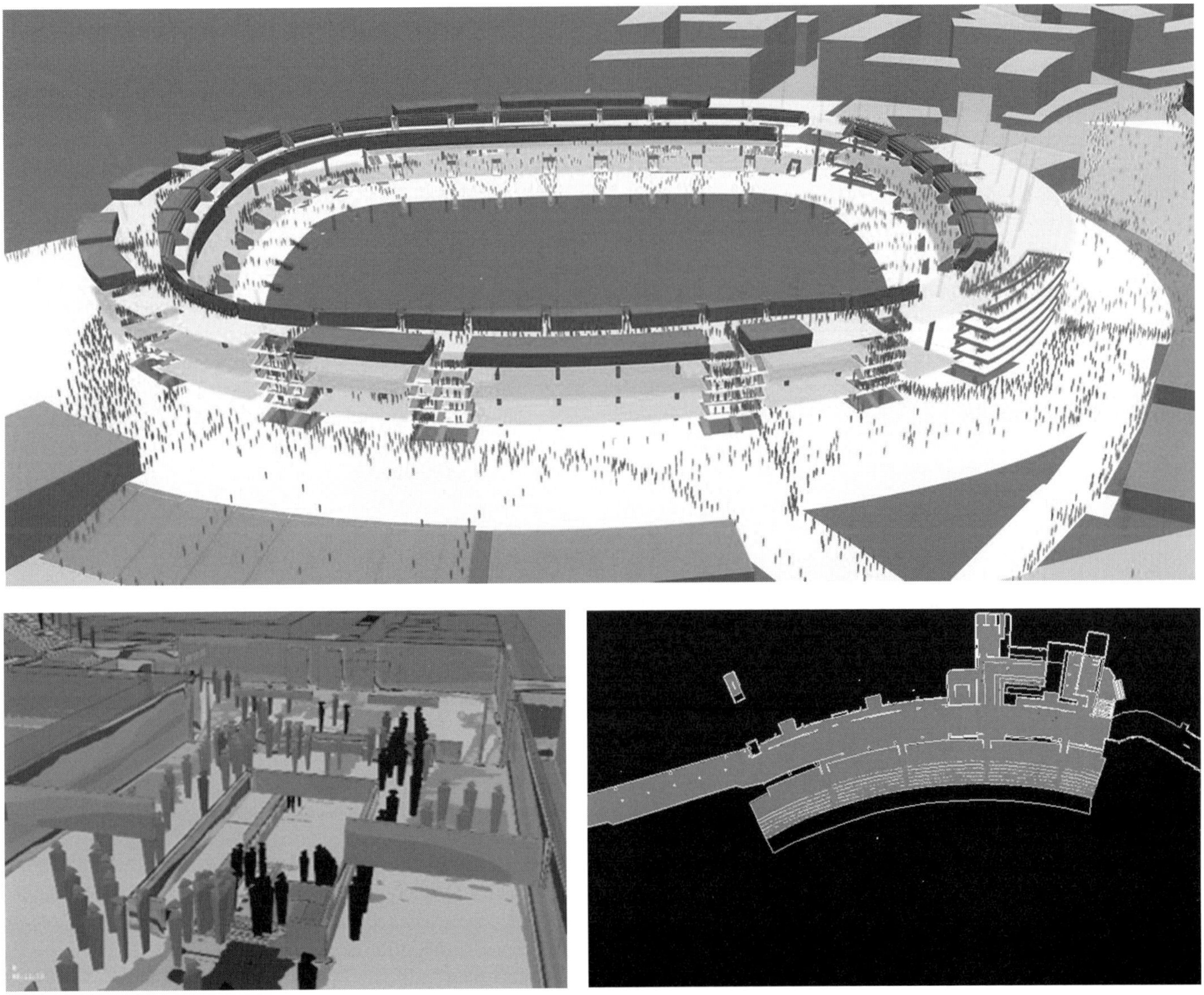

plan service/replacements can improve the operational efficiency of assets.

- *Reactive maintenance* and the interaction between building users and building operators can be improved by the adoption of social media and linking complaints and maintenance requests to components, which can in turn feed directly into the scheduled maintenance tools.
- *Dynamic maintenance* takes the operational data about how a component is behaving operationally and makes decisions to upgrade or replace based on the performance data in comparison with more efficient components.
- *Risk-based maintenance* compares operational data obtained through sensors with expected normal operation, and then makes a decision about whether to replace/maintain ahead of schedule.

Many of these techniques are well understood in safety-critical assets such as power stations. The decreasing cost of sensor technology and the increasing sophistication of computerised techniques will make them more commonplace. This is particularly important for complex assets such as large sport venues.

Building information modelling (BIM)

BIM is gaining popularity within the AEC industry and can be described as a 'coordinated digital dataset that contains

Figure 11.11 The three areas of BIM

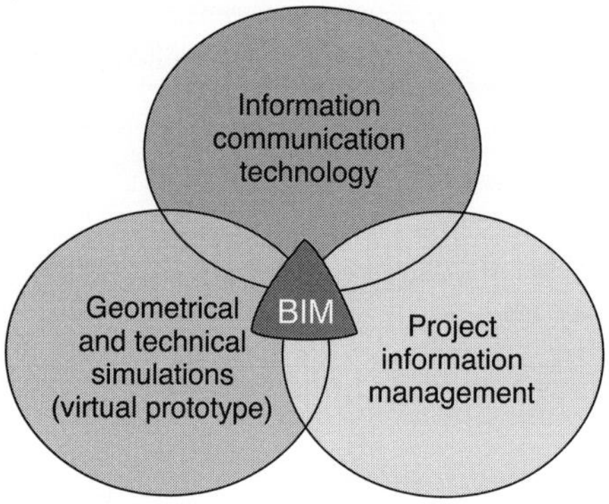

appropriate computable information necessary to design, build, operate and ultimately decommission an asset'.

From this definition we can identify that BIM is:

- not just applicable to buildings but to any asset
- appropriate for the whole project life cycle
- software agnostic (in fact BIM is not a tool, but a process enabled by technology).

Fundamentally, BIM is achieved by sharing structured information and allowing project stakeholders to collaborate in order to deliver the best possible project. It can effectively be summarised as 'the *right people* making the *right decisions* at the *right time* with the *right information*'.

BIM comprises three areas (Figure 11.11), which are equally important:

- *Geometrical and technical simulations*. The project team must have confidence that the solutions they propose will work in the real world. BIM allows the project team to build and test the project many times virtually before committing to building it in the real world.
- *Project information management*. Members of the project team must be able to share and reuse information produced by others in their own simulations. This is the process by which project team members know who is producing what information, when it will be available and what form the information will take.
- *Information communication technology*. ICT provides the backbone for BIM. It facilitates the storing, sharing and archiving of information. It gives confidence that everyone is working on the latest information and that there is only 'one version of the truth'.

Conclusion

This chapter has discussed many of the requirements and techniques for the computerised approach to the design, construction and operation of stadiums. It has focused on geometrical and technical simulations that comprise the virtual prototype. When combined with effective project information management and ICT, the application and adoption of BIM to stadiums and other sport venues can yield significant benefits to the client, project team and other stakeholders.

REFERENCES

Arup Sir O (1970) *The Key Speech*. http://www.arup.com/~/media/Files/PDF/Publications/Corporate/the_key_speech.ashx (accessed 12/02/2015).

Eastman C and Sacks R (2008) Relative productivity in the AEC industries in the US for onsite and off-site activities. *Journal of Construction Engineering and Management* **134:7(517)**: 517–526.

Flager F, Adya A and Haymaker J (2009) *AEC Multidisciplinary Design Optimization: Impact of High Performance Computing*. CIFE Technical Report TR186. Center for Integrated Facility Engineering, Stanford University, Stanford, CA, USA.

ISE (Institution of Structural Engineers) (2008) *Dynamic Performance Requirements for Permanent Grandstands Subject to Crowd Action*. ISE, London, UK.

Structural Safety (2012) *FC Twente Stadium Roof Collapse – Learning from the Fatal Consequences*. SCOSS Topic Paper. See http://www.structural-safety.org/media/363023/fc-twente-stadium-roof-collapse-final-1-oct-12-revised-july-2014.pdf (accessed 12/02/2015).

Stadium and Arena Design
ISBN 978-0-7277-5790-6

http://dx.doi.org/10.1680/saad.57906.133

Chapter 12
Concrete technology in stadium development

Ralph Belperio Aurecon, Adelaide, Australia

Forms of concrete used in stadiums

Concrete can be used in a myriad of ways in modern stadiums. In its simplest form, passively reinforced concrete can be used for most structural applications that may be encountered in stadium construction. With maturity of the local markets, more sophisticated forms of concrete construction are possible. These forms include:

- precast elements
- post-tensioned elements
- pre-tensioned elements
- proprietary products (such as hollowcore and the like)
- elements composite with formwork (steel or precast deck).

The use of concrete in stadium roofs is also relatively common, where mass is required to resist the uplift on cantilevered roofs and the introduction of tie-downs is to be avoided to maintain clear sightlines.

Site issues

Modern stadiums are constructed in a variety of locations, economies, communities and environments. From refurbishments of existing stadiums in bustling city centres, invariably with the need for continuing use during redevelopment, to brand new stadiums built on greenfield sites, complete with all the associated necessary infrastructure, the challenges facing the designers of these facilities differ significantly from project to project.

The vast majority of these developments are heavily time constrained, almost always having a requirement to be ready for a certain event or season. This is often a vital component of the business case in justification of the redevelopment.

These unique issues can have a significant impact not only on the selection of materials for certain applications but also on the type of a particular product chosen (e.g. whether to use in situ or precast concrete elements, passively reinforced or stressed).

There is a myriad of other factors that influence material selection, not the least of which is the maturity and capability of the market in which the stadium is to be constructed.

Design issues

Dynamics

Design for dynamics is critical in the performance of modern-day stadiums. The inherent mass of concrete is often beneficial in controlling the dynamic behaviour of structures. Such behaviour is particularly evident in grandstands, where groups of supporters may jump in a rhythmic motion, which can result in excitation of the vibration modes, with several famous examples. The advent of advanced modelling tools, including finite-element analysis, allows the dynamic performance of structures to be analysed in significant detail. Models that consider the interaction of the seating plat, the supporting raker and the concrete frame as a complete unit are now commonplace, and by adjusting the relative stiffnesses of the individual elements the dynamic performance of the structure can be optimised.

Seismicity

Design for seismicity varies in locations around the world. Significant decisions need to be taken about whether to lock structures together and design them to resist the seismic actions as a single unit, or to seismically separate the various elements of the structure. The latter invariably introduces complications of moving joints between isolated structures and accommodation for movement to occur between adjacent structures without causing distress to the structural fabric, or to services that necessarily need to bridge the movement joint. In our recent experience with stadiums in (seismic) New Zealand, grandstand structures were built as separate units to allow shrinkage movements to occur, but were locked together in their final configuration to act seismically as a unit. This removes the complexity of having to design services and other items to bridge the seismic joints, which may need to accommodate movements of up to, say, 200 mm (approximately 8 in.).

Durability

Design for durability is important in the context of the specific application within the structure, and the exposure conditions encountered. Concrete has the advantage that, when properly detailed and maintained, the life cycle can be matched to or beyond the design life of the stadium. While steel structures usually rely on protective coatings or galvanising technology to achieve a design life, the concrete design can account for the specific exposure condition or proximity to the coast of the individual element. Concrete surfaces are also durable as a trafficable wearing surface, and usually require less maintenance than other trafficable surface finishes.

Shrinkage and creep

When concrete is poured and allowed to set it shrinks, mainly due to the loss of water from the mix. The shrinkage of concrete can be detrimental to a structure, as it can cause widespread cracking. The shrinkage process continues for years after the concrete has been poured. However, the majority of this movement occurs in the early stages after pouring, particularly the first 1–2 months.

Another phenomenon that adds to the deformation caused by shrinkage is creep. Creep occurs when there is a load applied to the concrete, which then deforms over time and is largely irrecoverable. The deformation due to creep follows the same pattern as shrinkage; that is, when the load is applied, the majority of the deformation occurs early but also continues over time. Typically, the creep deformation is in the same direction as shrinkage, and therefore results in an overall shortening of the concrete unit.

Stadiums often contain long conjoined concourses over hundreds of metres. The summation of these movements can be a critical consideration for the design team. Consideration may need to be given to delaying the locking up of concrete elements to allow the majority of the expected movements to occur, thereby minimising the risk of permanent distress to the structure. In some cases, it may be necessary to make temporary arrangements to allow the building to go into service prior to the contractor returning to site to finally lock the building into its final configuration.

Strength

The availability of higher strength concrete has improved in recent years. A review of recently constructed stadiums indicated that concrete strengths utilised varied across projects and continents. This is due in part to the ability of the local market to supply concrete in the quantities required for large stadium projects. The case studies presented below highlight the differences in approaches for different projects. At the Melbourne Cricket Ground (MCG), concrete strengths were increased as necessary up to 80 MPa to maintain the consistency of structural sizes throughout different areas of the project. In contrast, all of the Peter Mokaba Stadium in South Africa was designed to use 30 MPa concrete, for ease of supply and in consideration of the need to site batch concrete for the project.

Fire performance

An advantage of structural concrete over other construction materials, when detailed appropriately, is its inherent fire resistance, which facilitates the ability to design for fire performance, without reliance on coatings or the high-end fire-engineering analysis that may be required for steel structures.

Aesthetics

Many contemporary stadiums utilise concrete as a practical, attractive and durable finish. The sculptural forms that are possible in concrete are often exploited by stadium designers as iconic, expressed features of the stadium.

Thermal mass

The ability of concrete to act as a heat sink can be beneficial in managing patron comfort in stadiums and in achieving sustainable outcomes. In warm climates with cooler evenings, concrete seating plats can act as a heat store from the radiant solar gain during the day, to assist in managing the ambient temperature in the bowl during cooler evening events. For day events in hot climates, the reverse can be also true, where the concrete structure can maintain a cooler environment in hot climates as a result of the structure being cooled by ambient conditions overnight.

Waterproofing

Ideally, concrete structures are designed and detailed to be waterproof in their own right. Practically, this is not always achievable. The propensity for concrete to shrink and crack results in the formation of paths for moisture ingress to occur. Pre- and post-tensioning can be effective in making concrete waterproof, as the tensioning process minimises the formation of cracks that occur in the concrete due to shrinkage, and can assist in maintaining a waterproof envelope.

Many admixtures and coatings are available to improve the waterproofness of concrete. The way in which each of these treatments improves the performance is specific, and requires diligence from the designers to understand the benefits and shortcomings of each proprietary system.

Concrete properly designed to remain waterproof should provide good service for many years, with a minimum of maintenance.

Construction issues

Many issues affect the ability to construct in concrete. Construction issues need to be carefully considered on a site-by-site basis.

Issues that need to be considered include:

- concrete availability (cement, aggregates, etc.)
- concrete performance
- proximity to site of ready mixed plant
- raw material suitability (particularly aggregates and effect on shrinkage performance)
- cost of labour
- skill of local labour and experience
- skill of local precast industry
- safety culture
- cranage (availability and access)
- supply logistics
- formwork quality and aesthetics
- programme
- pumping/height issues
- access to site
- existing upgrade versus greenfield construction.

These are but a few issues that can affect the selection of concrete as a construction material in stadiums. Designers should invest early in research to inform themselves of what is achievable considering the specific constraints of the project in question. Investing in this knowledge early in the project can reap the dividends of sound design and of minimising rework.

Summary

In this chapter we have endeavoured to provide an overview of some of the many issues that face stadium designers and influence the selection of materials in stadiums. We have focused on concrete as a material of choice for modern-day stadiums. Below are presented two case studies that highlight the differences in approach necessary on the basis of the criteria described above.

Case Study: Melbourne Cricket Ground, Melbourne, Australia

The Melbourne Cricket Ground (Figure 12.1) is one of Australia's premier sporting venues (Sheldon *et al.*, 2005). Built in 1853, the MCG has undergone continual upgrading to its current configuration with a capacity of in excess of 100 000 patrons. The most recent upgrade to the northern side of the MCG was completed in 2006 for the Melbourne Commonwealth Games, and comprises capacity for some 52 500 patrons, including 3000 standing.

The recent MCG redevelopment features many different examples of the uses for concrete in the construction of contemporary stadiums. Due to the enormity of the task, the construction and design team developed a series of different structural systems, each chosen to be the right solution for that particular set of circumstances.

Figure 12.1 Melbourne Cricket Ground, Melbourne, Australia

Figure 12.2 The new northern stand at MCG, indicative section

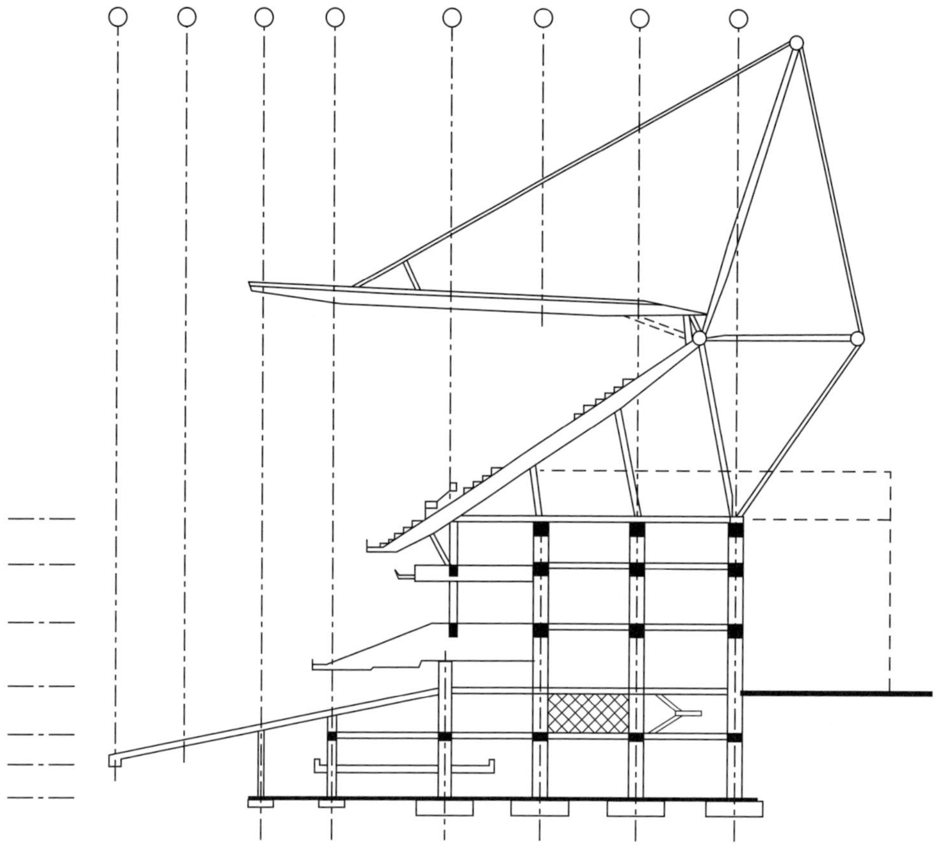

The structural frame

The northern stand is a 400 m (1312 ft) long seven-storey building, with two basement levels. The team looked at many alternative structural schemes before adopting a solution of what is effectively five buildings, separated by movement joints. Within each of these buildings, shear walls have been incorporated in the radial direction to provide stability and maximise ceiling zones for reticulation of services around the stand. In the circumferential direction, however, shear walls would hamper viewing areas, so beams have been provided, working with the columns as a sway frame (Figure 12.2). This assisted with support of the floors, as hollowcore planks span radially for the typical suspended floors of the stadium – a win–win situation.

Another aspect of the project, where concrete has been used extensively, is in the retention system to the double basement. Bored piers were drilled from the surface level and shotcrete panels sprayed between the bored piers progressively during excavation. Interestingly, wherever possible large-diameter bored piers are designed to cantilever out of the base in the temporary condition, rather than smaller-diameter piers restrained by temporary ground anchors.

The floor system

With over 10 ha (24.7 acres) of floors to construct, the contractor elected to use a number of different flooring systems throughout the project. Examples of the types used, and the reasons for selecting these systems, are summarised in Table 12.1.

The typical grandstand circumferential beams are post-tensioned and span approximately 16 m (52.5 ft). To maximise services zones, the design incorporated haunches in these beams, with depths of 900 mm (35.4 in.) at columns, stepping down to 600 mm (23.6 in.) at midspan. Beam depths were set to maximise the use of standard forms, recognising that to speed up construction the hollowcore planks were to be supported by the formwork prior to the beams being poured. Considerable coordination was undertaken by the project team, particularly at the many floor set-down areas, to achieve this goal (Figure 12.3). This enabled the hollowcore planks to be used as a working platform for installation of reinforcement and post-tensioning tendons in the circumferential beams. The structural topping to the hollowcore was poured simultaneously with the beams, to minimise construction time on site. In a

Table 12.1 Types of flooring system used at the Melbourne Cricket Ground

Structural system	Locations used
Conventional concrete construction (formwork, reinforcement and in situ concrete)	Areas with large penetrations or where the floorplate was curved in plan Cantilevered slab edges or slabs requiring cast in plates Main circumferential beams that support the hollowcore precast floor planks on levels 1–4 Inclined columns in the main entry atria Level 1 seating plats using bespoke jigs to form up the seating plats incorporating a prefabricated reinforcement system
Precast floor planks (hollowcore)	Typical level 1–4 suspended slab areas with minimal penetrations in the floor plate
Pre-tensioned precast concrete seating plats	All seating areas on levels 2, 3 and 5
Proprietary metal composite formwork slabs	Areas with very high soffits, such as the bar over the members' atrium Areas with tight floor-to-floor height, such as the long room mezzanine Areas with sloping soffits not suitable for hollowcore Non-typical areas where the architectural finish to the soffit was not a driving issue

cooperative approach, the floor system was designed in conjunction with the supplier as continuous construction. The supplier was also responsible for the design of the planks, with the design team responsible for the top flexural reinforcement over supports. Particular attention was paid to the shear capacity of the planks in zones of negative bending.

The seating tiers

A feature of the stadium evident to all spectators is the large cantilevering steel rakers that support the upper seating tiers (Figure 12.4). The largest of these cantilevers are approximately 11 m (36 ft) and are spaced approximately 12 m (39.4 ft) apart.

To support the folded precast concrete seating plats with crowd loading, a significant steel beam was required. These elements were designed for more stringent dynamic performance, to meet the required criteria.

The designers developed an analysis methodology that modelled the interaction of the steel beams, the concrete supporting frame and the precast seating plats. By adjusting the relative stiffnesses of the respective elements and developing a rational design criteria, the designers were able to save the client in excess of AUD$1 million compared with other stadiums designed using less sophisticated 'deemed to comply' approaches. Even with this reduction, steel beams up to 2.7 m (10.6 ft) deep with 80 mm (3.15 in.) thick flanges were required.

The upper level seating tier consists of 36 m (118 ft) long boxed steel girders supported on sloping tapered columns, and carries

Figure 12.3 The new northern stand at MCG under construction

Figure 12.4 Construction of the seating tiers at MCG. (This figure is reproduced in the colour-plate section as Plate XIV)

considerable forces from the roof down to the superstructure. Because of the large horizontal thrust forces, up to 18No. × 15.2 mm (0.6 in.) diameter post-tensioned strands are coupled to the steel's baseplates and traverse the top floor at each grid to counterbalance the opposing forces.

A similar challenge was how to connect the 2.7 m (8.86 ft) deep steel cantilever beams at level 2 back to the concrete frame. Typically, 36No. × 15.2 mm (0.6 in.) post-tensioned strands are anchored though a 3 m (9.8 ft) deep concrete backspan beam into the superstructure. These required careful detailing to avoid significant congestion of reinforcement at the junction with the columns, while maintaining the required structural behaviour.

Precast columns

In keeping with contractor's philosophy for the project of prefabrication for speed of erection, many of the columns are precast concrete. Strengths of up to 80 MPa were used to enable the sizes of columns to be standardised, reducing the number of moulds. The use of precast columns expedited site construction times, but required additional input from the design team in terms of coordination prior to fabrication.

Foundations

One of the areas of risk identified early in the design phase was potential delays associated with excavation for footings on a site with a history of previous buildings. It was also known that a dyke of weaker material, the extent of which was not fully known, traversed the site. As a result, the design and construction team decided to reduce the number of columns through the structure in the final design. This increased the spans of the circumferential beams and the stability moments in the columns, but reduced the number of footings. Allowable bearing pressures were increased on the basis that significant additional inspections be undertaken by the geotechnical engineer to validate the assumptions.

A novel use of concrete

The brief for the main roof over the new northern stand was for an elegant, light and transparent roof structure that maintained the MCG's iconic status.

The engineering team developed several options for the roof structure, and tested each against criteria, including architectural, cost and staged-construction requirements. The adopted solution involves an innovative tension structure with rafters suspended from cable stays running over individual masts.

Figure 12.5 The bowl of the MCG during construction

Utilising counterweight to resist uplift wind forces, the structure also makes use of a cable net to distribute peak wind loads determined from wind tunnel modelling. The engineering team was then able to engage a large portion of roof weight to resist uplift winds – which occur in peak zones – resulting in a very efficient and light structure. The cable net also serves the purpose of restraining the top of the individual masts, eliminating the need for a bulkier bracing system.

The primary roof structure comprises fabricated steel box-section rafters ranging in length from 37 to 42 m (121–138 ft). The rafters are concrete filled to provide counterweight against wind uplift, and to further assist this a front hoop beam, consisting of a 600 mm (23.6 in.) square fabricated steel box section, was located towards the front of the roof and is also concrete filled for ballast. The fabricated steel masts are tapered to give a further element of slenderness to their appearance, and vary in height from 13 m (42.7 ft) to 28 m (92 ft) (Figure 12.5).

Case Study: Peter Mokaba Stadium, Polokwane, South Africa

The Peter Mokaba Stadium (Figure 12.6), in Polokwane, in the Limpopo province of South Africa, was constructed for the 2010 FIFA World Cup. The stadium is named after Peter Mokaba (1959–2002), who was a hero of the struggle against apartheid and was imprisoned on Robben Island (Bokelman and Mentz, 2010).

The crisp, modern design was actually inspired by the baobab (*Adansonia digitata*), an ancient tree revered by the local population. At each corner of the steel and concrete structure is a squat, cylindrical structure that resembles the tree's muscular trunk (which can grow up to 28 m (92 ft) in girth).

In the 2010 FIFA World Cup, the stadium, which has a seating capacity of 45 000, proudly hosted four games.

In contrast to the MCG, this stadium is an example of a new stadium built essentially on a greenfield site. In this case study we will endeavour to highlight the use of concrete in this modern contemporary stadium.

The structural frame

The stadium comprises grandstands on all four sides of the rectangular pitch. The northern and southern stands are the shorter of the four stands, addressing the ends of the pitch. The eastern and western stands, while appearing similar, are different in that only the western stand is roofed. The four stands are structurally isolated from the radial structures in the four corners, which include the squat, cylindrical structures referred to above.

The western and eastern stands are approximately 140 m (460 ft) long, five-storey high buildings. The northern and southern stands are approximately 90 m (295 ft) long and of similar height to the western and eastern stands. The radial structures in the corners of the pitch, which join the four grandstand structures, include the circular structures that pay homage to the baobab, as described above. These corner structures provide the primary support to the long-span 175 m (574 ft) roof girder. The stadium was planned to have roofs over all the stands but, due to budget constraints, a roof was only provided for the main west grandstand. The corner structures also serve as the primary routes for vertical circulation to the various stands on the upper tier, with spiral walkways incorporated around the outside of the circular structures and lifts on the inside (Figure 12.7).

In contrast to the MCG, the design team for the Peter Mokaba Stadium had a different set of criteria and drivers that influenced the selection of material and solutions for the project. Key issues influencing design decisions included:

- an abundance of space on site to establish plant and layoff areas

Figure 12.6 Peter Mokaba Stadium, Polokwane, South Africa

- the need to site batch concrete for the works
- solutions that were within the capability of the local contractor and market
- minimum procurement of materials outside South Africa.

Figure 12.7 Joints and structural modules at the Peter Mokaba Stadium

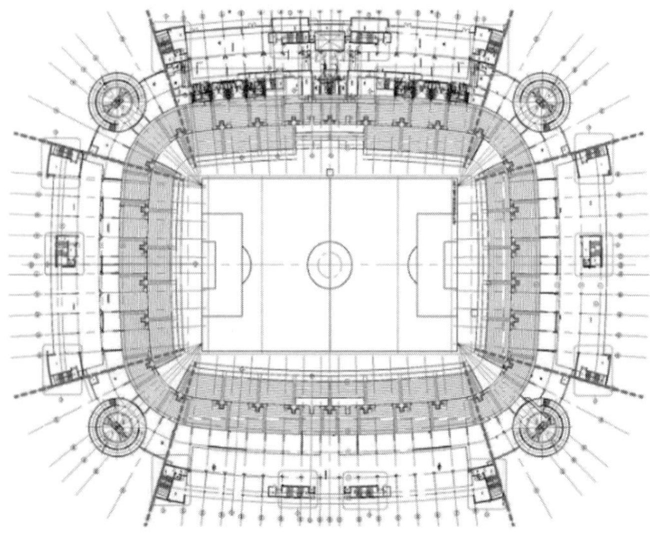

In response to these criteria, the structural framing is entirely of in situ concrete construction. The contractor invested heavily in very high-quality steel formwork (three for lower-tier raker beams and three for upper-tier raker beams, including the Y-columns) and these were reused repetitively throughout. The upper-tier raker beams and Y-columns (Figure 12.8) were cast

Figure 12.8 Y-columns at the Peter Mokaba Stadium. (This figure is reproduced in the colour-plate section as Plate XV)

in one pour to cut down time on erection of formwork. Although the vomitories and some of the associated elements were designed as precast elements, the contractor elected to cast these in situ to the same dimensions using high-quality formwork.

With the requirement to site batch concrete, the design strengths were relatively conventional with 30 MPa concrete used throughout for the in situ works.

The construction programme was under severe pressure from the onset, due to the late appointment of the design team and contractor, and this required various ways of speeding up the process. The structure was therefore designed such that a very high degree of repetitiveness could be achieved. This resulted in an acceleration of the construction progress, with the labour force becoming more proficient due to the constant repetition of the same process.

The floor systems

The floor systems on this project can be simply classified into two types. Examples of the types used, and the reasons for selecting the systems, are summarised in Table 12.2.

The typical grandstand raker beams are passively reinforced, and typically are nominally 1000 mm (39.37 in.) deep by 500 mm (19.685 in.) wide, spanning some 9.2 m (35.8 ft).

Floor-slab thicknesses generally vary between 250 and 300 mm (10–12 in.) and the slabs are of typical in situ concrete construction. Passively reinforced concrete beams are used along the column lines to provide frame action together with the raker beams. A number of shear walls are used perpendicular to the frames for lateral stability.

The eastern and western stand floor slabs were cast with two 4.5 m (17.7 ft) wide shrinkage strips to allow most of the initial shrinkage to occur before ‘locking up’ these very long structures of approximately 140 m (approximately 460 ft). One 4.5 m (14.8 ft) wide shrinkage strip was used for both the northern and southern stands. The shrinkage strips were cast between 60 and 75 days after the initial casting of the floor slabs, as approximately 85–90% of the initial shrinkage in the slabs had occurred by that time.

Table 12.2 Types of flooring system used at the Peter Mokaba Stadium

Structural system	Locations used
Conventional concrete construction (formwork, reinforcement and in situ concrete)	Used for all floor slabs throughout, except seating tiers
Precast concrete seating plats	All seating areas

Figure 12.9 Precast seating plats being constructed on site at the Peter Mokaba Stadium

The seating tiers

The precast elements for the seating plats were all cast on site (Figure 12.9). The contractor invested in a complete batching plant and curing area on site, where the completed elements were stored prior to installation. Although a design strength of 30 MPa was employed for the design of the plats, the contractor used 40 MPa for the precast elements in order to gain higher early strengths to allow the elements to be manoeuvred earlier from the casting facility to the storage facility on site.

From the stacking yard these elements were transported into the bowl using tractors and trailers. The elements were then lifted into position using a large 550 ton crane. Elements were designed and cast as single units (one row of seats between two raker beams) to decrease the size of crane required.

Columns

In keeping with contractor's philosophy for the project of simplicity in design and construction, all columns on the project are of cast in situ construction.

Foundations

On a greenfield site with well-known geotechnical conditions, pad foundations were deemed suitable. In some areas on the site, the founding rock was of a hardness such that blasting was required to create some of the foundation excavations (Figure 12.10).

Figure 12.10 Aerial view of the Peter Mokaba Stadium, October 2009

REFERENCES

Bokelman K and Mentz S (2010) Peter Mokaba Stadium – proudly South African. *International Association for Bridge and Structural Engineering (IABSE) days, April 2010, South Africa.*

Sheldon M, Guldager-Niesen E and Richardson S (2005) Concrete design and construction issues for the MCG redevelopment. *Proceedings, Concrete Institute of Australia Biennial Conference, Concrete 2005, Melbourne, Australia.*

FURTHER READING

Department for Culture, Media and Sport and Football Licensing Authority (2008) *Guide to Safety at Sports Grounds*, 5th edn (The Green Guide).The Stationery Office, London, UK.

Fatur E, Krebs J, Cummings G and Retzner J (2003) Architectural and structural precast/prestressed concrete renovates landmark Ohio stadium. *PCI Journal* **Jan/Feb**: 50–63.

Hardie MP, Miller G and Manley K (2005) Innovative connecting system for precast concrete planks on a major sports stadium. *International Group for Lean Construction*, IGLC-13, July 2005, Sydney, Australia (accessed 12/02/2015).

Korkosz W, Haris A and Andrews D (2004) Precast concrete transforms the University of Oregon's Autzen Stadium. *PCI Journal* **May/Jun**: 44–54.

Stadium and Arena Design
ISBN 978-0-7277-5790-6

http://dx.doi.org/10.1680/saad.57906.143

ice
Institution of Civil Engineers
publishing

Chapter 13
Formwork for precast/prestressed stadium components

Marianne Methven Hamilton Form, Fort Worth, Texas, USA

As stadium design and construction becomes increasingly more complex and competitive, the use of precast concrete has grown dramatically. Today, almost every stadium project makes use of some precast concrete elements. Structural and architectural precast components offer designers flexibility, durability and speed of construction. In addition, precast concrete is functionally resilient, providing protection from fire and flood, and can be economically designed to resist tornados, hurricanes and wind, all of which are great advantages in stadium construction.

Long spans and the ability to eliminate costly field formwork are key features that have contributed to the success of precast/prestressed concrete. Seating risers, raker beams, pedestrian ramps, mezzanine floors, stairs, suites, offices and locker-room areas can all be framed and constructed using precast/prestressed concrete. Architectural precast concrete delivers unlimited design potential for exteriors and wall panels.

Precast/prestressed concrete is produced in forms fabricated to hold rigorous dimensional tolerances, in a production facility, by trained personnel. This allows for a level of quality control only achievable in a production facility. Precast components can be produced in longer spans and larger panels to reduce installation time. Components that would require tall scaffolding towers to form in the field are simplified by being cast at a precast plant, delivered to the site and quickly lifted into place (Figure 13.1). Quality production, quick installation and long-lasting service make precast/prestressed concrete the most structurally sound and cost-effective option for stadiums and arena components.

Precast concrete stadium components

A number of different precast components are typically used in stadium and arena construction. Virtually all stadiums and arenas use precast concrete seating risers and raker beams, long, narrow, stepped pieces that support seating risers. Precast beams, double tees and wall panels are used as framing elements (Figure 13.2).

Seating risers can be made in single, double or even triple sections. Typically, double and triple risers are used because they go up quickly and reduce the number of crane picks and the number of joints required, making them extremely efficient to install and reducing facility maintenance. Risers are made in L shapes or a dropped stem or hammerhead shape. Walkways are typically produced by tub forms (Figure 13.3). Precast/prestressed products are plant cast, and can be stored at the plant and shipped to the site when ready for erection.

Just as the number of different precast/prestressed elements used in stadiums grows, so too does the demand for greater investment in the formwork that the components are cast in. Formwork must be fabricated to exacting tolerances and designed to resist the hydrostatic and compressive forces introduced when prestressing. Because formwork has a significant impact on the success of the project, it is important to secure the expertise of a company with proven experience in the field. Collaboration between the designer, construction manager, precaster and formwork supplier early in the process is the key to getting the most value from the form and the precast for the project.

Precast/prestressed formwork

Forms for casting precast concrete are typically designed and built for a specific stadium, and represent a significant investment. To be practical, the number of individual forms required for the project is determined by the number of different products specified, the quantity of each and the delivery schedule. To be profitable, the form supplier and precaster must determine which, if any, of the forms can do double, triple – or even multiple – duty. Occasionally, forms from a previous project can be used or modified for a new project. Even using one form from a previous project can save a significant amount of money.

Figure 13.1 Precast elements are plant-cast and delivered to the site when ready for erection. (Courtesy of Hamilton Form Company, Ltd)

While some stadiums call for relatively few forms, others may require as many as eight, 12 or even more unique forms. High-volume products may require a dedicated form, while other forms can be built to multitask. Multitasking, adjustable forms are relatively easy to build and can be adjusted by using inserts and fillers. Adjustable forms are especially cost-effective and practical for low-volume products.

The key to designing a stadium to get the most out of the formwork is to maximise the repetition of the components. Strategies such as keeping structural bays uniform and minimising the number of ovals and curves will increase the number of like products. Maximising repetition provides the most value from the form and set-up labour required to cast each piece. Although simple, repetitive patterns are very cost-effective, one of the benefits of precast concrete is its ability to accommodate virtually any product configuration. Both benefits are enhanced financially by the speed with which precast elements can be quickly set in place when they are delivered to the job site.

Figure 13.2 (a) Precast seating risers with raker-beam supports. (b) Raker beams supported by precast columns. (c) Precast concrete double tee floor systems are economical and durable, and allow for greater spans between beam and column supports. (Courtesy of the Precast/Prestressed Concrete Institute (PCI))

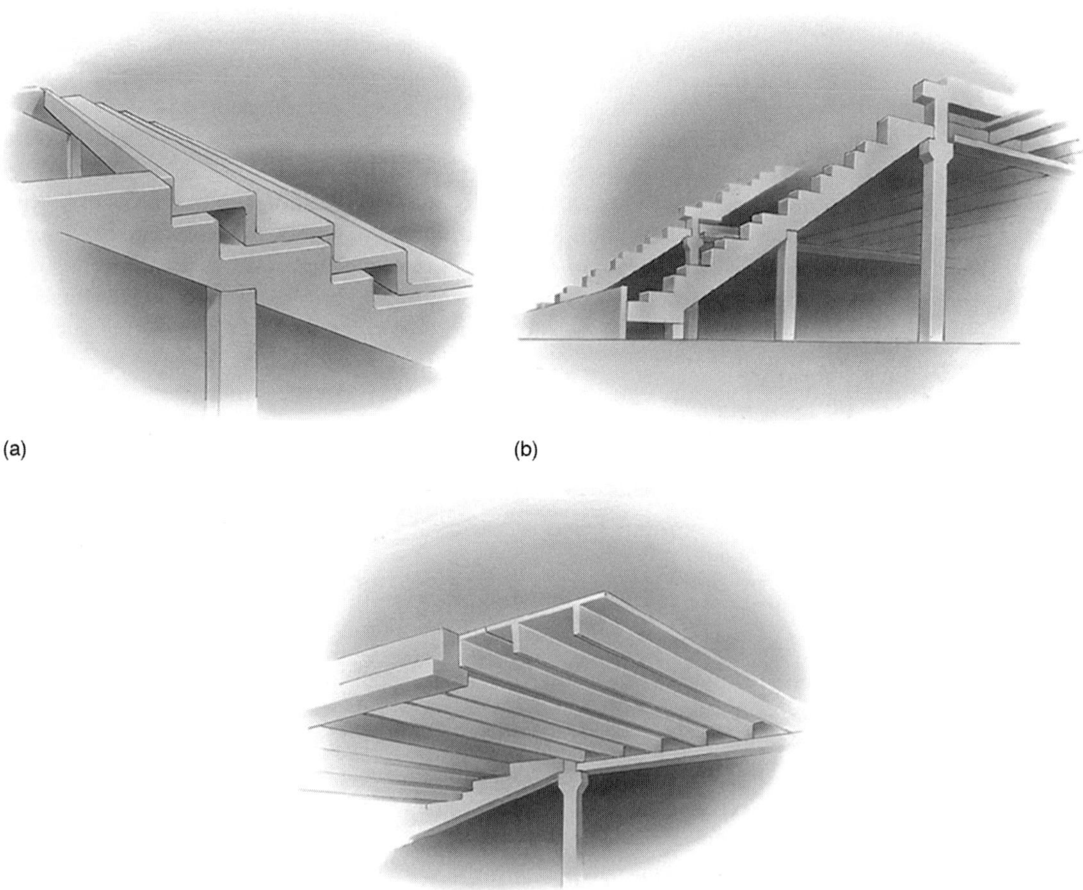

(a) (b) (c)

Figure 13.3 (a) Triple-L seating riser form. (b) Seating riser form with integral walkway. (Courtesy of Hamilton Form Company Ltd)

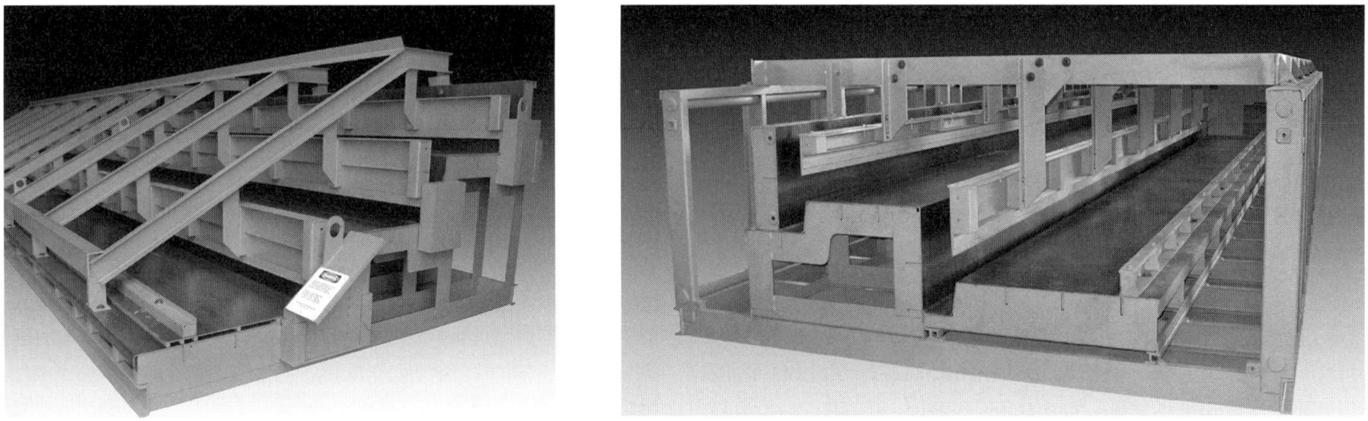

(a) (b)

Formwork for smaller stadiums can be especially challenging. Smaller stadiums usually have the same – or an even greater – number of different products than some of the larger stadiums. Because the investment in forms has to be applied over fewer pieces, the forms need to work harder. Smaller stadiums often utilise forms with a great deal of adjustability so they can cast a number of different products. Adjustments increase set-up time at the plant and make the forms more complicated, but can pay for themselves by reducing the overall cost of the project.

The best way to design precast/prestressed products for a stadium is with the help of the precaster, form manufacturer and an engineer that specialises in precast/prestress design. These specialists can help standardise components to take advantage of repeated form utilisation. Precast specialists can also help design load paths to simplify the structural design of the form.

Flexible formwork for small arenas

Hamilton Form Company is a custom precast form supplier in Fort Worth, Texas, that has designed and produced precast/prestressed concrete steel forms for many significant stadium and arena projects. An outstanding example of the use of Hamilton Form's modular formwork is the 12 000 seat INTRUST Bank Arena in Wichita, Kansas. The precaster was Prestressed Concrete, Inc. of Newton, Kansas.

To maximise the precaster's investment, Hamilton Form designed a pair of self-stressing double-L riser forms with different tread fillers and lifters to change riser depths and heights. Each form was used to cast about a dozen different products. Two forms with dropped stems, also known as 'hammerhead forms', were used for 11 more set-ups. A single form was designed to cast both the upper and lower tub sections. Providing base forms that could be easily modified resulted in needing only five forms to cast the entire arena. This practical approach limited the number of forms the precaster had to purchase, and helped trim capital expense for this relatively small arena (Figure 13.4).

Planning pays off

When precaster Coreslab Structures (Indianapolis) Inc. was awarded the precast contract for the Indianapolis Colts' football stadium, it called on Hamilton Form to develop the formwork. Working closely together, the different product configurations and quantities were determined. 'Building forms for a stadium requires a lot of planning', says Bill Daily, President of Hamilton Form. 'We've designed dozens of stadiums and always start by thinking through the most efficient and economical way to

Figure 13.4 Adjustable formwork allows several products to be cast in the same base form. Notice the multiple bolting locations used for adjusting the form. (Courtesy of Hamilton Form Company Ltd)

Figure 13.5 Large, thin brick-clad columns and wall panels are made off-site and quickly erected at the construction site. (Courtesy of Gage Brothers Concrete Products Inc.)

build the formwork. For the Colts' Stadium, many of the forms have fillers and adjustable headers to cast several products with the same base form. At the same time, because there were large quantities of several of the products, we didn't want to build forms so complex that too much time would be wasted in setting up and adjusting forms for different products.'

The stadium was built from the top down, with most of the precast erection taking place at night. Most of the products were double-L and triple-L seating risers designed with adjustable end plates to accommodate different strand patterns. The tub forms for the walkways had bolt-on hangers to make them easy to modify for various products. Matt Ballain, VP and General Manager at Coreslab Indianapolis said: 'This solution made change-overs quick and helped us remain ahead of a very demanding schedule.'

Modifications of old forms save big bucks

The AT&T Stadium, formerly known as the Cowboys Stadium, in Arlington, Texas, seats over 100 000. Formwork for this large stadium was made more economical by modifying forms used by the precaster for other stadium projects. Engineering firm CEG, precaster Heldenfels Enterprises and Hamilton Form met early in the project to discuss details. Hamilton Form sent a team of engineers to Heldenfels to inspect forms from previous stadium projects that the precaster had in inventory. Although several new forms were required for the project, several others were pulled from storage and modified for use on the Cowboys Stadium. In all, 1839 seating risers, 880 wall panels, 165 stairs and 109 flat slabs were required for the project. Use of some existing forms expedited delivery of forms to assist the quick turnaround that was required. New forms were delivered to Heldenfels to correspond with the casting schedule to keep the project on track.

The need for speed

When architectural firm Populous and engineering firm Walter P. Moore began design work for Target Field, the stadium built for the Minnesota Twins baseball team, it was quickly determined that both wanted to use precast elements throughout. 'Architecturally we wanted to use a very large, patterned natural stone on the exterior of the building. Those large pieces would have been difficult to set as a traditional, load-bearing masonry', said Bruce Miller, principal, Populous. Gage Brothers, the architectural precaster supplied over 1000 pieces of locally mined Quarry Creek vetter stone precast panels for the exterior of the stadium. Gage Brothers also supplied precast concrete, thin, brick-clad panels and columns (Figure 13.5). 'The precast solution allowed us to use large pieces and get them erected quickly', said Miller.

Slight shift, big savings

The structural precast products included columns, beams, raker beams, double tees and hollow core. To design the forms for the project, Hamilton Form worked closely with CEG, the engineering firm, and Mike Lewis of Hanson Structural Precast. As in most projects, the forms were designed so that simple adjustments could be made to cast several products in one form. One example is the tub forms used to cast the walkway sections. Each section included a sloping drain that shifts position from section to section, relative to the centre of the product. This was accomplished by building the forms in sufficient lengths to allow headers to shift position to accommodate the different drain locations. On the same form, CEG was able to relocate two strand positions to eliminate the need for extra steel-plate material at the top of the form, reducing the cost of the formwork.

Quick lift

Justin Barton, an associate with Walter P. Moore, served as project engineer on Target Field. According to Barton, the use of precast provided a number of advantages, including speed of erection, overall quality and the ability to span 13 m (42.5 ft) between raker bays. 'One of the things the precast erector used on the job was a suction or vacuum lift, as opposed to the typical embedded lifting devices', said Barton (Figure 13.6).

Figure 13.6 Vacuum lifting quickly positions precast components into place. (Courtesy of Hanson Structural Precast)

'That worked really well in terms of getting precast units up and less crane time per lift.' The vacuum lift allowed crews to lift and set as many as 30–50 pieces a day.

Arena design throws precast a curve

One of the benefits of precast concrete is that it can accommodate even complex designs. An example of an innovative formwork solution for a complex design is a front-faced adjustable double-riser stadium form that was built for an arena designed with seating sections with curved faces. To facilitate this design, the radius of each precast concrete section changes slightly as it curves around the arena. The radius also changes on each row from the lower to the higher sections.

Hamilton Form designed a form that could be modified for each curve change. Using hangers and steel shims, the radius adjusts in 1/16 in. (1.6 mm) increments (Figure 13.7). Risers and fillers were used to adjust the height of the product. The back pan also had both vertical and horizontal bolted adjustments. Without an adjustable form, it would not have been practical to use precast concrete for this project.

Figure 13.7 This riser form is designed so that the radius can be adjusted in 1/16 in. increments for curved seating. (Courtesy of Hamilton Form Company Ltd)

Architectural precast

In Blacksburg, Virginia, Virginia Polytechnic Institute and State University (Virginia Tech) wanted to do more than just match the existing buildings on campus with its new Hahn Hurst Basketball Practice Center: it wanted to impress.

Local precaster Smith-Midland hand-built custom wooden forms for exterior wall panels with the imprint of basketballs, the Virginia Tech logo and the school's seal cast directly into the

Figure 13.8 Architectural precast concrete can be designed to blend in with existing structures and add a personal element or accommodate almost any design. (Courtesy of Smith Midland Corporation) (This figure is reproduced in the colour-plate section as Plate XVI)

concrete 'The forms took a couple of weeks ... We did a mock-up and sent it to Virginia Tech to approve, because they obviously wanted to see what it looked like before they had basketballs and VT logos around the entire building.'

Smith-Midland was able to match the look of existing structures on campus. 'If you've ever been to Virginia Tech, they have what they call "Hokey Stone" all over campus', said Ashley B. Smith, president of Smith-Midland. 'The architect, Tae Jung, with Canon Design, incorporated the Hokey Stone into the practice facility and our precast panels blend right in.' The inspiration behind the design included incorporating the keys of a piano and the movement of a basketball into the panels, both of which have an affinity in terms of back-and-forth movement, along with the Virginia Tech logos and seal (Figure 13.8).

Meadowlands Stadium features open views with longer spans

Completed in 2010, the new \$1.7 billion Meadowlands stadium gives both the New York Jets and New York Giants a spectacular home field. The complex is the first specifically designed to house two National Football League teams.

The massive, 195 000 m^2 (2 100 000 ft^2) facility incorporates 41 340 m^2 (445 000 ft^2) of precast concrete. The seating bowl and sightlines were developed based on careful study of sections from other stadiums. The result is a stadium with 82 500 seats surrounding the entire field, providing excellent views even from the farthest seat.

The column-to-column spans are much larger than in most other stadiums. Typically, spans range from 11 m (36 ft) to 12.8 m (42 ft). In the New Meadowlands they are 15.2 m (50 ft) to 16.75 m (55 ft). The span increase helped reduce the number of beams that had to be erected, and the precast concrete was designed to minimise deflection, vibration and harmonics so that fans feel safe in the stands despite the larger spans.

Form meets function

Precast concrete is a structurally robust and durable material that offers several design advantages, including its ability to be used as spans and be quickly erected. The material is adaptable to almost any product configuration, and the design possibilities using architectural precast panels are virtually limitless. The quality of the precast concrete is directly dependent on the quality of the forms it is cast in.

FURTHER READING

D'Arcy T (2010) Large stadium projects using precast structural systems. *Proceedings of the 2010 Structures Congress*, Orlando, FL, USA, 12–15 May 2010 (Senapathi S, Casey K and Hoit M (eds)), pp. 2925–2936 See http://cedb.asce.org/cgi/WWWdisplay.cgi?263824 (accessed 12/02/2015).

Methven M (2008) Custom stadium formwork. *Concrete Plant International, North American Edition* **4**: 204–207.

PCI (Precast/Prestressed Concrete Institute) (2010) *PCI Design Handbook*, 7th edn. PCI, Chicago, IL, USA.

Stelsel K (2010) Raising the roof, precast takes sports stadiums to new levels. *Precast Solutions* **Fall**: 16–23.

Stadium and Arena Design
ISBN 978-0-7277-5790-6

http://dx.doi.org/10.1680/saad.57906.149

Chapter 14
Steel in stadium engineering

Peter Culley

How many differences can you see between the photographs of 21st century stadiums in this book and the stadium shown in Figure 14.1, photographed on 14 November 1914? The most obvious missing ingredients are roof structures and cladding – if it rained, spectators at Ann Arbor got wet. Other notable absentees include under-terrace facilities, crowd-safety structures, floodlighting and scoreboards. Then, thinking about movement into and around the stadium, you might say turnstiles, signs/sign gantries, gates, fences and vendor kiosks. All of these types of stadium feature have over the years come to be made of steel by design; unlike 100 years ago, the stadium visitor of today tends to notice steel elements of construction rather than timber or concrete.

Modern stadium design
Let us take a look at the final requirements of the design brief for the redevelopment of the north stand at Twickenham (Figure 14.2), home of the English Rugby Football Union:

- maximum number of seated spectators
- optimum sightlines and visibility
- unobstructed views
- appropriate protection from the weather
- a good standard of seating and range of facilities for both abled and disabled spectators
- maximum amount of hospitality boxes
- maximum volume of space within the stand for other accommodation
- low-maintenance regime
- minimal adverse effects on the condition of the pitch.

In this case, a 39 m (128 ft) steel cantilever truss roof was designed to provide optimum protection when taking into account the construction cost and the shadow effect of any roof on the condition of the pitch. Filon translucent glass reinforced polyester (GRP) roof sheeting provides natural light and is supported on Grade 43 rectangular hollow section (RHS) purlins at 2.3 m (7.5 ft) centres spanning beneath the lower booms of the trusses (to prevent pigeons perching on the underside of the roof). Because viewing quality is fundamental to any stadium design brief, I want – before looking more closely at steel construction – to consider sightlines.

Sightlines
A feature that sets a stadium apart from other sports facilities is the much greater amount of seating incorporated for spectators. The modern stadium has to offer excellent views from quality seating. To achieve this there must be no structural impediments to stadium sightlines, and the spectator must be able to see over the head of the person sitting in front. To make this happen, a line drawn from the spectator's eyes to the lowest point of the viewing area has to be at least 100 mm above the eyes of the spectator one row in front. This figure is arrived at by using the 'C value', a measurement (120 mm or 4.8 in.) of the distance between the centre of the eye and the top of the head. In some sporting events for which spectators wear hats (e.g. horseracing) the C value may be increased to 150 mm (6 in.) or even 200 mm (8 in.). At a cricket ground, where the action seldom comes close to the seating, a C value of 90 mm (3.5 in.) may be acceptable – this is because we tend to tilt our heads backwards slightly as the action moves closer towards us, reducing the distance between the centre of the eye and the top of the head to around 90 mm. At a football match the action moves to all parts of the playing surface, so, ideally, every football stadium would be designed to provide a C value of 120 mm to all parts of the playing surface:

$$C = \frac{D(N+R)}{D+T} - R$$

The calculation to determine the C value for sport to be viewed is, therefore,

$$\begin{aligned} C &= \frac{20\ 000(365+6000)}{20\ 000+800} - 6000 \text{ mm} \\ &= \frac{20\ 000 \times 6365}{20\ 800} - 6000 \text{ mm} \\ &= \frac{127\ 300\ 000}{20\ 800} - 6000 \text{ mm} \\ &= 120 \text{ mm} \end{aligned}$$

Figure 14.1 Michigan versus Cornell, Ann Arbor, MI, 1914. (© G. R. Swain)

where C is the viewing standard (i.e. the C value), D is the distance from the spectator's eye to the point of focus (typically the near touchline), N is the riser height, R is the height between the eye and the point of focus, and T is the tread depth (i.e. the depth of the seating row).

While this calculation is straightforward, it has to be made for every row of seating and from every variable that the stadium design throws up (e.g. rake or angle of stand, curvature of particular corner or height and depth of concrete treads and risers). A higher C value has consequences for the rake and height of the stand that are particularly challenging when designing a large, multi-tiered stadium. As a result, in some areas of a large stadium it may be difficult to achieve a C value greater than 60 mm.

Bringing the touchline closer to a stand makes it possible to maintain an optimum C value of 120 mm but affects the height of the stand. It increases the angle of rake, the measurement of how steeply or gently the stand or terrace slopes down towards the touchline. Getting as many spectators as possible as close to the action as possible results in very steeply raked stands. Italian

Figure 14.2 Rugby Football Union HQ, Twickenham. (© Corus Group)

codes of practice suggest that a stadium rake can be as steep as 41°. Rakes of more than 35° can be found in North American stadiums. In the UK, the rake angle is determined by safety limits for staircases, and the Green Guide (DCMS/FLA, 2008) recommends a rule of no more than 34°, which may be increased if compensatory measures are taken. Rakes exceeding 34° can induce vertigo, and the steeper Italian stadiums have handrails provided in front of each seat. Shallower rakes are used on lower tiers, with the upper decks of stands being steeper in order to accommodate more spectators closer to the playing surface, with an acceptable standard of view.

Structural steelwork

Structural steel can – very importantly – be used to frame roof structures that shelter stadium users while not impeding their view of the playing area. A list of the advantages of using structural steel in its various forms, in stadium engineering, could include:

- Long spans and clear floor areas can be easily achieved.
- Steel frames are fast to erect, making possible early or timely revenue returns when construction is completed, say, within a 'close season' or to meet the needs of a specific event to be hosted.
- The relatively low weight of a steel structure can produce savings in foundation costs, a particular advantage in poor ground conditions.
- Steel is a material of guaranteed high quality, with the structural elements prefabricated in a factory environment.
- Delivery and erection of elements often require little space, time or inconvenience, and noise is seldom a problem.
- Because of the ductility of steel, it is beneficial for long-term integrity if subjected to dynamic loading.
- The framework affords a light, strong material for fixings, and will accept a wide range of cladding materials.
- The clean lines and slender elegance of steel construction are aesthetically appealing to stadium users.

Because of its reliability and consistency of manufacturing quality, steel is one of the most versatile of the traditional construction materials available today. It offers a high strength/weight ratio, and can span long distances gracefully and economically. These advantages, allied with a reduced need for internal supporting columns, create flexibility of future occupancy in stadium construction. Using modern corrosion-protection systems, steel can provide a structure with a relatively long life span. Also, a significant residual value can be achieved – when the life of the structure is at an end, the steel frame can be recycled. In stadium projects, in particular, the frame can be prefabricated off-site while, at the same time, the site foundation work is being carried out. This means shorter construction periods, leading to earlier completion dates.

Structural steel is a material of wide capability (Tables 14.1 and 14.2). It is compatible with, and can be joined efficiently with, almost all other construction materials. For example, a combination that can be particularly well exploited in grandstands is the use of steel with precast concrete elements. Significant economies can be achieved through the interaction of these materials by designers who use the advantages of both, while suppressing their disadvantages. With both materials having virtually identical coefficients of thermal expansion, differential movements

Table 14.1 Structural steel: nominal values of the yield strength (f_y) and ultimate tensile strength (f_u)

Steel grade and subgrade	f_y: N/mm^2 Nominal thickness of element, t: mm				f_u: N/mm^2 Nominal thickness t: mm
	$t \leq 16$	$16 < t \leq 40$	$40 < t \leq 63$	$63 < t \leq 80$	$3 \leq t \leq 100$
S275JR					
S275J0	275	265	255	245	410
S275J2					
S355JR	355	345	335	325	470
S355J0					
S355J2					
S355K2					
S355JOH	355	345	335	325	470
S355J2H					
S355K2H					

Notes:
As stated in the National Annex (NA 2.4) to BS EN 1993-1-1:2005 (BSI, 2005), the ultimate strength f_u should be taken as the lowest value of the range given (in the product standard). This minimum value is quoted above
Although not stated in the Eurocodes, for rolled sections, t may be taken as the flange thickness

Table 14.2 Summary of leading properties for steels to BS EN 10210:2006 (BSI, 2006a,b)

Grade	Design strength: N/mm	Charpy test (27 J)
S275	275	−20°C
S355	355	−20°C

Other properties of steel:

Modulus of elasticity

$E = 205 \times 10$ N/mm^3 (205 kg/mm) (BS EN 1993-1-1:2005)

Coefficient of thermal expansion

0.12×10^{-4} per °C

Density or mass

7850 kg/mm^3 (7.85 tonne/m or 78.5 kN/m)

Elongation to fracture (200 mm gauge length)

Grade 275: 20%

Grade 355: 18%

due to changes in temperature are not a major problem when the two materials are combined.

Steel for structural use is normally hot-rolled from billets in the form of plates, flats or sections – in various shapes – at the mills. Fabrication of sections can involve operations such as sawing, shearing, punching, grinding, drilling, bending and welding. Therefore, the steel must be suitable for all these processes without degradation of its required properties. The sections being used must possess reliable and predictable strength properties in order that the structures can be safely designed to carry any specified loading. It is generally economic to use high-grade steel (S355J2) due to its favourable cost/strength ratio, compared with lower grade steel (S275J2). This choice leads typically to a 20% advantage. Where deflection limitations are the criteria, it may be more economical to use the lower grade steel, which can also be advantageous for very small projects or where the weight in a particular size is less than 5 tonnes (in which case the material can be obtained from a stockholder at short notice).

Structural sections

Hot-rolled sections are produced in the form of universal beams (UBs), universal columns (UCs), asymmetric beams (ASBs), channels and rolled steel angles to BS4 and BS EN 10056 (BSI, 1993, 1999). Less frequently used are structural tees, cut from universal beams and columns.

Structural hollow sections are produced in circular (CHS), square and rectangular (RHS) and oval shapes. They are used as columns and struts, and where clean appearance is a criterion, such as in exposed structures. As wind resistance is less on hollow sections than on open sections, an advantage is gained in open-braced structures such as long-span roof trusses on stadiums.

The range of universal beams and columns offers a number of section weights within each serial size (depth D and breadth B). Heavier sections are produced with the finishing rolls further apart such that the overall depth and breadth increases, but with the clear distance between the flanges remaining constant (Table 14.3). Structural hollow sections are also produced hot-rolled, in a number of different thicknesses. However, while the overall external dimensions remain the same for each serial size, designers can benefit from using the range of wall thicknesses available for each serial size. There is also available an extensive range of European, North American and Japanese section sizes.

Many other structural sections can be fabricated using the above sections, together with plates. These include plate girders, box girders and composite sections, using both steel and concrete.

Wide-span solutions

The steel shapes described in this chapter are available in different steel strengths. High-strength steel was developed by British Steel in the 1960s specifically for the roof of the BOAC 747 01 Hangar at Heathrow Airport (Figure 14.3) – a direct, necessary and urgent response to a global demand created by the introduction of the new wide-bodied generation of passenger aircraft. The 1200 tonne structure was the biggest of its kind in the world, designed to house two Boeing 747 jets. British Steel's structural marketing manager Ron Taylor said that from now on architects would be constrained only by the limitations of their imagination. He encouraged architects to think wide span, and then wider and wider still. One of the first sports manifestations of high-strength steel was the 9000 seat Celtic Football Club Grandstand, which was completed in an 18-week contract period, April to August 1971, in readiness for the 1971–1972 playing season. In association with Ron and his colleagues, architect James Cunning and structural engineer Vivian Rossi reached the conclusion that it would be technically feasible to give the grandstand's full complement of spectators an uninterrupted view of the playing area from their seats. The proposal which made this possible was for a roof supported by an enormous central spine girder, 97.5 m (320 ft) long × 5.4 m (17.7 ft) deep, fabricated from CHS to BS 4360:1986 (now BS EN 10025 (BSI, 2004a–f)) in Grades 55C, 50C and 43C. The main chords of the twin top and bottom booms are 406.4 mm (16 in.) outside diameter in Grade 55C in all but the end booms. This was the largest tubular steel girder of its kind in Europe.

Wide-span and closing-roof solutions

The 80 000 seat AT&T Stadium (formerly known as the Cowboys Stadium) in Arlington, Texas (Figure 14.4), is the

Table 14.3 Details of structural shapes

Shape		UK size range: mm	Use
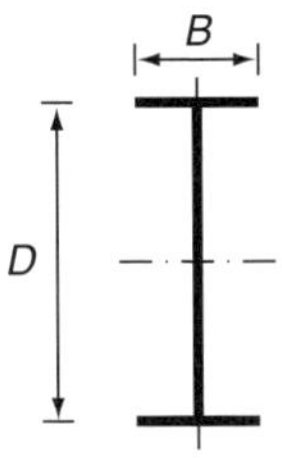	Universal beam (UB)	$B \times D \times$ kg/m 127 × 76 × 13 to 1016 × 305 × 487	Beams
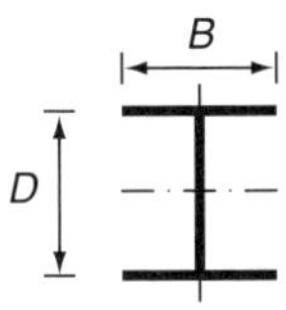	Universal column (UC) H-pile Bearing piles	152 × 152 × 23 to 356 × 406 × 634	Columns Shallow beams Heavy truss members
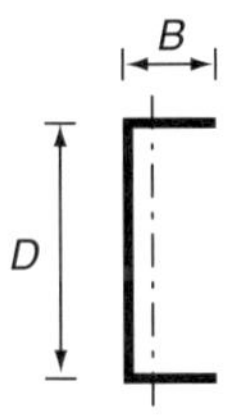	Parallel flange channel (PFC)	100 × 50 × 10 to 430 × 100 × 64	Bracings, ties Small beams
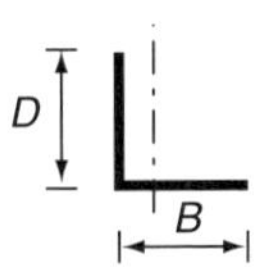	Equal angle (UKA)	$B \times D \times t$ 90 × 90 × 7 to 200 × 200 × 24	Bracings Truss members Purlins Sheeting rails
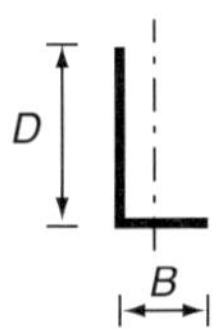	Unequal angle (A)	100 × 65 × 7 to 200 × 150 × 18	
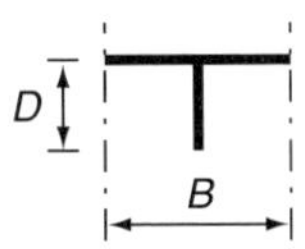	Structural tee (T)	$B \times T \times$ kg/m (UB) 133 × 102 × 13 to 178 × 102 × 37 (UC) 152 × 76 × 12 to 305 × 152 × 79	Truss chords Plate stiffeners
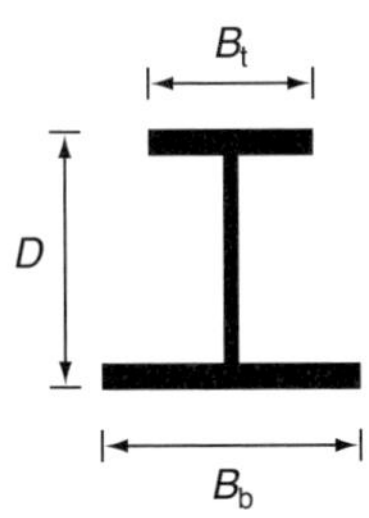	Asymmetric Slimflor beam (ASB)	280 ASB 74 to 300 ASB(FE) 249	Structural floors
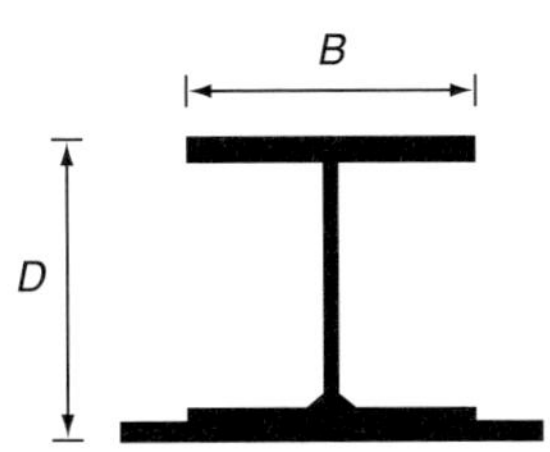	Slimflor beam (SFB)	152 × 152 × 64.4 to 354 × 406 × 707.4 kg/m	Structural floors

Table 14.3 Continued

Shape		UK size range: mm	Use
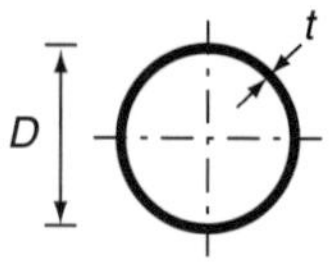	Circular hollow sections (CHS) and tubes	$D \times t$ 21.3 × 3.2 to 508 × 16 Tubes up to 2134 × 22:2	Columns Truss members Bracings, piles
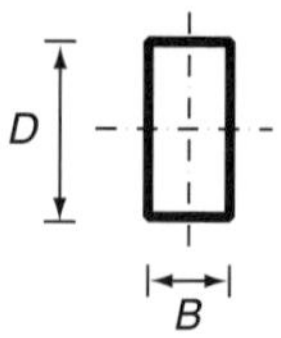	Rectangular hollow sections (RHS)	$D \times B \times t$ 50 × 30 × 3.0 to 500 × 300 × 20	Columns Warren girders Vierendeel girders
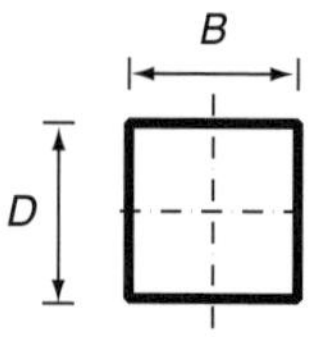	Square hollow sections (SHS)	40 × 40 × 3:0 to 800 × 800 × 60	As above
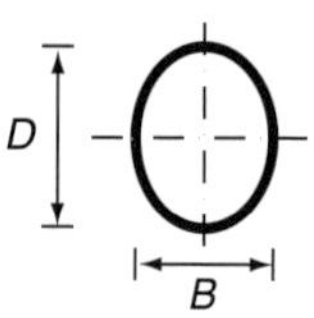	Oval sections	H × B × t 150 × 75 × 4.0 to 500 × 250 × 16	Architectural features

All the above hollow section members are hot-finished BS EN 10210:2006 (BSI, 2006a,b)
Similar sizes of hollow section members can be obtained cold formed to BS EN 10219 (BSI, 2006c,d)

Figure 14.3 BOAC-01 Hangar, Heathrow Airport. (© British Steel Corporation, Tubes Division)

largest domed stadium in the world and has the world's largest column-free interior. Two monumental arches rise to a height of 91 m (320 ft) above the playing surface and support a retractable roof. Each boxed arch is 5.2 m wide × 10.7 m deep (17 ft × 35 ft), weighs 3255 tons and spans 393 m (1290 ft). As one of the first components to be constructed, the arches form a distinctive silhouette on the Arlington landscape as sunlight reflects against the metallic surface during the day (a tailored lighting system reflects the form at night). When closed, the stadium's 61 316 m^2 (661 000 ft^2) retractable roof encompasses 2.95 million m^3 (104 million ft^3) of volume. The roof is designed to allow views inward and outward. Each panel can open or close in 12 minutes. From directly above, the entire playing field can be seen, and the opening itself is visible from an elevation of 8 km (5 miles). Each retractable end-zone door has a five-leaf, clear opening measuring 36.6 m high × 55 m wide (120 ft × 180 ft). The five 11.6 m (38 ft) panels can be opened in 18 minutes. Through the clear glass door panels spectators experience panoramic vistas from their seated positions within the bowl and when circulating through the stadium concourses. The building design allows for improved air circulation during events. A very large (1070 m^2, 11,520 ft^2) high-definition video screen is suspended approximately 33.5 m (110 ft) above the field, from the roof structure, and spans between the 20 yd lines.

Figure 14.4 Chelsea versus Club America for the World Football Challenge 2009, Cowboys Stadium, Arlington, TX. (© Mahanga, 2009)

It features four individual boards: two face the sidelines and two face the end zones. The sideline boards measure 54.9 m wide × 15.2 m tall (180 ft × 50 ft), while those facing the end zones measure 14.6 m × 8.2 m (48 ft × 27 ft). One of the stadium's finest architectural features is its 26.2 m (86 ft) high canted glass exterior wall. The wall slopes outward at a 14° angle to create a luminescent glow. A fritted glass system transitions up the elevation to create a dynamic, ever-changing aesthetic depending on the time of day. At night, a series of internal lights washes the glass wall to create a glow across the façade.

The New English National Stadium at Wembley, London, is the tallest stadium in the world. With a seating capacity of 90 000, it is also the second largest stadium in Europe and the world's largest football (soccer) stadium with every seat under cover. The stadium is signposted from miles around by the iconic 315 m (1033 ft) Wembley Arch, a 138 m (452.7 ft) high, 7 m (23 ft) diameter unclad lattice structure (Figure 14.5). The arch is formed of 457CHS (18 in. outside diameter (od)) longitudinal chords with diaphragms at approximately 11 m (36.1 ft) centres. Alternate diaphragms are primary and support the roof stays. Steel grades are S355 JO or J2 to BS EN 10025 (BSI, 2004a–f). Rolled hollow sections are S355 J2H to BS EN 10210:2006 (BSI, 2006a,b). Protection is 400dft micron epoxy primer/buildcoat and a 75 ft micron finish coat, over a blast clean surface to Sa 2.5 of BS 7079:2009 (BSI, 2009). Access to the arch, through its centre, permits structural inspection, lighting maintenance/replacement, repainting (30-year intervals), festivity/celebration (e.g. pyrotechnics) and dressing the arch with flags or banners. The stays are spiral strand galvanised wires grade 1570. The roof plate main structure runs north–south. The roofing material is a mixture of standing seam aluminium

Figure 14.5 New English National Stadium, Wembley. (© Simon Atkinson) (This figure is reproduced in the colour-plate section as Plate XVII)

Figure 14.6 Millennium Stadium, Cardiff. (© W. S. Atkins, 2000)

and 30% translucent polycarbonate sheeting (the latter allowing diffused light through the roof leading edges). A moving roof over the whole of the southern side of the stadium was required to provide maximum covered seating for spectators while providing daylight for the turf. The permanent roof structure running north–south provides the runway beams supporting the track for the panels. The area of roof that moves is split into five bays. The middle section extends the 135 m (443 ft) length of the pitch, and there are two bays at each end which cover the end stands. Each of the two end-bay panels is subdivided to enable double-stacking on top of the fixed roof, without projecting over the southern edge of the building. The roof panels are framed by fabricated box sections up to 3 m (9.8 ft) deep (for the central large cantilever panel), which are connected to the running bogies. Secondary framing UB sections are used with full diagonal rod bracing for each panel, to ensure that racking of the panel does not occur. A full cycle for the roof to open or close takes 20 minutes.

The Millennium Stadium, Cardiff (Figure 14.6), was the first closing-roof stadium to be built in the UK (Table 14.4). It needed to be of the order of 50 m (164 ft) larger than the pitch in all directions in order to accommodate the 72 500 seats required, and the opening had to be at least the size of the pitch. This gave roof dimensions of the order of 220 m long × 180 m wide (720 ft × 590 ft) with an opening of approximately 120 m × 80 m (390 ft × 260 ft). At an early stage of design it was decided that the following criteria should be adopted: to keep the roof as low as practicable to reduce the stadium's impact on adjoining buildings; to keep the structure and edge of the opening as low as possible to reduce the extent of shading on the pitch; and to make the track, for the retractable roof to move along, as near to flat as possible to assist with making the retractable roof mechanism simple and, therefore, more cost-effective and less problematic. These criteria were met in a structural solution that incorporates continuous primary plane lattice trusses over the full 220 m length of the stadium, rising 35 m (115 ft) above the pitch.

Aluminium roof coverings

Steel structures may be covered in steel but can also be covered using aluminium (Figure 14.9). The most common source of aluminium is the ore called bauxite, which was discovered in 1821 near the village of Les Baux in France. Aluminium on its own is too soft for structural purposes, but it can be alloyed effectively with copper, magnesium, silicon, nickel and zinc. Alloying can produce tensile strength equivalent to that of mild steel. Because such alloys weigh about 2800 kg/m^3 (175 lb/ft^3),

Table 14.4 Selected projects – stadiums with closing roofs

Name	City	Country	Open	Capacity	Owner	Architect	Engineer
Singapore National Stadium	Kallang	Singapore	2014	55 000	Singapore Sports Hub Consortium	DP Architects	Arup Sport
Stade Borne de l'Espoir	Lille	France	2012	50 186	Urban Community of Lille Métropole (ELISA design-and-build contract)	Valode & Pistre et Atelier d'architecture Pierre Ferret	Eiffel (steel roof structure)
Friends Arena	Solna	Sweden	2012	50 000	SFF, Solna Municipality, PEAB, Fabege, Jernhusen	Pierre Ferret	SWECO
Marlins Park	Miami, FL	USA	2012	37 000		Populous	Bliss and Nyitra, Inc. (bowl and track), Walter P Moore (roof)
AT&T Stadium	Arlington, TX	USA	2009	80 000	Arlington, TX	HKS Inc.	Walter P Moore
Aslantepe Türk Telekom Stadyumu	Istanbul	Turkey	2009	52 647	Galatasaray SK	Mete Arat	Schlaich Bergermann und Partner
Lucas Oil Stadium	Indianapolis	USA	2008	70 000	Indiana Stadium and Convention Building Authority	HKS Inc.	Walter P Moore/Fink Roberts & Petrie
New English National Stadium, Wembley	London	England, UK	2007	90 000	The Football Association	World Stadium Team (Foster and Partners + HOK Sport)	Mott Stadium Consortium (Mott MacDonald, Aurecon and Jacobs)
University of Phoenix Stadium	Glendale	USA	2006	73 719	Arizona Sports and Tourism Authority	Peter Eisenman/HOK Sport	Walter P Moore, Buro Happold, TLCP Structural, Inc.
Commerzbank-Arena (Waldstadion)	Frankfurt	Germany	1925 (rebuilt 2005)	52 300	Waldstadion Frankfurt Gessellschaft für Projektwicklung	Gerkan, Marg ünd Partner	Schlaich Bergermann und Partner, KreBS und Kiefer
Esprit Arena (LTU Arena)	Düsseldorf	Germany	2004	54 500	City of Düsseldorf	Hascher Jehle and Associates	Inros Lachner AG
Reliant Stadium	Houston, TX	USA	2002	71 500	Harris County	HOK Sport	Walter P Moore
Veltins Arena	Gelsenkirchen	Germany	2001	61 482	Schalke 04	Hentrich-Petschnigg ünd Partner	HBG Engineering
Oita Stadium	Oita	Japan	2001	40 000	Oita Prefecture	Kisho Kurokawa	Takenaka
Toyota Stadium	Toyota	Japan	2001	45 000	Toyota City	Kisho Kurokawa	Arup

Table 14.4 Continued

Name	City	Country	Open	Capacity	Owner	Architect	Engineer
Miller Park (Figures 14.7 and 14.8)	Milwaukee, WI	USA	2001	42 200	Southeast Wisconsin Professional Baseball District	HKS, Inc + NBBJ + Eppstein Uhen Architects	Arup + Flad Structural Engineers
Minute Maid Park (formerly Enron Field)	Houston, TX	USA	2000	40 950	Harris County – Houston Sports Authority	HOK Sport, Molina & Associates	Walter P Moore
Etihad Stadium (formerly Telstra Dome, Docklands Stadium, Victoria Stadium, Colonial Stadium)	Melbourne	Australia	2000	53 355	James Fielding Funds Management	Daryl Jackson Architects + HOK Sport	
Safeco Field	Seattle, WA	USA	1999	47 116	Washington-King County Stadium Authority	NBBJ + 360 Architecture	Skilling Ward Magnusson Barkshire (roof), Stephen Tipping + Associates (bowl)
Millennium Stadium	Cardiff	Wales, UK	1999	74 500	Welsh Rugby Union	HOK + LOBB Partnership	W. S. Atkins
Chase Field (formerly Bank One Ballpark)	Phoenix	USA	1998	49 033	Maricopa County, Arizona	Ellerbe Becket	Hatch Associates Ltd
Gelredome	Arnhem	Netherlands	1998	32 500		Kjell Kosberg	
Amsterdam Arena	Amsterdam	Netherlands	1996	51 628	Gemeente Amsterdam Stadion Amsterdam NV	Bouwcombinatie	
Fukuoka Yahoo! Japan Dome (formerly Fukuoka Dome)	Fukuoka	Japan	1993	35 695	Hawks Town	Takenaka Corporation + Maeda Corporation	
Rogers Centre – formerly Skydome	Toronto	Canada	1989	31 074	Rogers Communications	Rod Robbie	Michael Allen
Montreal Olympic Stadium	Quebec	Canada	1976 (roof 1987)		Régie des Installations Olympiques	Roger Taillibert	

Selected projects have a spectator capacity of 30 000 or more
Spectator capacities quoted are for the principal sport mode (higher capacities may be achieved for hosting sports with reduced playing areas and for hosting concerts)

Figure 14.7 Miller Park, Milwaukee, WI: roof closed. (© Tim Griffiths, 2001) (This figure is reproduced in the colour-plate section as Plate XVIII)

Figure 14.8 Miller Park, Milwaukee, WI: roof open. (© Tim Griffiths, 2001) (This figure is reproduced in the colour-plate section as Plate XIX)

Figure 14.9 Aluminium roof covering, New English National Stadium, Wembley. (© Multiplex, 2006)

compared with around 7850 kg/m^3 (490 lb/ft^3) for steel, lighter structures can be designed in aluminium alloys than in mild steel.

More significantly, perhaps, aluminium is regarded as a non-combustible material. It will melt at about 620°C (1148°F), but it does not burn, ignite, add to the fire load or spread surface flame. Its thermal conductivity is four times that of steel, and its specific heat is twice that of steel. Because heat is conducted away more quickly in aluminium than in steel, a greater heat input is necessary to bring aluminium up to a given temperature. These qualities make it a leading choice for roof coverings, including roof coverings for stadiums.

Aluminium sheeting is transported to site in coils and is passed continuously through a machine to form standing seams in situ. This 'long-strip' system removes the need to form joints transverse to the standing seams up to a maximum of 7 m (23 ft). The standard thickness for long-strip aluminium roofing is 0.8 mm (0.03 in.), and the recommended maximum width of 450 mm (17.7 in.) produces standing seams at 375 mm (14.8 in.) centres. A minimum fall of 1.5° is recommended.

Façades – sheet-steel cladding

One of the simplest, shortest and best definitions of a façade is 'the exterior front or face of a building'. This definition incorporates cladding, which creates weatherproof enclosures by spanning between the elements of structure of a building, and facing, which requires continuous background structure. Cladding carries its own weight and thereby eliminates the need for continuous background structure. Examples include precast concrete panels and profiled sheeting in plastics or metals. I like sheet-steel cladding best.

The durability of prefinished sheet-steel products is dependent on the metallic coating type, pretreatment chemistry, primer quality, topcoat paint quality and thickness, and manufacturing process quality and consistency. Environmental factors that may act or interact to degrade the external surfaces of prefinished sheet-steel products include water, oxygen and salts, sunlight, chemicals and pollutants, and abrasives such as wind-blown sand and salts. In specifying prefinished steel cladding, it is important to choose the cladding product that best matches the durability requirements of the specific project. Tata Steel's Colorcoat HPS200 Ultra and Colorcoat Prisma have been used on many stadium projects because of their combination of aesthetic appearance and durability in performance (Tata Steel, 2015). Tata Steel's Colorcoat Celestia was specified for the Millennium Stadium in Cardiff (see Figure 14.6) because it conferred an appropriate superior quality to the UK's first closing-roof stadium.

Steel and brick façades

Steel is also used effectively with other façade materials. Bricks are available in a wide range of colours. This, together with the availability of different mortar colours, textures and laying patterns, creates a wide diversity of design opportunities. Brick is often chosen to complement and integrate with surrounding buildings. It combines aesthetic appeal with impact resistance, inertness, sound-insulation properties, low maintenance requirements, and resistance to water, wind and fire.

Glass and steel façades

Glass (Figure 14.10) is, surprisingly to many people, stronger than structural steel but it has relatively low fracture toughness. It can, however, be toughened by heating to about 700°C (1292°F) and then cooling the surfaces rapidly while the centre is still viscous. This hardens the surfaces, which are put into compression as the centre cools and contracts. Walk-on glass demonstrates the possibilities in a spectacular steel and glass building design solution (Figure 14.11). Glass façades can be fixed to steel structures using a variety of steel fixings, including

Figure 14.10 North atrium, New English National Stadium, Wembley. (© Multiplex, 2006)

Figure 14.11 National Glass Centre, Sunderland. (© Space Decks)

bolts, plates, countersunk bolts and stud assemblies. Proprietary systems include the Pilkington Planar system, which allows the panes of glass to move in relation to their supporting structure, while maintaining a smooth outer surface appearance.

Peter Rice once wrote: 'The techniques of fixing glass and exploiting its structural capacity ... allow structural elements to be eliminated from the glass plane. This has a particular potential in architectural expression, because until recently the transparent or luminous surface has been dominated by the mullions or glazing bars that have been necessary to hold the glass in place. The flimsiest steel or aluminium profiles, and even glass stiffening fins, detract from the pure planar nature of the glass surface, giving it a relief standing out from the reflective or transparent planes. Removing them creates a new architectural device – a pure planar surface, perhaps 20 or 30 m high of transparent, reflective or luminous glass' (Rice and Dutton, 1995, p. 16).

Polycarbonate and steel façades

Polycarbonates are used in façades for their strength, light weight and translucency. In combination with metal fixings, they present opportunities for creative design/detailing and efficient assembly/installation. PQE Technology Ltd of Cootehill, County Cavan, Ireland, manufactured 2000 unique types of component for the Aviva Stadium, Dublin, which hosted the 2011 UEFA Europa League Final. More than 500 000 of these components were used to secure the external transparent polycarbonate cladding to the main structure (Figure 14.12) and to support the extensive array of gutters throughout the stadium. Various grades of aluminium, mild steel and stainless steel were specified and fabricated using computer numerical

Figure 14.12 Aviva Stadium, Dublin. (© 2014, PQE)

Figure 14.13 Plunge bath, Emirates Stadium, London. (© Terry Francis, 2014)

control (CNC) milling, CNC turning, saw cutting, plasma profiling, laser cutting, folding, shot blasting, painting (including powder coat painting), galvanising, zinc plating and anodising.

Stainless steels

Stainless steels are popular for a host of stadium applications ranging from façade fixings to wet areas (Figure 14.13), food handling and hygiene (Figure 14.14) and people–movement applications. The escalator cladding (Figure 14.15) at Paul Brown Stadium, Cincinnati, OH, is in Gauge 16, type 316L stainless steel by Langdon Inc. Sheet Metal Fabrication. Stainless steels possess enhanced corrosion resistance because of the addition of chromium to alloys of iron and carbon. In comparison with mild steel, they have greater corrosion resistance, cryogenic toughness, work hardening rate, strength, hot strength, hardness and ductility. They also have a more attractive appearance and lower maintenance requirements. Worldwide demand for stainless steels is growing at approximately 5% per annum, and new applications are continuously being discovered or invented for them. Stainless steels are available to the construction industry in the forms of plates, bars, sections, sheet strip and tubes. They are widely used by architects and engineers in North America, Japan and Western Europe (but traditionally less so in the UK).

Stainless steels are commonly divided into five groups: martensitic, ferritic, austenitic, duplex (ferro-austenitic) and precipitation-hardening. For exterior applications, the most appropriate grades

Figure 14.14 FA dining – kitchen area, New English National Stadium, Wembley. (© Multiplex, 2006)

Figure 14.15 Paul Brown Stadium, Cincinnati, OH. (© Langdon Inc.)

of austenitic steels are type 304 and the molybdenum-bearing types 315 and 316. For interior applications, ferritic stainless, which contains little or no nickel, may be used (commonly as types 430 and 434). Standard finishes for architectural applications range from semi-dull to mirror.

Stadium and arena seating

Seating (Figure 14.16) is fundamental. The Emirates Stadium (Figure 14.17) was designed with more legroom than any other stadium in the UK. In their recommendations for new stadiums, FIFA/UEFA state: 'All spectators must be seated in individually numbered, shaped seats with backs. Benches are not acceptable. Standing viewing areas should not be provided. There must be unrestricted view of the playing area from each seat, bearing in mind that advertising boards may be erected around the boundary lines.' (Note: The expression 'shaped seats with backs' does not exclude the use of certain types of bucket- or tractor-type seating.)

Seat dimensions must be considered in combination with seating-row depth and the measurement, called the 'seatway', between the foremost projection of the seat (if a tip-up seat, then in the tipped-up position) and the back of the seat in front. In the UK, The Green Guide states that the seating depth should be at least 700 mm for new construction and that the seat width should be a minimum of 460 mm (40 mm greater if fitted with arms). The Green Guide recommends a minimum 305 mm seatway and, for non-tip-up seats, the Football Stadium Advisory Design Council recommends a minimum of 400 mm. If installing seats with armrests, the projection of the armrests should not reduce the 305 mm minimum seatway measure. In Scotland these dimensions may need to be increased to meet the Scottish

Figure 14.16 Scottish Rugby Union, Murrayfield, Edinburgh. (John Clarke, © Corus) (This figure is reproduced in the colour-plate section as Plate XX)

Figure 14.17 Emirates Stadium, London. (© Terry Francis 2014)

Building Regulations (E2.63). Seating-row depth should be at least 760 mm for new constructions.

Different models of tip-up seats have different dimensions, affecting the resultant seatway dimension and useful seat dimension (the horizontal distance between the back of the seat, measured at seat height, and the rear of the back of the seat in front). In the UK, an average seated male measures 614 mm from front of the knee to the rear of the buttocks. The British Association of Spectator Equipment Suppliers (BASES) recommends, therefore, a minimum useful seat dimension of 660 mm. The seat width should be a minimum of 460 mm (or 490 mm with arms).

Providing wider and deeper seating increases spectator comfort but increases stadium cost while reducing revenue potential. The seating consideration is key to answering the question: 'Why did the English National Stadium at Wembley not have a 100 000 capacity?' At Wembley the architects adopted a useful seat dimension of 800 mm rather than 660 mm. Seat width was also increased from 440 mm, in the old stadium, to 500 mm in the new stadium.

In the UK, The Green Guide states that the number of seats in a row should not normally exceed 14, where there is a gangway at one end only, or 28 where there is a gangway at both ends. BASES suggests, in addition, that BS 5588-6:1991 (BSI, 1991) should be the guide where escape-route distances are taken into consideration (in Scotland the number of seats between aisles has a direct bearing on the minimum seatway measurement). The figure of 28 seats per row in the UK compares with 22 in the USA and Australia, 48 in Italy and 96 in Germany (in English theatres the maximum is 22 seats per row).

The above paragraphs have flagged up some of the main principles relating to seating in stadiums, together with some of the differences in approach. For readers of this book, there is a highly developed industry out there to advise on the details of stadium seating. Many suppliers offer steel posts and brackets as primary fixings to riser or tread. Seats may be supported on a steel frame. The frame fixings are often in stainless steel, with non-metallic bushes and sleeves to eliminate metal-to-metal contact.

Suites and club-type facilities

Stadium VIP and hospitality facilities (Figure 14.18) have benefitted from advances in glass and steel technologies. In the UK, examples range from the directors' box at the Emirates, located on the halfway line in the middle tier of the west side of the ground (Figure 14.19) to the eight-person hospitality box at Wembley (Figure 14.20). Suites have, historically, 12–14 seats on two or three risers in front of the box between the lower and upper seating decks. The stadium or arena geometry limits the number of suites possible in this configuration. Variations

Figure 14.18 Dodger Stadium, Los Angeles: club facilities. (© Cherie DTLA, southparkinla, 2010)

Figure 14.19 Directors' box, Emirates Stadium, London. (© Terry Francis, 2014)

include stacking the suites in two or three layers at midlevel, adding side suites within the lower seating bowl (to augment the midlevel ring) and tucking bunker suites under the seating deck (these do not usually have a view of the action). In some arenas the suites are all on one side and are stacked, because that way there is no end zone or bad location. Suite concourses were originally little more than circulatory corridors. They have become lighter and more airy as they have become dedicated social spaces. Current designs feature upscale concessions and merchandising points of sale, as well as lounges and bars. Club seats, typically located in the bowl, cost less than suites but offer access to high-quality restaurants and bar privileges attached to the best seats. The loge box is a recent development. A typical loge has seating in the bowl with flat-screen TVs in front of the seats. The chairs are usually roll-around rather than fixed, with cabinets behind the seats containing a minibar, under-counter refrigerator and serving space. Loges are usually located at the front edge of an upper seating deck or on a separate deck away from standard seats. Priced between a suite and a club seat, the loge has become very popular, very quickly. Statutory authorities and sports governing bodies include suites and club-type facilities in their directives and guidance on stadium development (e.g. the Football Rugby League in the UK requires Super League clubs to have the demonstrable capacity to entertain a minimum of 500 people in hospitality boxes, banqueting suites and restaurants).

Temporary structures

Considered in the widest sense of the term, 'temporary structures' may be of limited design life, installed for specific stadium

Figure 14.20 Eight-person hospitality box, New English National Stadium, Wembley. (© Multiplex, 2006)

Figure 14.21 Club concourse drink stands, New English National Stadium, Wembley. (© Multiplex, 2006)

and arena events, or retrofitted to enhance existing facilities. They range from the self-supporting concourse drinks stand at Wembley (Figure 14.21) to the 13.5 m × 2.8 m (44 ft × 9 ft) players' tunnel at the Emirates Stadium (Figure 14.22).

Stadium safety structures

Structural steel is used in all sorts of crowd and pedestrian movement control systems at sports grounds. The application that visitors may encounter first is the turnstile. The full-height Tornado (Figure 14.23) is manufactured by Boom Gate Systems (Pty) Ltd of Roodepoort and Cape Town, South Africa. Turnstiles were for many years a mechanical application of what is a structural steel section. But nowadays turnstiles are more of an electromechanical application. The Tornado, for example, can interface with any biometric or card reader system, remote control or keypad. Stadium safety structures such as crush barriers, segregation and perimeter fences, handrails, guardrails and gates are usually steel by design (Figures 14.24 to 14.26). Large quantities of steel go into these applications. The Emirates Stadium, for example, incorporates 4500 m (14 764 ft) of handrailing. Structural hollow sections (SHS) are particularly popular steel sections for safety structures because of their high strength/weight ratio, torsional rigidity, buckling resistance, enclosed shape (with no sharp edges), smooth surfaces (again no sharp edges) and reduced maintenance requirement (because of their reduced surface area in comparison with open steel

Figure 14.22 Players' tunnel, Emirates Stadium, London. (© Terry Francis, 2014)

Figure 14.23 Tornado turnstile. (© Boom Gate Systems (Pty) Ltd)

Figure 14.24 UK Football Stadium post-Taylor Report: gated enclosure. (John Clarke, © Corus)

Figure 14.25 Crowd Safety Barriers, Avenue Stadium, Dorchester Town FC, Dorset. (© Peter Culley, 1990)

Figure 14.26 Aisle steps protection, Avenue Stadium, Dorchester Town FC, Dorset. (© Peter Culley, 1990)

sections). Figure 14.27 communicates these attributes better than words can.

When I was a structural advisory engineer with British Steel, I would use the following types of calculation for designing stadium safety structures such as tubular steel crush barriers and tubular steel stairway handrails:

- Calculations for unbraced crush barrier on stepped terrace for safety considerations when the barrier shall be designed to resist a horizontal load of 5.0 kN/m, using Grade 43C steel (now Grade S275JO steel) to BS 4360:1986 (now BS EN 10025 (BSI, 2004a–f)):

  Height of step: 180 mm

  Height of rail above nosing: 1.1 m

  Post centres: 2.0 m

  Horizontal load at top rail level uniformly distributed: 5.0 kN/m

Figure 14.27 Elephants at Knowsley Safari Park, Prescot, Merseyside (John Clarke, © Corus, 1994)

Design of rails
Maximum bending moment $M = WL/8$

$$= (5.0 \times 2 \times 2)/8$$

$$= 2.5 \text{ kN m}$$

Section modulus $Z = M/f$

Permissible stress $f = 180 \text{ N/mm}^2$

$Z = 2.5 \times 10^3 \text{ cm}^3/180 \text{ N/mm}^2$

$= 13.89 \text{ cm}^3$

Nearest size of section = 60 × 60 × 4.0 RHS*

Or, alternatively (circular) = 76.1 diameter × 4.0 CHS*

*The examples give the size of section necessary to achieve suitable strength; research has demonstrated that a rail of 100 mm vertical depth has advantages.

Design of posts
(As a cantilever with a point load acting at 90° to the free end.)

Horizontal force at top = 2.0 × 5.0 kN = 10 kN

Maximum bending moment $M = WL = 10 \times (1.1 + 0.18)$

$$= 12.8 \text{ kN m}$$

Section modulus required $Z = 12.8 \times 10^3 \text{ cm}^3/180 \text{ N/mm}^2$

$$= 71.11 \text{ cm}^3$$

Nearest size of section = 120 × 120 × 5.0 RHS

Or, alternatively (circular) = 139.7 diameter × 6.3 CHS

The anchorage of the posts should be capable of resisting 2.0 times the maximum overturning moment calculated from the loadings specified.

Example in this case = 2.0 × 12.9 kN m = 25.8 kN m

- Calculations for handrails parallel to the direction of flow – when handrails are provided parallel to the direction of flow they should be capable of withstanding a force of 2.2 kN/m, using Grade 43C steel (now Grade S275JO steel) to BS 4360:1986 (now BS EN 10025(BSI, 2004a–f)):

Height of top rail: 1.1 m

Post centres: 2.0 m

Horizontal load at top rail level, uniformly distributed: 2.2 kN/m

Design of rails
Maximum bending moment $M = WL/8$

$$= 2.2 \times 2.2 \times 2.0/8$$

$$= 1.21 \text{ kN m}$$

Section modulus $Z = M/f$

Permissible stress $f/Z = 180 \text{ N/mm}^2$

$$= 1.21 \times 10^3 \text{ cm}^3/180$$

$$= 6.72 \text{ cm}^3$$

Nearest size of section = 50 × 50 × 2.5 RHS*

Or, alternatively (circular) = 60.3 diameter × 3.2 CHS*

*The examples give the size of section necessary to achieve suitable strength; research has demonstrated that a rail of 100 mm vertical depth has advantages.

Design of posts
(As a cantilever with a point load acting at 90° to the free end.)

Horizontal force at top = 2 × 2.2 kN = 4.4 kN

Maximum bending moment $M = 4.4 \times 1.1 = 4.84$ kN m

Section modulus required $Z = 4.84 \times 10^3 \text{ cm}^3/180$

$$= 26.89 \text{ cm}^3$$

Nearest size of section = 80 × 80 × 5.0 RHS

Or, alternatively (circular) = 114.3 diameter × 3.6 CHS

The anchorage of the posts should be capable of resisting 2.0 times the maximum overturning moment calculated from the loadings specified.

Example in this case = 2.0 × 4.84 kN m = 9.68 kN m

Responsibility for spectator safety at sports grounds lies with stadium management, but The Green Guide states that: 'specialist advice, from an appropriately qualified engineer belonging to a recognised institution should be sought to assess the safety and strength of loadbearing elements in buildings and installations such as crush barriers'. The qualified engineer is usually appointed to undertake an annual inspection of the general structural fabric of the stadium to assess the current condition and make recommendations regarding maintenance or improvements necessary to maintain safety.

There is an additional requirement for the annual inspection and testing of specific items such as crush barriers and handrails. Essentially, and I oversimplify, an evenly distributed horizontal force is applied in increments to the top rail of the barrier and the resulting deflections of the rail and posts are recorded during the application and upon removal of the load. The test comprises a bedding-in cycle up to the design load and two proof test cycles to 1.2 times the design load. Each proof test load is held for 5 minutes and the recovery recorded. The bedding-in cycle must achieve at least 75% recovery or less than 2 mm permanent deflection, and the proof cycle must achieve at least 95% recovery or less than 2 mm permanent deflection, measured over the two proof cycles.

The curious case of the Chelsea Football Club floodlighting towers

Engineers have for many years been confident in the freedom from internal corrosion of sealed steel hollow sections. Other professionals have, from time to time, sought reassurance and, to meet this need, a variety of case study evidence was collected by the international steel community. In the 1970s, Chelsea Football Club commissioned one of the first new stands in the UK to be designed in accordance with the Wheatley proposals for crowd safety. Building the new stand necessitated taking down two floodlighting towers that had been erected in 1954. Sections from these redundant towers were despatched to British Steel Tubes Division's Research Centre at Corby.

Sample 1 was of a 139.7 mm od CHS. This was cut lengthwise to expose the internal surfaces (Figure 14.28). Little evidence of internal corrosion was found other than a discolouration of the surface caused by the oxygen and moisture in the entrapped air – much of the mill scale was still visible. A light rust in the centre of the sample was recent, having occurred after the tube was cut open for examination. Sample 2 of the 139.7 mm od CHS incorporated an intermediate flanged joint (Figure 14.29). The internal flange face had been enclosed – and thus hermetically sealed – by welding to the CHS, which was still comparatively bright, with the original marking-off lines clearly visible. Additional samples of 139.7 and 114.3 mm od CHS were examined and showed an almost total absence of corrosion (Dawson, 1977).

Non-stadium examples of freedom from internal corrosion of sealed hollow sections include the case of the davits from the four-funnelled ocean-going Cunard liner *S.S. Aquitania*, which was built in 1913–1914 and scrapped in 1950. No internal corrosion was found in these davits despite 37 years' service in a salt-water environment.

Lighting continued

Floodlighting towers have been a feature of English football since the 1950s. However, the advent of tiered all-seater stadiums in the late 20th century presented the opportunity to fix lighting instead to the leading edge of the stadium roof structure. This had not been feasible when many grounds were open-terrace because the roof lines on their covered stands were too low. Although eliminating floodlighting towers reduces an application for steel, the new types of lighting gantry (Figure 14.30) are also of steel construction.

Figure 14.28 Chelsea floodlighting tower sample 1 – section of 139.7 mm CHS. (© British Steel)

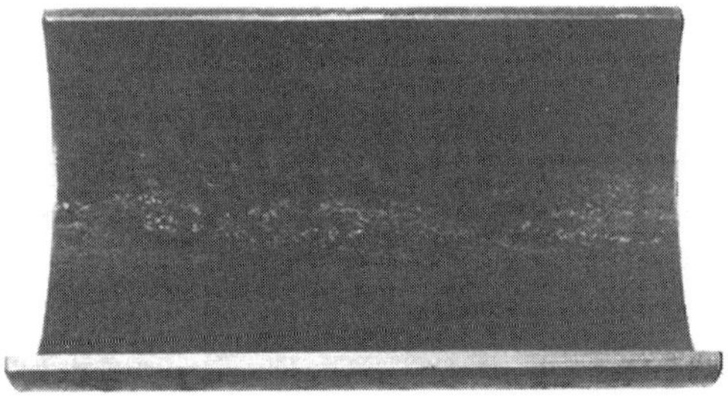

Figure 14.29 Chelsea floodlighting tower sample 2 – section of 139.7 mm CHS welded to an intermediate flange. (© British Steel)

Other lighting that you may see at and around stadium and arena developments includes the tubular-steel lighting columns and sign gantries on stadium approaches and perimeter roadways, walkways and car parks. For these types of tubular steel application the structure acts also as a services conduit, conferring exceptional protection and security to enclosed wiring or ducting.

Under-terrace accommodation

Under-terrace accommodation (Figure 14.31) is big business. One of the advantages of using long-span steel construction is that it enables stadium operators to eliminate columns beneath the terraces. For example, the 5400 spectator capacity Dolman Stand at Bristol City Football Club, UK, was conceived in the 1960s as having columns in the main structure and multiple columns beneath the terraces. Rethinking led to a main roof girder spanning the full length, approximately 100 m (328 ft), of the pitch. This eliminated the internal roof support columns and provided a clear, uninterrupted view of the playing surface for all 5400 spectators. This solution also eliminated most of the columns beneath the terraces, enabling the football club to use the space for the provision of general social facilities and a bowling green. Such clear space beneath terraces has the flexibility to be adapted to suit changing trends in indoor sport and entertainment. Examples include an 85 m (279 ft) indoor sprint straight with fitness, training, physiotherapy and TV facilities (Don Valley Athletics Stadium), hospitality boxes, shops, bars,

Figure 14.30 Pitch lighting gantry (north), New English National Stadium, Wembley. (© Multiplex, 2006)

restaurants, fast-food outlets, a museum, a national fitness centre, changing rooms and a medical centre for players and match officials (Rugby Union Stadium Redevelopment, Twickenham – Figure 14.32), a purpose-built cinema (Queen's Stand, Epsom Race Course) and a jockeys' weighing room (Goodwood Race Course).

Ken Rosewall Arena, Sydney Olympic Park Tennis Centre, Australia

This is a very simple building – the type I most like to write about. The bowl-shaped structure is of precast concrete terrace units and steel raking beams (Figure 14.33). These components are held together by a tent-style tension ring at roof level,

Figure 14.31 Manchester United Football Club redevelopment: under-terrace. (John Clarke, © Corus) (This figure is reproduced in the colour-plate section as Plate XXI)

Figure 14.32 Rugby Football Union HQ, Twickenham: under-terrace. (John Clarke, © British Steel, 1996)

Figure 14.34 Ken Rosewall Arena, seating and access. (© Christine Goodman, 2014) (This figure is reproduced in the colour-plate section as Plate XXII)

working in tandem with an even more effective tension ring at the top of the masts. The result is exceptional simplicity and economy of steelwork in an elegant circular form. The tennis centre was originally briefed at 7000 permanent seats and 3000 temporary seats. By exploiting the circular structure, the team provided 10 300 permanent seats (Figure 14.34) plus a roof at almost no additional cost to the Olympics authority. Essentially, the combined tension rings allowed the structure to support much more seating with less vertical support. A circular structure is stable and strong when symmetrically loaded, but what about when it is loaded unevenly? The team designed this structure for the most destabilising loading condition possible – when one half of the stadium is completely full and the other

Figure 14.33 Ken Rosewall Arena, Sydney Olympic Park Tennis Centre. (© Christine Goodman, 2014)

Figure 14.35 Saitama Super Arena, Japan. (© Ellerbe Becket) (This figure is reproduced in the colour-plate section as Plate XXIII)

half empty. Robert Pugh, the project structural engineer, noted that during the opening ceremony this most unlikely of occurrences actually happened. So there was very early proof that the structure, when loaded to its full extent, would hold up well.

Saitama Super Arena, Japan

This arena (Figure 14.35) opened up exciting new possibilities when architects Nikken Sekkei (MAS.2000 Design Team) and Ellerbe Becket, with consulting engineers Flack + Kurtz, responded to their client's wish for something with 'the functional diversity and flexibility of the Swiss army knife, offering a wide range of features and combinations'. A holistic response to the brief resulted in the world's first 'smart' arena, with the capability of converting from a concert venue for a string quartet to a full-scale stadium within 30 minutes. This versatility is achieved by a 41.5 m (136 ft) tall moving block – incorporating the spectator seating, shops and facilities – which weighs 15 000 tonnes. Moving 70 m (230 ft) horizontally, the block adapts the space to seat between 22 000 and 36 500 spectators. The 130 m × 130 m (426.5 ft × 426.5 ft) stadium roof is supported by large fan-shaped beams. Cable rails supply electricity and connect and reconnect ducts as the block moves. Additional retractable seats at the sides, a vertically moving floor, and movable partitions and ceiling panels can add more capacity, and an adjustable ceiling renders the acoustics appropriate for each configuration.

REFERENCES

BSI (British Standards Institution) (1991) BS 5588-6:1991. Fire precautions in the design, construction and use of buildings. Code of practice for places of assembly. BSI, London, UK.

BSI (1993) BS EN 10056-2:1993. Specification for structural steel equal and unequal angles. Tolerances on shape and dimensions. BSI, London, UK.

BSI (1999) BS EN 10056-1:1999. Specification for structural steel equal and unequal angles. Dimensions. BSI, London, UK.

BSI (2004a) BS EN 10025-1:2004. Hot rolled products of structural steels. General technical delivery conditions. BSI, London, UK.

BSI (2004b) BS EN 10025-2:2004. Hot rolled products of structural steels. Technical delivery conditions for non-alloy structural steels. BSI, London, UK.

BSI (2004c) BS EN 10025-3:2004. Hot rolled products of structural steels. Technical delivery conditions for normalized/normalized rolled weldable fine grain structural steels. BSI, London, UK.

BSI (2004d) BS EN 10025-4:2004. Hot rolled products of structural steels. Technical delivery conditions for thermomechanical rolled weldable fine grain structural steels. BSI, London, UK.

BSI (2004e) BS EN 10025-5:2004. Hot rolled products of structural steels. Technical delivery conditions for structural steels with improved atmospheric corrosion resistance. BSI, London, UK.

BSI (2004f) BS EN 10025-6:2004 + A1:2009. Hot rolled products of structural steels. Technical delivery conditions for flat products of high yield strength structural steels in the quenched and tempered condition. BSI, London, UK.

BSI (2005) BS EN 1993-1-1:2005. Eurocode 3. Design of steel structures. General rules and rules for buildings. BSI, London, UK.

BSI (2006a) BS EN 10210-1:2006. Hot finished structural hollow sections of non-alloy and fine grain steels. Technical delivery requirements. BSI, London, UK.

BSI (2006b) BS EN 10210-2:2006. Hot finished structural hollow sections of non-alloy and fine grain steels. Tolerances, dimensions and sectional properties. BSI, London, UK.

BSI (2006c) BS EN 10219-1:2006. Cold formed welded structural hollow sections of non-alloy and fine grain steels. Technical delivery requirements. BSI, London, UK.

BSI (2006d) BS EN 10219-2:2006. Cold formed welded structural hollow sections of non-alloy and fine grain steels. Tolerances, dimensions and sectional properties. BSI, London, UK.

BSI (2009) BS 7079:2009. General introduction to standards for preparation of steel substrates before application of paints and related products. BSI, London, UK.

Dawson C (1977) Chelsea Football Club floodlighting towers. In Michel Tournay (ed.) *La Résistance à la Corrosion de l'Intérieur des Profils Creux en Acier*, Chambre Syndicale des Fabricants de Tubes d'Acier (CSFTA), Notice 1059, & Comité International pour le Développement et l'Etude de la Construction Tubulaire (CIDECT), Rapport final 10 b 78/3, Octobre 1978.

DCMS/FLA (Department for Culture, Media and Sport and Football Licensing Authority) (2008) *Guide to Safety at Sports Grounds*, 5th edn (The Green Guide). The Stationery Office, London, UK.

Tata Steel (2015) Colorcoat-online. See http://www.colorcoat-online.com/en (accessed 12/02/2015).

FURTHER READING

Ancon Building Products (2015) See http://www.ancon.co.uk (accessed 12/02/2015).

Culley P (1990) *Action Steps for Sports Ground Safety*. British Steel Tubes Division, TD 336/10E/90.

Goppert K (2011) New retractable roof solutions for sports stadia. *35th Annual Symposium of IABSE/52nd Annual Symposium of IASS/6th International Conference on Space Structures*, London, UK.

Inglis S (1991) *Seating – Sightlines Conversion of Terracing Seat Types*. Football Stadia Advisory Design Council, London, UK.

Innovative Brick Systems, LLC (2015) *Sporting and Event Center Construction Projects*. See http://www.mbrick.com/sports.html (accessed 12/02/2015).

ISE (Institution of Structural Engineers) (2010) *Appraisal of Existing Structures*, 3rd edn. ISE, London, UK.

Kalzip (2015) Metal standing seam roof and wall cladding systems. See http://www.kalzip.com/kalzip/uk/home/default.aspx (accessed 12/02/2015).

New Zealand Steel (2015) Zincalume. See http://www.nzsteel.co.nz/products/zincalume%C2%AE (accessed 12/02/2015).

Pilkington (2015) *Structural Glazing System*. See http://www.pilkington.com/europe/uk + and + ireland/english/products/bp/bybenefit/glasssystems/planar/default.htm (accessed 12/02/2015).

Rice P and Dutton H (1995) *Structural Glass*. Spon, London, UK.

Simpson JL and Webster DA (1995) The redevelopment of Twickenham for the Rugby Football Union. Presented at the Institution of Structural Engineers, 7 December 1995.

Society of Facade Engineering (2015) See http://www.facadeengineeringsociety.org (accessed 12/02/2015).

SCI/BCSA (Steel Construction Institute and British Constructional Steelwork Association) (2015) *National Structural Steelwork Specification for Building Construction*, 5th edn. See https://www.steelconstruction.org (accessed 12/02/2015).

Stadium and Arena Design
ISBN 978-0-7277-5790-6

http://dx.doi.org/10.1680/saad.57906.175

Chapter 15
Natural and artificial pitches for stadiums

Andy Cole Director of Consultancy, Sports Turf Research Institute (STRI)
Marke Jennings-Temple Senior Consultant, Sports Turf Research Institute (STRI)

Introduction

Different sports demand different playing surface characteristics, and it is important to consider these when designing and constructing pitches. Individual sports combined with the multi-use of the stadium will also influence whether natural or artificial turf is chosen for the pitch design.

For natural turf it is perhaps surprising to learn that little attention is given to the pitch in the planning and design of new stadiums, even though it is one of the most important aspects of use within the stadium. Some modern stadiums struggle to maintain high-quality pitches, where the design has resulted in large sections of the pitch covered in almost permanent shade; natural grass pitches demand direct light and airflow for the grass to grow effectively and stay healthy (Figure 15.1).

Light and air movement are obviously not issues where artificial pitches are being installed, and this is one of the reasons why artificial turf has become more popular in recent years.

Specialist pitch consultants can provide advice on the design of both natural and artificial pitches for stadiums, and effectively manage the installation process to ensure that specifications and quality standards for the playing surface are met.

Multi-purpose stadiums

Designing stadiums to host sporting and entertainment events will increase their utilisation and improve financial viability. Stadiums are not limiting their sporting activities to a single code but expanding to include a variety of sports, including football, rugby, cricket and American football, to maximise spectator interest and the return on investment. Stadiums may also want to host entertainment events, including concerts, festivals, monster truck shows, theatrical extravaganzas and trade/consumer shows (Figure 15.2). Some of these events take place on natural turf which is covered for a short period of time, but artificial turf allows faster and less costly conversions from sport use and back again.

Some of the key factors to be taken into consideration when deciding what uses can be accommodated within the stadium include: ease of access to the pitch for vehicles, materials and machinery required for the conversion, additional dressing rooms for athletes and performers and additional pitch-level storage. Also ensuring that there are adequate infrastructural services, including additional power supply and water recirculation, further enhance the possibilities of multiple use.

To accommodate these different uses, it is important not to change the stadium to an extent that has a negative impact on its primary purpose. For example, making the pitch considerably larger for another sport or adding a running track around the pitch can result in spectators being much further from the pitch and removed from the action. This reduces their sense of involvement and engagement with the game, and diminishes their excitement. Clever architecture can accommodate a running track below temporary seating, which helps retain that closeness of spectators to the action, which would otherwise be lost, yet once again maximising the space available within the bowl.

Pitch orientation

Great care must be taken in respect of the pitch orientation in relation to both the sun and prevailing weather conditions. Match participants, spectators and media representatives must be protected as much as possible from the glare of the sun and be able to watch the sport in relative comfort against the elements. However, consideration must also be given to the negative effect the roof will have on the quality of the grass being grown inside the stadium. Grass requires adequate light, air, moisture, heat and nutrition to develop healthy growth, and limitations to these five requirements will result in inferior turf, which will have a negative impact on the quality and sustainability of the grass being grown. Ideally, all sides of the pitch would receive a reasonable amount of direct sunlight.

A north–south pitch orientation is often considered ideal but more sophisticated analysis has led stadium designers to choose

Figure 15.1 Forsyth Barr Stadium in Dunedin, New Zealand is an enclosed stadium with a natural turf pitch, the stadium incorporates a natural air system design to ensure good air circulation

Figure 15.2 The multi-use stadium is a popular model for many new stadiums – Wembley Stadium in England being transformed into a racetrack for motor racing

an angle equal to the average direction of the sun at half-time in an afternoon game. Stadium designers and architects are becoming more aware of the need to select materials to maximise light availability for the turf. That said, the diffused light created may still require supplementing with artificial grow lights, which have revolutionised the growing of grass within stadium environments. Light studies can be carried out at either the design stage or on completion of construction to determine the number of lighting rigs required and the deployment strategy to maximise growth within the shaded environment.

Pitch performance

The pitch must be absolutely smooth and level. It should consist of natural grass or artificial turf, and it needs to provide consistency and quality across the whole surface. High-quality pitch construction is essential to accommodate the various sporting codes and additional use for entertainment to fit the business model for the stadium. The profile will support a functional drainage system and, where events demand, a vacuum and ventilation system incorporated in the sub-base. Natural grass can only withstand so much wear and tear in any given period, and additional profile/turf reinforcement products are now essential to achieve greater levels of stability, durability and wearability at any given time of the year. Natural grass should have an efficient watering system to keep the grass in good order and to be able to improve the speed of the surface prior to use. In cold climates, the pitch should be equipped with an underground heating system to prevent it from freezing in extreme winter conditions.

When constructing a new pitch it is vitally important to use experts who have proven experience in laying pitches in a particular locality. The site will have been surveyed for topographical and geological interest to determine any level adjustments necessary, as well as the most suitable platform on which to build the new facility. The specialist will design the pitch to provide the client with the best possible playing surface to suit the business model and to deal with any local climatic issues, which would influence playing quality if not addressed during the construction phase.

Keeping the pitch in peak condition at all times requires a detailed maintenance plan, and resources that include trained personnel, suitable mechanical equipment, grow lights, fertilisers, pesticides and testing equipment.

Natural turf pitches

Stadium environment

When modern stadiums are developed, consideration needs to be given to how the pitch interacts with the stadium and the micro-environment it creates within the bowl. In particular, the effect of shade has a negative effect on the growth of the grass. Hemi-view shade analysis will determine the extent of the problem and the required solution, and should be carried out at an early stage of design (Figure 15.3). Depending on the grass species specified, consideration will be given to the use of high-pressure sodium lights or, in the future, light-emitting diode (LED) lights. If supplementary pitch-level lighting is needed

Figure 15.3 Hemi-view image of Generali Arena, used by Sparta Praha in the Czech Republic. Hemi-view provides an analysis of light and shade across the pitch for all times throughout the year

to help the pitch survive, it is important to allow for storage space and pitch-side power supply at the design stage. Wind and air movements are also important functions to consider, and both should be modelled before proceeding with a full stadium design. The pitch and stadium micro-environment will also influence the pitch design and the approach taken to its development, depending on the client requirements.

The effect of climate and weather on the pitch must also be considered and will strongly influence species selection for each individual project – this is covered in more detail later in this chapter.

Business planning

A review of the business plan of the facility and its potential effects on the pitch is an important part of the planning process and will influence design and management options. Pitch-protection systems must also be considered alongside options for removable pitches or highly reinforced options.

Pitch management

The ability of the stadium to maintain the pitch to a high standard before, during and after the event is essential to the success or failure of the project. From an early stage of planning, the budget must be in place to have appropriate staff numbers, training, equipment and an annual renovation programme to keep the pitch quality at a high standard. Ongoing independent performance testing and independent agronomic advice should also be factored into this equation to ensure optimum playing-performance characteristics for each discipline. Maintenance contracting and advice should be considered separately to ensure that the best procedures are implemented.

Design requirements

The design approach to pitches is site-specific and must be considered on a site-by-site basis. The majority of new natural turf pitch developments require specifically designed drainage and soil profiles to ensure satisfactory playing conditions throughout the season which function correctly in the proposed environment. Laboratory testing, through an approved and accredited laboratory, is essential to ensure selection of the correct materials and, most importantly, quality control throughout the installation process (Figure 15.4). Physical and chemical characteristics must be measured to achieve optimum performance.

Tender criteria

Only suitably experienced contractors should be considered for the work. Designing the pitch, building the pitch and supplying the materials should each be considered as separate entities. The criteria for qualifying contractors should be based on:

- Experience: a portfolio of stadium pitch construction in a variety of different areas to indicate a depth of knowledge and capability to deliver at the highest level.

Figure 15.4 Material testing in accredited laboratory ensures quality and durability of materials, which directly influences the playing performance on the pitch

- Resources: adequate and correct machinery in order to complete the installation efficiently and to the required standard.
- Technical backup: a labour force that mirrors the level of experience within the company to deliver to the highest standard. A working knowledge of drainage, soils, irrigation, nutrition and growing in are essential traits for the successful contractor.
- The company's financial ability to complete the contract: the company must have a solid financial background and proven track record to be short-listed.

It is sensible to pre-qualify all contractors that wish to bid on work by using the above criteria before issuing tenders.

Level adjustments

For minor level changes (± 50 mm), cultivation and laser grading within the soil profile may be acceptable, depending on the situation. Where greater adjustments are needed, a civil engineering type cut-and-fill approach may be necessary, based on a 3D topographical survey of the site. Geological testing is also essential on new-builds to determine the nature and stability of the sub-base, which will govern any specific remediation before construction.

Drainage

Suspended water table rootzone pitch

In the majority of new stadium pitches this type of construction is recommended (Figure 15.5). The selection of material used to build this type of pitch should consider the climate and also stability, drainage, reinforcements, grasses to be used and water retention for grass growth. All materials used should be pre-approved by an expert following detailed laboratory analysis.

Figure 15.5 Cut-out diagram showing the construction of a suspended water table pitch with reinforcement

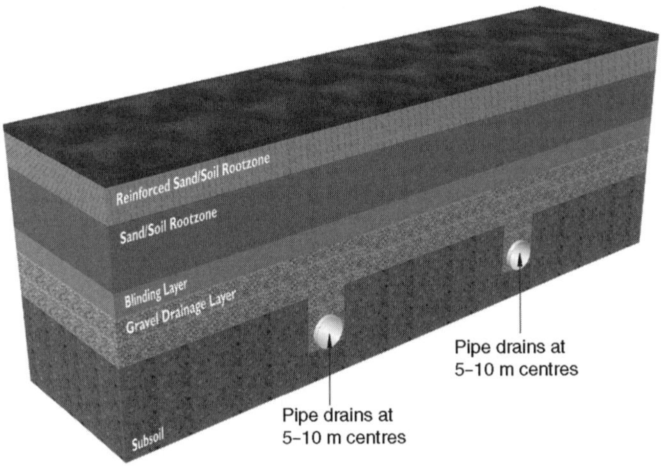

Where climatic conditions indicate periods of sudden and heavy rainfall, it would be prudent to consider the installation of a vacuum and ventilation system, installed in the sub-base alongside the drainage system. This system will mitigate against the unexpected periods of heavy rain that could disrupt or even lead to cancellation of a fixture if not available. The ventilation function is extremely beneficial for gaseous exchange, as well as enhancing any heating or cooling elements within the profile.

Soil over drainage layer

This option can also be used as a design proposal for high-level/high-use training pitches. The approach involves installation of a drainage system in the subgrade, followed by the spreading of a carefully selected layer of gravel and an intermediate layer of coarse grit. The design, spacing and depths of the materials are subject to various tests and recommendations based on soil analysis, usage levels and climate. Particular attention should be given to receiving professional advice when constructing this type of pitch.

In situ, palletised or tray systems

The stadium and pitch design will also include consideration of an in situ pitch system, where all elements remain in place for either sporting or entertainment events. In this situation, the playing surface must be protected to limit turf damage.

Palletised systems provide for a drainage medium; rootzone and turf, with or without reinforcement, which can be moved into or out of the stadium depending on the use of the bowl at the time. The system requires a suitable outdoor storage area where the pitch can be 'grown' while not in play in readiness for its return to the stadium for the next event.

Sliding pitch systems require an area of land outside the immediate environs of the pitch, and this needs to be big enough to house a full-sized pitch. The area is used for growing and storing the turf while the bowl is being used for alternative events. The main limiting factor to this system is the availability of land surrounding the stadium, where land value is often at a premium and, therefore, land is limited.

Pitch reinforcements

Reinforcements allow the surface to remain safe and playable when grass cover is lost but also reduce the level of divoting and improve the durability of the surface. The selection of the reinforcement must consider the intended use of the pitch over a whole season, which will determine the need, or not, for sod/turf replacement.

Hybrid systems involve stitching fibres into the rootzone (Figure 15.6), which in turn adds strength to the surface and profile, giving a green appearance amongst the natural grass. Alternative hybrid systems involve the laying of a synthetic carpet-like material, into which the rootzone is added, prior to establishing the grass cover. This system can be harvested and reinstalled up to a maximum of three times if suitable harvesting and storage facilities are in place.

Rootzone mixed-fibre or fibre plus elastic systems involve the amelioration and blending of elastic and polypropylene fibres into the sand rootzone, which improves shear strength to the rootzone and reduces divoting and increases usage levels.

Figure 15.6 A hybrid grass system includes stitching fibres into the rootzone to strengthen the surface and profile

Figure 15.7 Re-turf operation at Croke Park in Ireland following the U2 concert

Construction materials

The selection of sand, gravel and soil for pitch construction and drainage is a specialist task requiring analysis of materials in an accredited laboratory. Internationally recognised grading curves for pitch construction must be used to determine suitability against climate data, species composition, water availability and quality, and reinforcement selection.

Seeding the pitch

Establishing grass from seed is the most cost-effective and, in most cases, the best way to establish a new pitch. It does, however, take time and at least 12 weeks of good growing weather, and an intensive maintenance period is generally required.

When choosing the grass type, the following should be considered:

- climate and geographical location
- macro- and micro-environments within the stadium
- status of the venue and games to be played/usage levels
- performance of grass species for specific sport
- variations in performance and aesthetics with seasons
- legacy and maintenance requirements
- rootzone reinforcement used
- time available for grow-in.

Turfing the pitch

Where establishment time is limited, turfing may be considered as an option for providing a playing surface in minimal time (Figure 15.7). Compatibility of rootzone and imported material brought in with the new turf can cause some issues, which can be reduced by the use of bespoke material or washed turf. The turf must be tested for compatibility with the underlying materials. Turfed pitches often require intensive maintenance once laid to ensure that optimum playing conditions are achieved.

Turf nurseries are beneficial where localised repair and renovation is likely through the season. Appropriate harvesting and laying equipment may be necessary in this situation.

Irrigation systems

Fully automatic pop-up irrigation is a must on new pitch constructions where a high sand content is used to provide good surface drainage. The irrigation system is required for grass growth but also for match preparation. The system should provide valve-in-head control to allow perimeter or centre watering to be carried out independently; this is especially important during match-day preparation. Off-pitch water cannons are not ideal, as the controlled application of water is not as direct as with a fully automatic system, and keeping spectators and media dry can prove challenging.

The following should be considered:

- the safety of players and officials
- spacing of irrigation heads and 'head-to-head' coverage
- location of heads to ensure the system works when advertising hoardings are in place
- location and number of heads and speed of rotation to allow watering of the central area within 10 minutes
- type of irrigation heads with turf cups preferable over solid sprinklers to improve safety
- automatic control system for night watering
- emergency shut-off
- link to weather station
- tank storage capacity engineered to water supply and usage requirements
- correctly sized pipes and pumps to guarantee full system pressure when the stadium is operational.

Irrigation heads should not pose a safety risk for players, and should be set 3 cm below the surface so that players do not come into contact with the solid part of the head.

Undersoil heating

Where the risk of frost is present, an undersoil heating system is recommended. As well as preventing frost, the undersoil heating system, in combination with supplementary pitch lighting, allows the grass to grow all year round. The undersoil heating system should not interfere with or prevent deep aeration practices to 250 mm on the pitch, and hence all components must be designed to operate effectively below this level.

Cooling/ventilation systems

The ability to ventilate the pitch from the base as well as the surface provides an advantage in certain environments. It can also improve drainage rates of the pitch during high rainfall events by creating a vacuum within the rootzone. Where very heavy rain is expected, such systems should be considered (Figure 15.8). Cooling technology is currently in its infancy, but is certainly a topic for the future in tropical countries attempting to grow cool-season species.

Pitch covers

Depending on pitch design and climate, it may be necessary to have rain or snow covers to manage the elements, especially in Eastern Europe, where winter temperatures can fall as low as −40°C. Where vacuum systems have not been installed, rain covers can protect the surface from heavy rainfall events prior to kick off, but these are obviously limited to pre-and post-match use only.

The frost covers can also double up as germination blankets, which can be extremely beneficial during the establishment phase of seeding, to accelerate the process and reduce moisture loss as well as preserving heat.

Figure 15.8 Installation of an in situ air sparging system (ISASS) at Metalist Stadium in Kharkiv, Ukraine. The system can be used for warming, cooling, ventilation and drainage

Environmental and sustainability

Environmental improvements can be instigated from an early stage in design and carried through to maintenance. These include:

- grass-clipping composting
- use of recycled products in pitch construction
- use of biodiesel in equipment
- cultural practice in disease control
- water use and management.

Post-construction pitch testing

Following construction, it is essential to identify the performance characteristics that were identified in the design stage to ensure client and player satisfaction. Performance measurements would include firmness, moisture, traction and hydraulic conductivity, measured using standard operating procedures for repeatability. Ongoing testing during the playing season and pre- and post-events can assist the groundsman in guiding maintenance requirements based on changes in performance figures.

Artificial turf pitches

Background

Artificial turf has become an acceptable playing surface for most sports as continued development of the turf carpet has resulted in high-quality options specific to each sport (Figure 15.9). Artificial playing surfaces are now available that enable players to perform both dynamically and safely.

Figure 15.9 Stade Marcel-Picot in Nancy, one of the few French stadiums with an artificial pitch

The advantages of an artificial turf playing surface are numerous. Artificial turf remains green throughout the year and in different climates. Players may need a short period of time to adapt to playing on artificial turf but the evenness and consistency of the playing surface make for a precise game that encourages the skills of technical and physical players.

In addition, artificial turf pitches can be used much more than natural pitches, increasing utilisation of both the pitch and the facility. Artificial turf pitches do not suffer damage due to inclement weather (e.g. rain, snow), whereas these elements can significantly affect the use of natural pitches. Regular maintenance, which is critical to keeping artificial turf pitches in top shape, may be less expensive than maintaining a good-quality natural turf pitch, depending on the nature and type of use the artificial turf pitch receives.

For these reasons, artificial turf is a viable and attractive alternative to a natural turf pitch. To make the right decision regarding artificial turf, the following should be taken into consideration.

Specialist advice

Building regulations and local conditions must be fully respected, and architects or engineers with experience of building sporting facilities should be consulted when planning pitch development. Manufacturers of artificial turf systems can play a supporting and informative role with regard to the choice of a suitable turf carpet but independent advice may assist with navigating through the marketing hype. Construction work should only be awarded to firms with a successful track record and specialist equipment. Supervision of the building work by the planner or designer and monitoring by external inspection bodies are crucial to the successful completion of such projects.

Artificial turf projects

The work should only be awarded to firms that specialise in artificial sports pitch construction. If no such firms are available, then local engineering firms that are able to carry out ground works and have experience in placing and compacting aggregates to fine tolerances and installing concrete formwork should be contracted. The work can be carried out by an engineering contractor, with the artificial turf suppliers as subcontractors, or vice versa. Whichever model is chosen, it is important that a single party is responsible for guaranteeing both the base and the artificial turf.

Products

The choice of artificial turf depends on the intended use of the surface (i.e. football, rugby etc.), as each sport has its own specific requirements.

In summary, the turf systems may have either fibrillated or monofilament fibres (or a combination of both), with the choice of yarn length being dependent on the sport and whether or not a shockpad is to be installed. Infill types vary, and can include crumbed truck tyres (SBR), ethylene propylene diene monomer (EPDM) manufactured from raw ingredients, and natural infill options such as cork and coconut fibre. There are systems available with and without a shockpad. Each available turf system (the carpet, the infill and the shockpad, where installed) has a range of advantages and disadvantages, and the final choice must consider the local environment, the amount of maintenance likely to be carried out and the predominant use of the surface.

Laboratory tests

Products used in artificial turf pitches must first pass laboratory tests to determine their composition, and then must be tested for durability, joint strength, climatic resistance, player–surface interaction and ball–surface interaction, all specific for the given sport and grade of sport. If the turf passes all these tests, this completes the first stage of the process towards the award of the certification. Where a pitch is going to be certified for a particular sport (after construction), it is recommended that the specified turf product has been through this initial laboratory test procedure.

Planning and design

Sub-base

In the event that geological information is not already available from the planning of the stadium, soil-drilling tests should be conducted to determine how to prepare the sub-base and build the base. The stability of the sub-base is critical to the long-term success of the surface.

Specific characteristics of the site, such as water-table depth and bearing strength, must be checked prior to the feasibility study to ensure that the sub-base is adequately prepared, either through drainage and/or stabilisation and/or the inclusion of a geotextile between the sub-base and the base.

Base

The final choice of construction materials for the base will depend on the site characteristics, but may also be influenced by the building materials available locally. It is inadvisable to have too many layers in the base; a great deal of precision is required to achieve flatness and evenness criteria.

Base layers for vertically-draining pitches will be formed using clean, highly-specified aggregates (i.e. gravel). Large-scale roading-type machinery is required for the installation of these layers. The base should be tested before the artificial turf system is laid, and strength tests should also be conducted during the building work.

Drainage

Drainage design will be required for the vertical movement of water through the profile, with a provision for surfacewater collection under high rainfall events. Therefore, local rainfall conditions and storm intensity/duration data are critical when designing the drainage system. In practice, the dimensions of the drainage system can vary depending on the amount of rainfall the surface may receive, the extent and direction(s) of surface slope and the type of pipe network preferred.

Two drainage systems may be required – one for the artificial turf and one for groundwater control in the sub-base – although one system can typically cater for both, depending on soil conditions. Water collected off surrounding areas (i.e. footpaths) should also be taken into account in the overall design. The vertical drainage rate of the immediate surface is determined by the permeability of the turf itself, which is a function of the perforations (per m^2) in the backing material.

Under heavy rainfall events on vertically-draining surfaces, surfacewater may shed off the surface and will need to be able to enter the drainage system by way of a gutter or surface drain. Inspection points should be designed into the system to monitor the drainage.

Edging

Where pitches are installed with single or multiple surface gradients, the low sides of the pitch should feature drainage channels for collecting surfacewater, and the high sides of the pitch should feature kerb stones (or similar). Typically the kerbstones would be positioned such that they finish approximately 50 mm above the maximum height of the turf. In locations with low rainfall, the gutter can be dispensed with, except where the anticipated volume of a single rainfall event is large and can fall in a short period of time.

If the pitch is surrounded by a running track, a gutter or surface drain for each may be preferable, although these would discharge into a shared drainage system. It is advisable to make the surfaces of the running track and the pitch flush for safety, particularly when players take corner kicks. Edging and pitch-edge drainage are important factors in the coordination of the project, as there are options for forming the drains and edging on-site during the works or purchasing preformed materials.

Depending on the intended sport, artificial turf is usually laid on a completed base with a maximum gradient of 1%, and very often less. It is useful if the sub-base has been formed to the same gradient, as this makes it easier to install layers of even thickness and achieve the required compaction criteria.

However, foundations for the goalposts, spectator fencing and corner flags should be installed before the artificial turf is laid.

If possible, the artificial turf should be laid after the physical work has been completed.

Watering

Watering is preferable in conditions with consistently high temperatures, and is generally advisable for all turf surfaces. Artificial turf and certain infill types absorb heat and, for the players' comfort, it is advisable to cool the surface. Having a watering system will also assist with surface cleansing.

In order to keep water consumption to a minimum on grounds of cost, automatic sprinklers are advisable, even if they are only used occasionally. Sprinklers should be placed outside the pitch of play. It is useful to install quick coupler valves to enable localised hand-watering (i.e. for small spills) should it be required.

Cable ducts

Any underground ducts (for TV etc.) should be installed outside the playing surface and during the formation of the base.

Installation

The materials required for artificial turf are voluminous, and it is therefore necessary to ensure there is an access road to the site that is capable of tolerating heavy vehicles, and that adequate material storage space exists. Laying artificial turf depends on the weather and requires the programme of works to be adapted to local conditions. Ideally, during turf installation the weather should be dry, the temperature should be within the preferred range for the seam glue being used and humidity should be low.

Interruptions to the physical works should be avoided, and the forward work programme adapted to the weather forecast. The artificial turf should be laid by qualified staff. The project manager or project engineer should check whether the materials delivered for the base and the artificial turf match those specified in the plans and manufacturers' guidelines.

Post-construction pitch testing

After construction, an installed pitch must be tested on site by an accredited testing organisation 2 or 3 months after the turf surface is completed. Specialised pitch-testing equipment measures the performance characteristics of the surface to assess whether the ball and surface–player interaction requirements of the sport have been met. The first certification tests also confirm that the installed product matches what was specified (Figure 15.10).

Figure 15.10 Allianz Park, home of Saracens Rugby Club, in London, UK, was the first professional rugby union club in the world to play on entirely artificial turf with a 3G synthetic pitch

Ongoing certification tests are required, and the frequency of retesting is dependent on the sport and the grade of sport the surface is being used for.

Safety

A cautionary note needs to be sounded regarding safety at a stadium with an artificial turf surface. Artificial pitches are installed using a variety of materials, some of which may be flammable under extreme conditions. Items such as firecrackers or flares are occasionally thrown onto the pitch during a match. These items can burn at temperatures of up to 2000°C (3632°F), which is more than sufficient to ignite some synthetic materials and damage the pitch. Such items are easily extinguished with the use of water or sand, and it is recommended that pitch staff be trained and on hand to ensure that a rapid response is made to extinguish these items if thrown on to the pitch. This will limit the amount of damage the pitch is exposed to.

Ongoing maintenance

Ongoing maintenance must be carried out immediately after the pitch starts to be used. Unlike natural turf, an artificial turf pitch cannot recover from damage, so the role of maintenance is to slow the decline in quality of the pitch and maintain the pitch within the performance criteria set by the governing body of the sport(s) the surface is being used for.

Maintenance schedules should be specific to the use the surface is receiving, and be flexible enough to enable high-wear areas to be targeted. In general, maintenance activities should include grooming or brushing (to redistribute infill and stand the fibres upright), aerating (to decompact the infill and maintain drainage rates), raking (to loosen the upper infill layer) and sweeping (to pick up debris and leaf litter and keep the surface clean). There is a huge range of machinery options available, and the warranty documentation that comes with turf will indicate the preferred types of machine and frequency of operations.

Stadium and Arena Design
ISBN 978-0-7277-5790-6

http://dx.doi.org/10.1680/saad.57906.187

Chapter 16
Retractable roof solutions for sports stadiums and arenas

Knut Göppert Schlaich Bergermann und Partner, Germany
Christoph Paech Schlaich Bergermann und Partner, Germany

Retractable roof solutions for sports stadiums and arenas

The worldwide growing demand for multifunctional sport stadiums creates great opportunities for innovative engineering solutions. In general, smart engineered roof and façade structures are currently very popular, and interest in these solutions within the architectural community is expanding (Figures 16.1 to 16.3). In addition, we recognise a growing demand for adaptable building envelopes that can react to changing climatic conditions and functional demands. We now have the necessary tools to make architecture a dynamic machine, as Le Corbusier proposed: 'A house is a machine for living in' (Ishii, 2000).

This trend is being served by traditional building engineering services but also increasingly so by movable structures and new methods of climate engineering, where basically the airflow is being created and/or directed without the implementation of mechanical fans or similar equipment.

One additional requirement, resulting from the important aspect of sustainability, is to minimise grey energy and energy consumption during the operation of the building. In many cases this points directly to lightweight structures and membrane roof solutions. Schlaich Bergermann und Partner are leaders in this field and work from a strong engineering basis, including civil, structural, mechanical, electrical and aeronautic engineers among their team. Their portfolio consists of 10 implemented large retractable roof structures for sports arenas, and three more are currently on the drawing boards (Pawlitschko *et al.*, 2011).

Movable structures – adaptivity

Generally, adaptivity is characterised by the modification of functionality. The roof of a convertible car, for instance, can be closed to protect the driver from the sun and the rain, or it can be opened for the driver to feel the breeze. Comparably, adaptivity often defines a modification of visual appearance based on the geometry, lighting conditions or the colour.

In the field of architecture and engineering, the use of convertible/adaptive elements has enormous potential and has been used more and more frequently in recent years. Regarding the aesthetics and the utilisation of a building, the ability to react to changing requirements, in combination with a varying appearance, offers many advantages.

In engineering and design, adaptivity usually means that individual components move within the space. However, intelligent, adaptive structures can also convert their load-bearing system, their geometry or their properties to react to, for instance, changing load conditions or solar radiation. In this way, energy-efficient systems can be produced at a time of energy shortage, when resources are becoming more and more scarce. Such systems either save, for example, on building materials, or they achieve higher yields of energy, thanks to an automated adjustment of individual elements. (Schlaich, 2004).

History

Certainly, at a small scale, adaptivity has already been used in architecture and engineering for centuries (e.g. the drawbridge of a medieval castle). However, until recently, the requirements for these elements were fairly limited. The principles of motion were very simple (e.g. rotation around one axis), as were the drives. It was only in the second half of the last century that the value of transformation was introduced. Engineers pushed the boundaries of design, and with the developments by Frei Otto, Felix Escric, Jörg Schlaich, Rudolf Bergermann and Bodo Rasch the engineered transformable structure entered the architectural world (Bergermann *et al.*, 2004).

Figure 16.1 National Stadium, Warsaw. (© Marcus Bredt)

Movable construction and adaptivity

Movable, and also convertible, structures mechanically react to the conditions of the outer framework. The entire motion process consists of a series of fine-tuned motion sequences. There are many examples known from other disciplines (e.g. vehicle construction); although these are on a smaller scale, they are very often produced in larger quantity.

The use of movable components increases with the simplification of mechanical construction methods; with the aim

Figure 16.2 BC Place Stadium, Vancouver. (© Michael Elkan)

Figure 16.3 National Stadium, Bucharest. (© Marcus Bredt)

to generate durable, reliable and economic systems. The smaller the component that has to be moved, and the lower the movable weight, the lower is the energy consumed by the movement. At the same time, maintenance intervals increase for the lightweight solutions.

It is also possible to design a load-bearing structure in such a way that it reacts to external loads. Adaptive load-bearing structures are characterised by a minimal use of material on the one hand, and more sophisticated drive and controller technology on the other. This idea, however, is mostly used to limit deflections, vibration and acceleration. The active damper systems used in the demanding design of structures (not only for roofs, but also for high-rise towers) in areas with high earthquake activity are well known (Göppert *et al.*, 2011).

Principles of movable structures

We can differentiate between two different principles for movable components.

Stiff elements

Stiff elements are shifted or rotated in space. In addition, individual components can be connected by hinges. It is possible to combine translations with rotations. The movement will always be a 3D rigid-body rotation, or a shift.

Starting with the large domes in Toronto and, mainly, Japan in the 1980s, about 30 stadiums were built with rails to guide the roof elements (Figures 16.4 and 16.5).

Figure 16.4 Movable roof of the casting house at Landschaftspark, Duisburg-Nord. (© Thomas Eicken, Mühltal/planinghaus architekten bda)

Figure 16.5 Esprit Arena (formerly LTU Arena), Düsseldorf. (© euroluftbild.de)

Three different driving technologies are typically used:

- gear drive/rack and a pinion drive
- adhesion drive with an electric motor
- wire rope winch.

Important parameters for this type of roof are the driving speed and the track slope, and the energy required for the driving operation depends mainly on these two parameters. Roof elements of up to 1000 tonne need to be moved, and it can easily be seen that the inclination has a major effect on the choice of technology (a friction/adhesion drive has a limit) and the energy used.

Nevertheless, although they represent a major investment and involve high energy consumption, these technologies are seen as robust and are regarded as 'proven technology'. In fact, the major shortfall of these concepts, apart from the energy requirement, is the limited flexibility in the architectural design. Large mega-girders in a parallel arrangement and bulky retractable roof elements, parked on top of the fixed roof, are not the first choice in contemporary stadium design. Therefore, designers have increasingly looked for new possibilities derived from lightweight systems.

Flexible and pliable elements

Flexible and pliable materials, such as cables or fabric membranes, are folded or rolled up during operation. These systems are characterised by their low weight and, therefore, simplified driving technology.

Basically, a 2D membrane of a mechanically pretensioned structure can be folded if, during the driving operation, the distance between the two connection points never exceeds the distance of the mechanically pretensioned geometry. The application of this principle leads to interesting results, for example:

- The membrane is driven in a parallel way: the distance between the connection points remains constant during the operation.
- The membrane is driven radially from the outer rim to the centre: the direct distance gradually decreases.

In both cases rectangular or triangular membrane fields are suitable for the transmission of tensile forces by means of point loads. Options can be expanded beyond the exactly radial arrangement if the primary structure (the direct support structure of the membrane) is designed in such a way that its geometry can also adapt. Modifications like these even facilitate a narrower arrangement of the membrane fields in parking position than during the driving operation. For most applications the pretension is generated by means of hydraulic cylinders. These are controlled according to the required load transmission and the distance. In specific cases, if the membrane is cushion-shaped, the pretension depends on the internal air pressure. The opening and closing operation of the membrane is executed by means of electric winches and a continuous drive cable. This technology is fast, simple and strong enough to guide the membrane edges to the hydraulic cylinders for the tensioning process. Besides the actual drive system, the roof has to be equipped with sensors in order to achieve

Figure 16.6 Tensioning and drive technology on the National Stadium, Warsaw. (© Schlaich Bergermann und Partner)

synchronisation and to prevent excessive strain (Figures 16.6 and 16.7).

Interdisciplinary design and planning

Adaptive structures are challenging and, in most cases, complex buildings. Apart from the architectural design and the engineering of the load-bearing structure, other components such as the drive technology and control technology have to be designed. Therefore, the design and planning of convertible structures requires expert knowledge from various fields. Close collaboration is necessary in the first design stages between architects, civil engineers, mechanical engineers, electrical engineers and experts from other disciplines.

Figure 16.7 Drive technology for the vertically movable roof on the Vista Alegre, Madrid. (© Skyspan)

Figure 16.8 Simulation of the folding process. (© Schlaich Bergermann und Partner)

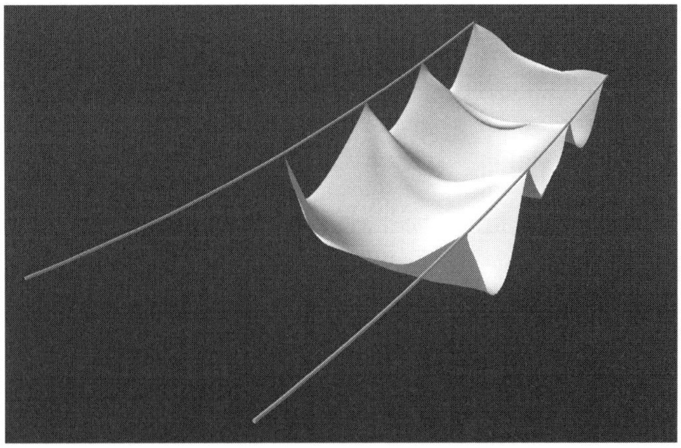

Movable structures are designed at the computer, using high-performance 3D drawing programs, finite-element software and special optimisation tools. Physical models are frequently employed to visualise complex motion and folding patterns. The functionality of the system (or of individual components) is verified upfront, using prototypes and test series to guarantee a smooth and successful installation and start-up on site (Figure 16.8).

Basic requirements of structural design for movable roofs

The following points should be considered and determined in order to ensure the safety and serviceability of structures equipped with retractable roofs and façades (Ishii, 2000):

- *Design life of the structure*. The design life of the retractable structure should be determined and the structure should be designed based on this definition.
- *Structural characteristics*. The designers should clarify and define all structural characteristics, such as strength, stiffness and geometry.
- *Structural safety*. The retractable structure design must be structurally safe against predicted loads and external forces in its open and closed positions, and also during the opening and closing process. Special care has to be taken to include within the design process the interaction between fixed and retractable structural elements.
- *Structural analysis and testing*. Beside all elements that are subject to standard or non-standard structural analysis,

experiments and tests for elements or details might be required.

- *Serviceability*. The retractable structure must be designed to provide the desired function adequately under defined usage conditions. This must be clearly agreed and communicated between the client, operator and design team.
- *Operation safety*. All intermediate stages (during the retraction process) must be safe and controllable. If restricting conditions, such as wind-speed limits, are agreed, they must be safely implemented within the operation procedures and be outlined in the operation manual.

Concept for lightweight retractable roofs

The concept for lightweight adaptable roofs can be characterised by:

- the use of low-mass materials, mainly fabrics and membranes
- the application of folding patterns to reduce the size of the roof from fully deployed to the storage position (typical reduction factors in plan are seen in a range from 1/20 to 1/100)
- the structural concept is reduced to tension members and the application of prestress in the structural members
- the differentiation of driving technology and stressing technology. This is a key point in keeping the mechanical system simple, reliable and economical. Long distances for travelling require small forces, whereas short stressing lengths require significantly higher forces.

Materials for folding membrane structures

The selection of the best suited material for an application where the material is subject to folding cycles is dependent on the following main aspects:

- environmental conditions (indoor/outdoor/climate)
- loading conditions (wind/snow/rain/hail/temperature)
- number of envisaged cycles within the lifespan of the material
- required protection level (water/wind/sun/temperature).

At the current state of knowledge, the most suitable fabrics for the membranes are materials made from polyester (PES) or polytetrafluoroethylene (PTFE) fibres. PES fibres need to be protected from ultraviolet (UV) light, for which the state of the art would be either PVC or silicone. It is important to mention that if polyvinyl chloride (PVC) is used it must be ensured that softeners are used in the PVC to avoid embrittlement of the coating over time. Fluoropolymer-coated woven PTFE can be used with and without PTFE coating. The latter is the watertight version of a highly translucent membrane, with excellent self-cleaning properties and good folding behaviour. For current projects, we have developed special testing procedures to help us determine the long-term performance of the materials in the actual application (Balz and Paech, 2013; Balz *et al.*, 2013).

Geometrical arrangements and stressing strategies for retractable membrane structures

In general, a continuous membrane is only able to fold if, during the retraction process, the distance between two supporting points will never be bigger than the distance in the final deployed geometry. Applying this rule results in some interesting findings:

- the membrane will be retracted parallel, keeping the distance between the supports constant during the process
- the membrane will be retracted radially, deploying from the centre to the outer boundary.

In both cases, the rectangular or triangular shaped membrane bays, respectively, are suitable for a reasonable introduction of prestress by single point stressing units. Aside from the exact radial arrangement, the possibilities could be extended if the primary structure (the membrane supporting structure) is designed such that it will change its geometry as well. This allows panels that are narrower in the deployed position than while in motion.

For most of the application the prestress is introduced by means of hydraulic cylinders, which are force and distance controlled. In the special cases where the membrane is cushion shaped, the prestress is applied by means of internal air pressure.

The retraction of the membrane into its parking position, and vice versa, is typically achieved through the use of electrical winches and an endless actuation cable. This technology is fast, simple and strong enough to hand over the membrane to the hydraulic jacks for final stressing. Beside the actual driving systems, the roof needs to be equipped with sensors for synchronisation and to avoid overstressing.

Case studies – retractable roof structures

To explain and illustrate the above guidelines in more detail, some of the latest applications of retractable roof structures are described below.

Retractable membrane roof above the courtyard of the Vienna City Hall, Austria

The 34 m × 32 m (111.5 ft × 105 ft) retractable membrane roof on the Arkadenhof of the Vienna City Hall serves as a rain shelter and sun protection for various events (Figure 16.9). Rails are located at the longitudinal sides of the roof, supported on the existing building. Ridge cables are tensioned perpendicular to the rails and can be moved in parallel by means of sliding carriages. Between any pair of ridge cables the membrane is

Figure 16.9 The parallel retraction process at Vienna City Hall. (© Schlaich Bergermann und Partner)

weighed down in a line to generate pretension and create a linear folding pattern of the membrane during operation. To avoid the transmission of tensile forces into the existing structure the resultant tensile force from the membrane and the cables is short-circuited by means of four trussed girders. Each girder on both sides is supported on rollers, driven by synchronised gear motors, pinions and racks that are mounted parallel to the rails. The drive control coordinates and monitors the motion of the individual cross-beams (Bergermann *et al.*, 2004).

The casting house at the Landschaftspark, Duisburg, Germany

To be fit for purpose as a venue for the city's cultural festival Triennale, the quality of the former casting house of blast furnace 1 at the 'Duisburg Landschaftspark had to be enhanced. Pneumatically supported foil cushions make up the retractable roof, which rides on wave-like rails. This wave gracefully undulates from the casting house across the existing pipes, to cover the stand in the case of bad weather (Figure 16.10). The distance of approximately 20 m between the two wave-like rails is bridged by trussed girders. Each pair of girders is linked and equipped with four rollers. Frames are pin-jointed between

Figure 16.10 The retractable roof over the casting house in the Landschaftspark, Duisburg, in operation. (© Thomas Eicken, Mühltal/planinghaus architekten bda)

these 'carts'. In combination the carts and frames cover an area of approximately 29 m × 20 m (95 ft × 65.6 ft). Each of the four frames features an ETFE-foil cushion infill.

National Stadium, Warsaw, Poland

A 55 000 seat multifunctional arena was built on the earth wall of an ancient open-air arena close to the city centre of Warsaw to host the opening game of the European Soccer Championship in 2012. The grandstands are covered by a fixed roof, whereas a retractable roof above the green field allows it to be used as an indoor arena during the winter season or bad weather (Figures 16.11 and 16.12).

The seating area is permanently covered by 54 000 m^2 (581 250 ft^2) of PTFE-coated glass-fibre membrane supported by membrane arches. A 10 m (32.8 ft) wide glass roof clad with 4000 m^2 (43 056 ft^2) heat-strengthened glass forms the inner edge of the permanent roof, and provides the overlap between the permanent roof and the 11 000 m^2 (118 400 ft^2) PVC–polyester fabric of the inner retractable roof. Altogether this leads to a covered surface of nearly 70 000 m^2 (753 470 ft^2), with main spans of 280 m (919 ft) and 245 m (804 ft).

The structure is carried by a series of 72 columns founded around the bowl structure. A single compression ring rests on top of the columns and short-cuts the tension forces of

Figure 16.11 National Stadium in Warsaw, closed inner roof. (© Marcus Bredt)

the primary cable structure. In general, for spoke-and-wheel structures two basic strategies can be used: one compression ring and two tension rings kept at a distance by a series of flying masts; or two compression rings in combination with one tension ring. Each strategy will provide the required inclination of the two layers of radial cables that allow the structure to carry varying vertical loads. As an answer to the high roof span and loads, the cable structure for the National Stadium Warsaw is a combination of the two principles. Subsequently, the radial cables need to cross each other and thereby obtain an increased inclination, leading to increased stiffness and reduced cable forces. Instead of an upper compression ring, the upper radial cables are secured by an inclined strut-and-tie arrangement activating the lower and single compression ring and the foundations. The outer radial cables are connected to two tension rings, roughly following the inner edge of the seating area in plan view. A series of flying masts keeps the two tension rings at a distance and at the same time allows the fixing of a

Figure 16.12 National Stadium Warsaw, section. (© Schlaich Bergermann und Partner)

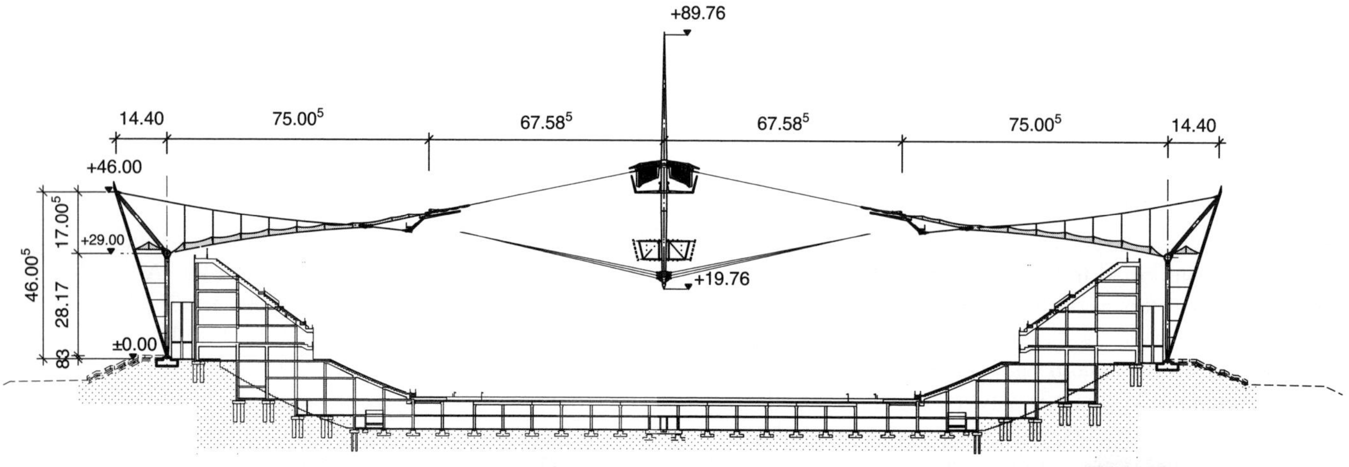

Figure 16.13 The inner roof on BC Place, Vancouver, closed. (© Schlaich Bergermann und Partner)

10 m (32.8 ft) wide glass roof cantilevering towards the pitch. For the inner part of the roof, carrying the retractable membrane, the lower radial cables have been concentrated to only four sets of three cables, each arranged along the diagonals of the green field. The retractable roof membrane itself moves along a series of 60 upper radial cables spanning between a central hub and the upper tension ring. This leads to a minimum number of structural elements protruding into the inside volume of the arena and provides a relatively organised bottom view. The compression member between the lower and upper central node has been elongated above the roof structure and placed as a widely visible central spire.

The retractable roof of the National Stadium in Warsaw has an additional requirement – usability in winter. A comparably high inclination and prestress level of the single layer of cables was chosen in order to limit deflections and to ensure dewatering of the retractable membrane. In addition, the span of the membrane was reduced. An alternating arrangement of the driving carriages with decreasing distance from the centre to the outer edge provides a roughly constant density of supports for the total surface. Instead of sliding carriages made of steel with sliding pads, the entire sliding carriage is made of polyamide and assembled with a stainless steel clamp, reducing the weight of the components (Jaeger and Göppert, 2012).

BC Place Stadium, Vancouver, Canada

After the Olympic Winter Games in Vancouver in 2010, the existing air-supported dome arena was replaced by an innovative new roof structure to meet the demands of a state-of-the-art sport and multipurpose facility, and to regain an iconic architectural image for the city (Figure 16.13).

The loading conditions for structures in British Columbia are characterised by extremely high snow loads as well as reasonable seismic loading. According to the local codes, a ground snow load of 1.75 kN/m^2 (36.5 lb/ft^2) has to be considered for Vancouver. A self-stabilising lightweight roof structure has been designed to reduce as much as possible the resulting support forces on the existing concrete bowl. The new roof provides clear spans of 227 m × 186 m (741 ft × 610 ft), and is designed to carry approximately 7000 tons of snow. Thirty-six radial aligned cable trusses form the primary structure of the roof. Due to the high structural demands the lower and upper cables of each truss are realised as a pair of cables (Figure 16.14). The cables are post-tensioned between a central hub and 36 perimeter masts that rise 47.5 m (156 ft) above the concrete structure. The offset moment about the base of the existing concrete structure is balanced by a pair of horizontal forces, one acting in the elevated compression ring and the other in a tension ring located at the base of the masts.

The roof envelope is composed of three elements, all of which are connected to the lower set of radial cables. The outer, fixed portion of the roof is covered with a PTFE-coated glass-fibre fabric, tensioned on steel arches, creating a double-curved membrane geometry. The second element forms the transition zone of the fixed outer roof and the retractable inner roof, and is realised as a glass roof that is supported by a steel ring truss (Figure 16.15).

Figure 16.14 The folded membrane on BC Place, Vancouver, in operation. (© Schlaich Bergermann und Partner)

The inner, retractable roof extends from the central hub to the glass-roof perimeter. It can be opened in summer to provide an open-air atmosphere for various events. During the winter or for special events requiring it, the inner roof is closed. With its inflated sealing system at the glass-roof edge, the retractable roof transforms the open stadium into a weathertight indoor arena.

The inner roof is created by pneumatically stabilised membrane cushions, which provide sufficient inclination and stiffness to prevent ponding and to limit the membrane stresses even under severe snow conditions. For the 8500 m^2 (91 500 ft^2) retractable roof, a total of 36 inflated cells (with an individual volume of approximately 105 m^3 (1130 ft^2)) are suspended from the primary cable structure. For the retraction process the individual cushions are first deflated and then folded and moved back to the central hub.

Along the radial axes the adjacent cushions are connected and supported by two radial polyester belts, each with a breaking strength of 540 kN (121 397 lbf), forming the valleys of the deployed retractable roof structure. The radial belts are mechanically prestressed by the hydraulic stressing units located at the perimeter of the inner roof. The cushions are made out of fluoropolymer-coated woven PTFE fabric (Tenara), which has a translucency of 40% and thus maximises the natural illumination of the interior.

To fully protect the interior against wind and precipitation in the deployed condition, an inflated closure is connected to the lower membrane layer, forming a continuous seal between the glass roof and the inner roof. Using the same inflation pressure and chamber as the cushion, the closure is permanently pressed onto the smooth surface of the glass roof. The patterning of the closure takes into account the changing geometry of the cushion. An internal suspension mechanism with elastic cords lifts the closure during retraction to avoid the membrane catching on the glass roof.

The inflation pressure of the inner cushions is adaptable to the permanent loading on the roof, making it both an economic and a sustainable system. The standard inflation pressure under typical daily loading is 500 Pa (10.44 psf). If the roof is subject to snow loads the inflation pressure is increased in several steps up to 2000 Pa (41.76 psf). The variation in the inflation pressure is controlled by the roof-control system that evaluates the temperature and humidity conditions and the load on the roof. The required load information is collected by 72 magnetic load sensors connected to the cable truss hangers. The inflation units, consisting of fans, exhausters, filters and air dehumidifiers, are located within the membrane garage. For the retraction process the cushions need to be actively deflated. One of the biggest concerns with regard to the retraction process of the inflatable cushions is that the cushions are not fully deflated prior to the

Figure 16.15 Section of the roof on BC Place, Vancouver. (© Schlaich Bergermann und Partner)

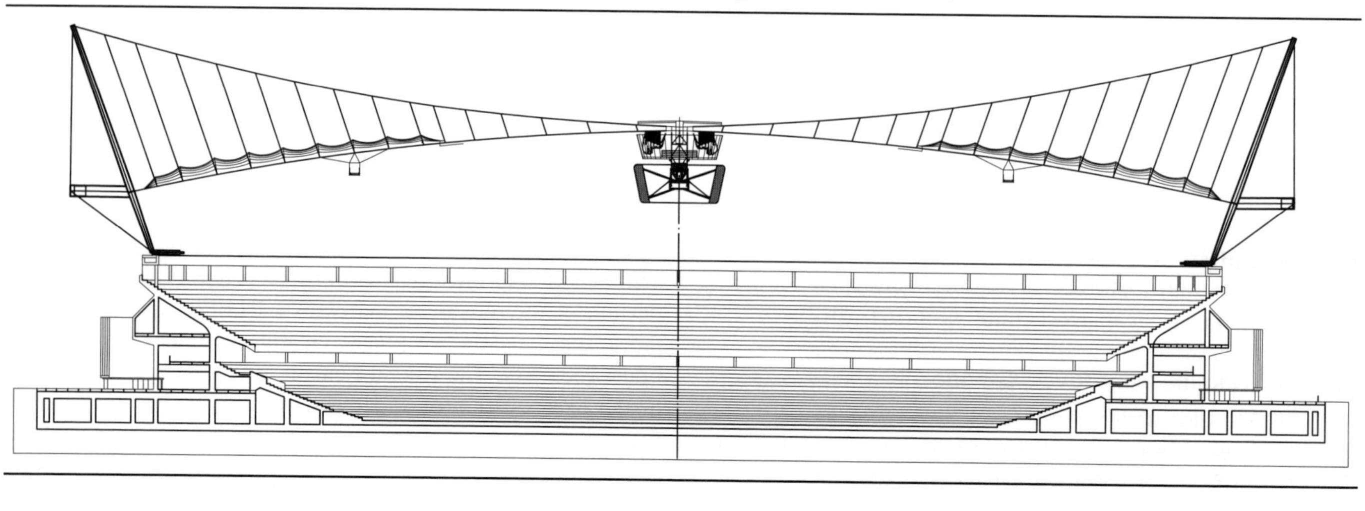

Figure 16.16 The folding process of the bridge at Kiel. (© Schlaich Bergermann und Partner)

moving process, or that air inclusions are generated during the retraction process. Both effects would prevent successful folding of the cushion to a minimal size. In order to fully remove air along the entire length of the chamber, a flexible, perforated tube is located in an open membrane pocket connected to the upper membrane layer. The 70 mm (2.75 in.) diameter tube starts directly at the exhauster and runs all the way to the outer seal. The performance of this arrangement and the folding ability of the complete inner roof, including the edge closure, was tested and verified in a full-scale mock-up of two bays.

Case studies – other movable structures

To illustrate the requirement for interdisciplinary design of movable structures, and the variety of solutions, some other structures designed by Schlaich Bergermann und Partner are described below.

Folding Bridge, Kiel, Germany

Located at the harbour side in Kiel, the design of the bridge adopts the themes of the surrounding vessels and container cranes. In a lowered position, the structure resembles a classic cable-stayed bridge. The bridge deck is suspended on both sides from cables. Each cable is deflected by way of its own mast portal before it is anchored in the foundation of the jetty. The three bays of the deck are connected by means of rotational axes, which allow it to fold up. This not only demonstrates a fascinating motion sequence, but it significantly reduces the surface area exposed to wind loads when the sections of the bridge are retracted. The folding process, which is operated by hydraulically driven winches, takes approximately 2 minutes (Figure 16.16). To ensure synchronised operation, the winches on each side of the bridge are connected by a shaft.

Harbour Bridge, Duisburg, Germany

The footbridge that crosses a harbour basin in Duisburg connects the historic city park with the opposite bank (Figure 16.17). In order not to impede the passage of larger vessels, the deck can be raised from its normal position to a clearing height of 10.6 m (34.75 ft) above the highest water level. Designed as a suspension bridge with end stays, the structure is 3.5 m (11.5 ft) wide and spans a distance of roughly 73 m (239.5 ft). The masts can pivot outwards by means of hydraulic cylinders in the stay cables. Thanks to the changed position of the support and suspension cables, the superstructure and the deck, which comprise an articulated chain, are drawn upwards into an arched form. The individual precast concrete elements are lined by steel frames. At the corners they are connected by way of eyes and pivot pins to the adjacent elements. Due to the spatial displacement curve, the suspension cables also required articulated connection points. During the lifting process, which

Figure 16.17 The Harbour Bridge at Duisburg in the maximum lifted position

takes about 5 minutes, the superstructure is continuously increased in length by additional elements, which rest in the parking position underneath the approach slabs.

Parabolic trough power plant, Andasol, Spain

Andasol I, II and III in the south of Spain are solar thermal power plants based on the Euro-Trough collector with a rated output of 50 MW each (Figure 16.18). The collector field of each plant covers a surface of around 1300 m × 1500 m (4265 ft × 4921 ft) and consists of 7296 individual trough collectors connected to 152 loops.

Each individual collector element has a width of 5.8 m (19 ft) and a length of 12 m (39.4 ft), and consists of parabolic mirrors, which focus the incoming solar radiation onto an absorber pipe. Temperature-resistant synthetic oil circulates inside the pipe. To keep the focal line steady on the absorber, the collectors must be adjusted to follow the course of the sun along a horizontal north–south axis. A 145 m (476 ft) long collector unit requires two hydraulic cylinders to rotate it by up to 180°. Solar-radiation sensors and an electronic control guarantee precise positioning. The complete power plant produces approximately 179 GW h annually, providing up to 200 000 people with electricity.

HelioFocus

A dish solar concentrator featuring a 500 m^2 (5382 ft^2) mirror surface has been developed for HelioFocus, a company based in Israel. This system comprises 219 mirror facets arranged in a Fresnel-like array, all pointing to one common focal point. Hence, a very high solar concentration can be achieved at the focal point. A heat exchanger (receiver) installed at this point generates hot air of up to 1000°C (1832°F). This air is collected in a number of concentrators to be fed into a conventional fossil-fuel power plant.

The structure of the concentrator is based on a stiff trussed girder with seven cantilever arms. These support the mirrors of curved glass by way of purlins. By means of a horizontal axis (elevation), the concentrator is pivoted on the turntable, which can be rotated about the vertical axis (azimuth) by way of four double wheels and a circular crane rail (Figure 16.19). Both axes are driven hydraulically, whereby the so-called 'back-stepping' approach is applied at the azimuth.

Summary

As large cable-supported roof structures provide structural and architectural advantages, the demand for solutions with integrated retractable structural elements is increasing. The combination of cable structures and membrane material for protection against all weather conditions is the first choice for lightweight tension structures. The examples presented here are specific solutions to the individual conditions. Extremely high snow loads in Vancouver required a pneumatic stabilised retractable roof and related solutions with regard to inflation of h cushions, and especially deflation for retraction. With the

Figure 16.18 The Andasol I power plant

Figure 16.19 Prototype Dimona, Israel

requirements of being able to move along a single layer of cables spanning up to 70 m (229.7 ft) without stabilisation and of winter use, the retractable roof on the National Stadium in Warsaw is a milestone for single-layer retractable roofs.

Conclusion

The growing functional requirements of the buildings of the future represent challenges for architects and engineers alike. An important aspect in responding to these demands will be the adaptability of the building envelope. The possible solutions need to take into account various aspects, including the required energy consumption, the recyclability of all materials used and the financial investment.

Retractable roofs for sport stadiums and multiuse arenas are good examples of the application of adaptable systems, and will, in many parts of the world, become more popular, not only to keep up with the increasing comfort requirements but also to react to climatic changes caused by global warming.

REFERENCES

Balz M and Paech C (2013) Use of textile membranes for adaptive structures. *Proceedings of the 2013 Tensinet Symposium*, Istanbul, Turkey.

Balz M, Göppert K and Paech C (2013) *Faltung von textilen Membranen bei Leichtbaukonstruktionen*. Bautechnik, Berlin, Germany.

Bergermann R, Gugeler J and Keck T (2004) Wandelbares Membrandach im Innenhof des Wiener Rathauses. *Stahlbau* **73(6)**: 373–380.

Göppert K, Haspel L. and Paech C. (2011) New retractable roof solutions for sports stadia. *Proceedings of the 2011 IASS Symposium*, London, UK.

Ishii K (2000) *Structural Design of Retractable Roof Structures*. Wit Press Southhampton, UK.

Jaeger F and Göppert K (2012) *Next 3 Stadia. Warsaw Bucharest Kiev*. Jovis Verlag, Berlin, UK.

Pawlitschko R *et al.* (2011) *Schlaich Bergermann und Partner*. DETAIL engineering 1, Edition DETAIL.

Schlaich M (2004) Aktive und wandelbare Ingenieurbauten. *Stahlbau* **73(12)**: 1001–1009.

Part IV

Case studies

Stadium and Arena Design
ISBN 978-0-7277-5790-6

http://dx.doi.org/10.1680/saad.57906.203

Chapter 17
Hazza Bin Zayed Stadium

Dipesh Patel and Jon-Scott Kohli

Overview

In early 2011, the Abu Dhabi Crown Prince Court invited Pattern Design Ltd to be the lead designer for a new football stadium for the Al Ain Football Club (AAFC). Al Ain is the home city of the Abu Dhabi royal family, and the AAFC is a favoured team within the United Arab Emirates (UAE) Pro League. The client's aspiration for the Hazza Bin Zayed (HBZ) Stadium was a spectacular sports venue within a mixed use development including a hotel, training centre, homes and offices.

The client's initial view was to replace AAFC's existing 12 000 seat stadium with a 40 000 seat, air-conditioned, retractable-roof multi-purpose venue. Through the feasibility stage, the designers identified that the energy and environmental impact of this brief would be immense in the desert climate. Furthermore, as current matches typically attract just a few thousand spectators, there was considerable risk that matches with small crowds would lack atmosphere in such a large stadium.

The designers responded to these challenges with an alternative proposal: a passively shaded, 25 000 seat stadium that would look and feel like a much larger stadium when full, but create a great atmosphere for small crowds (Figure 17.1). This proposal was ultimately built within a rapid 18-month construction programme, and this case study details the key design developments and technical solutions that made this innovative concept a reality.

Context and climate analysis

Cultural and climatic context was the starting point for the approach to the project brief. Al Ain is an oasis city located within the emirate of Abu Dhabi, 140 km (87 miles) inland from Abu Dhabi city close to the Omani border. The city was originally founded on the site of five oases. The oases have been in cultivation since ancient times, and Al Ain has long been a place of trade and farming in the region. The falaj irrigation system there is thought to be over 3000 years old. The palm tree is an important symbol of the oasis and the city, and is enshrined in its history.

Al Ain sits in an arid desert climate with a brief rainy season in March and April. The UAE Pro League season runs from late August to early June, with kick-off times generally between 16:00 and 18:00. Although average mid-summer high temperatures reach 45°C (113°F), the mean daytime highs during the season are 35°C (95°F). Thus, actual game-time conditions are typically hot and dry, suggesting that shielding spectators and players from harsh sun is more important than protecting them from rain.

Such conditions are in stark contrast to the temperate climate of European grounds. Standard European football stadiums typically feature a 'drip-line' roof that protects spectators from the rain. A review of regional precedents revealed that the drip-line typology is often exported to the Middle East. However, in light of the climate analysis the designers felt that a drip-line roof would not be appropriate in the desert conditions. Instead, passively controlling shade and airflow became the principal design driver for key aspects of the design.

Bowl and accommodation

Passive shading for the stadium begins with the massing of the bowl and main spectator accommodation (Figure 17.2). Preliminary solar analysis indicated that a north–south oriented pitch with a high west stand would provide the best passive shading for spectators and the field of play (Figure 17.3). This strategy was coordinated with other design drivers for the bowl and accommodation, namely the hospitality offer that is essential to the stadium's long-term success.

The designers undertook extensive analysis of international and regional precedents to determine the correct hospitality provision and arrangement for the HBZ Stadium. The designers found that, internationally, new stadiums are typically built with hospitality provision for 5–15% of the total capacity; generally large premier and national stadiums with capacities in excess of 50 000 seats are at the high end of this range. In the Gulf region, however, the provision is higher, even for small stadiums, with hospitality comprising 8–20% of total capacities. In Al Ain, specialist research

Figure 17.1 Hazza Bin Zayed Stadium from the south-east. (© Dennis Gilbert, courtesy of Pattern Design Ltd)

Figure 17.2 The bowl interior and the sweeping parasol roof. (© Dennis Gilbert, courtesy of Pattern Design Ltd) (This figure is reproduced in the colour-plate section as Plate XXIV)

Figure 17.3 Evaluation of typical stand arrangements. (© Pattern Design Ltd)

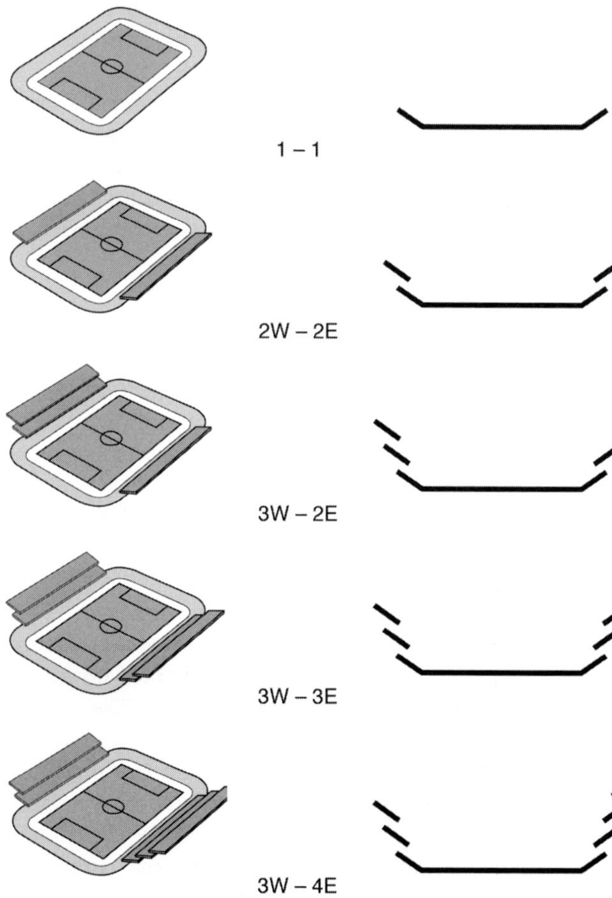

Figure 17.4 Perspective cross-section demonstrating the shade criteria. (© Pattern Design Ltd)

confirmed that the market demand would support a hospitality population of 16%.

The designers evaluated typical seating arrangements and found that a hybrid bowl–tier configuration, with three tiers on the west and two on the east, scored highest on a matrix of criteria that included atmosphere, suitability for hospitality and ease of shading. This unique arrangement for 25 000 seats features small tiers, and allows multiple tier overlaps to create a compact, high bowl. This is best for small-crowd atmosphere, as the size of a small crowd distributed through small tiers is perceived to be larger than the same crowd in a single large stand. It also increases the proximity of spectators to the field of play, creating the exciting sensation of being close to the action.

As the previous two AAFC grounds had athletic tracks, spectator proximity was a very important client aspiration. Modern stadiums have larger distances between the first row and touchline (the front row offset) than in historic examples. This is largely to accommodate modern broadcast and match protocol. However, it can be very detrimental to atmosphere, particularly in small stadiums. Thus, while Fédération Internationale de Football Association (FIFA) generally requires an 8.5 m (28 ft) front row offset (FIFA, 2011, p. 64), the designers agreed with FIFA representatives a smaller auxiliary zone that is appropriate to league play and midlevel events such as the U-17 World Cup.

Overall proximity was optimised by cutting the back of the seating bowl using an elliptical geometry to ensure that all seats lay within the industry standard 90 m (295 ft) optimum viewing distance. In elevation, cutting the bowl in this way opens up the corners (as in traditional football grounds), allowing increased airflow for comfort and good grass growth (Figure 17.4).

Within this bowl configuration, a balanced arrangement of hospitality options is provided, including a VIP lounge, club lounge, sports lounge, private boxes and outdoor terraces. The terraces are an innovative addition to the hospitality programme, and are located overlooking the pitch in the open corners of the bowl. Designed for flexible arrangements of loose furniture, they provide a relaxed and informal experience, made possible by the hot climate.

All the hospitality and supporting accommodation is stacked vertically on the west stand, generating a high west side that maximises passive shading at kick-off time. The concentration of all the catered spaces in one stand facilitates an efficient food and beverage operation. This illustrates the designers' overall approach to the stadium: addressing the complex and varied requirements of climatic response, spectator experience and market viability in an elegant and integrated solution.

Roof

Massing and orientation are critical but the roof is the primary element providing shade at HBZ Stadium. On the annual shading design day of 21 August, the roof achieves shade to the majority of seats by 16:00, complete pitch shading at 16:20, and full shade to every spectator by 17:45. The design day was determined as the earliest likely start date of the Pro League calendar, while the design time balances the contradictory needs of grass growth, cost and spectator comfort. The geometry required to achieve the shading criteria generates an exceptionally high west

Figure 17.5 Roof plan. (© Pattern Design Ltd)

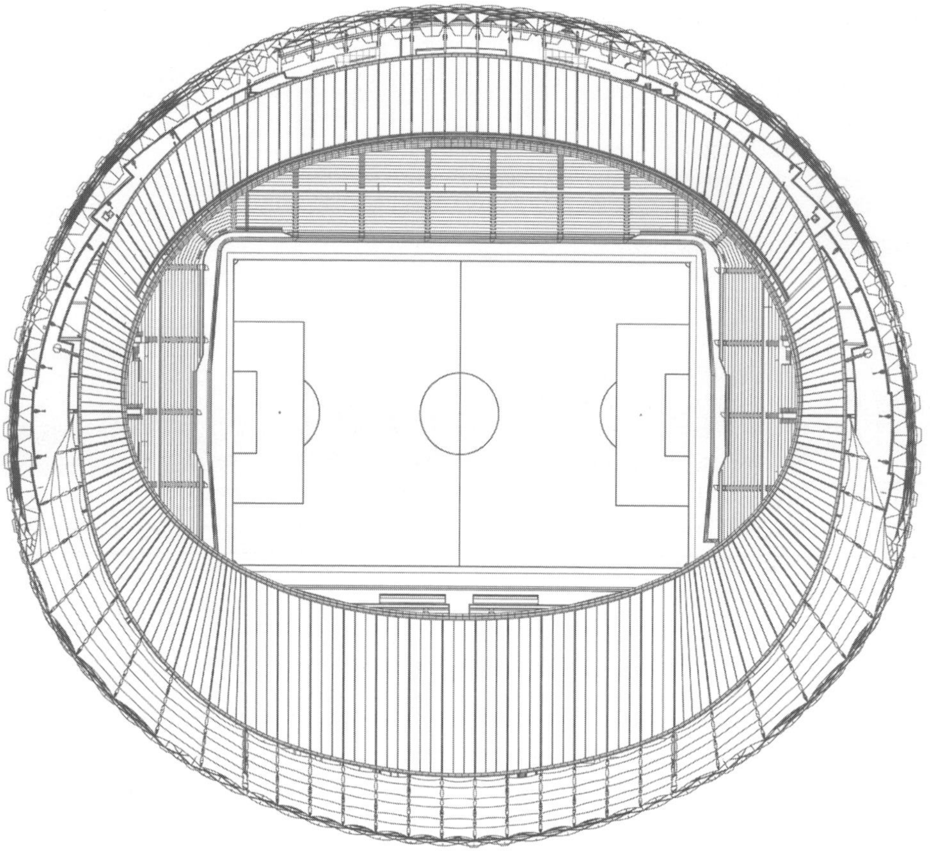

stand roof, reaching a peak of 50.7 m (166 ft) above the pitch, with an inverted low roof on the east, just 24.4 m (80 ft) above the pitch.

The designers conceptualised the roof as a piece fabric twisting in the wind, evocative of an Arabic headdress. The simple shading plane on section creates a complex 3D curvilinear form. The roof is a single continuously twisting surface, with inside and outside edges that vary in height at different rates. To resolve this highly complex 3D geometry, the roof form was generated using bespoke parametric scripts that defined the relationship of all parts of the roof construction, including primary surface, primary structure centre-lines, cladding, secondary structure, gantries, gutters and floodlighting. Using these scripts, the complex geometries were broken down into rational geometries for panels, gridlines and structural centre-lines, forming the basis of coordination between the engineers and architects.

This complex roof form is constructed using simple cantilevered triangular planar trusses with a back prop (Figure 17.5). This allowed the simple and common on-site process of assembling the trusses on the ground and lifting them into place. The top surface of the roof is clad with a white polyvinylchloride-III (PVC-III) membrane using a concealed kader tension detail. An acoustically-transparent PVC mesh clads the bottom surface of the roof, shielding access gantries, services and public address speakers from view. Together, the PVC fabrics have a translucency that allows a degree of natural light through the roof to ensure that the appearance of the roof is light and airy, in keeping with the architectural concept.

The designers intentionally set back the outside edge of the roof from the top edge of the façade in order to allow a 14 m (46 ft) clear area for air-handling plant ventilation on the top deck, immediately behind the back wall of the seating bowl. However, solar analysis demonstrated that in the low west light of the setting sun the gap created between the façade and roof required louvres on the western edge of the roof to achieve the pitch shading. A multidisciplinary solution was required, and the designers developed the fabric ‘sky louvres’. These are tensioned with alternating convex and concave curved booms between steel trusses at roof gridlines. The louvres allow the required ventilation while still achieving the passive shading criteria.

The extreme and varying pitch of the roof posed a particular technical challenge for the placement of sports floodlighting. In order to meet FIFA lighting requirements, the designers integrated a bespoke gantry and lighting rig into the underside of the west-stand roof and above the east-stand roof. Through modifying the parametric script, the designers mapped the largely uniform level of the lighting against the constantly changing roof height. The final effect evokes a string of pearls – an unbroken chain of lights along a sinuous curve that emphasises the dynamic twisting form of the roof.

Façade

The HBZ Stadium façade is a unique response to the cultural context of Al Ain and various technical requirements. Solar analysis revealed that the setting sun during matches required the façade to block low westerly light from reaching players and spectators. Natural ventilation through the façade was also a design driver, both for healthy pitch growth and passive spectator cooling. Internal conditioned spaces, such as lounges and boxes, required a high-performance thermal barrier to reduce energy use from air-conditioning. Thus, the façade had to create shade, allow airflow, and provide a thermal separation for internal spaces.

To achieve these criteria, the designers employed a double-façade system: an outer porous primary façade, and a secondary inner façade forming the thermal barrier. This approach has cost and buildability advantages: instead of one very complex façade system to solve all issues, the two simple systems each addressed specific requirements and could be installed faster and at less cost (Figures 17.6 and 17.7).

The secondary inner façade is of a standard system construction and built only where required to house internal conditioned areas and support spaces. It comprises painted and rendered blockwork walls, metal sand-trap louvres and reflective glazed curtain walling. The designers performed energy modelling to ensure that the wall assembly performance met the requirements of the Abu Dhabi Estidama 2-Pearl rating.

Figure 17.7 West façade and VVIP entrance. (© Dennis Gilbert, courtesy of Pattern Design Ltd)

The primary façade wraps around the inner façade and stadium bowl to shade the pitch and spectator seating. It is offset from the inner façade by 3–6 m (9.8–19.6 ft), creating a shaded colonnade around the stadium. As a continuous element it unifies all four stands and gives the stadium its elegant,

Figure 17.6 South elevation. (© Pattern Design Ltd)

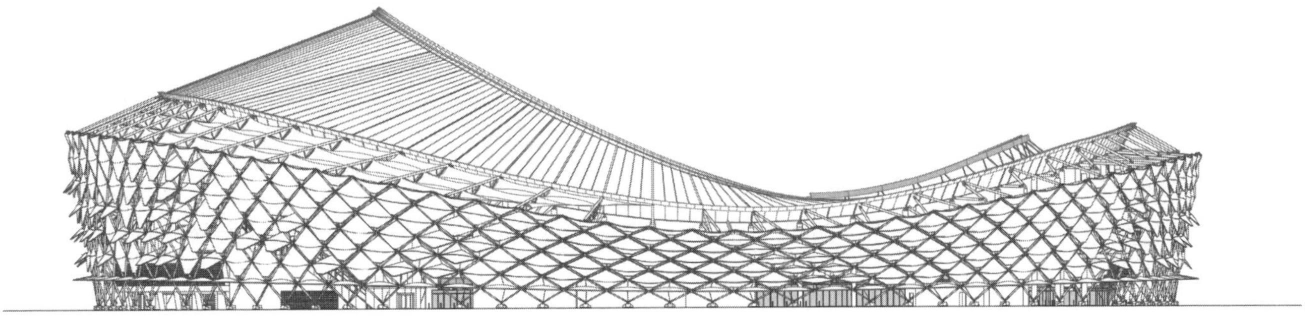

Figure 17.8 Palm bole and fronds.
(© Alfonso de Tomas)

Figure 17.9 Façade panels, diagrid and sky louvres.
(© Dennis Gilbert, courtesy of Pattern Design Ltd)

sweeping appearance. As the primary façade is the most publicly visible element of the stadium, the designers looked to local context to ensure the stadium was an expression of the culture and place. Given the importance of the date palm as a visual reference to the place and historical status of the Al Ain, it is a potent symbol for a building about sport, health and well-being (Figure 17.8).

Palm trees naturally follow an efficient formal pattern. As a palm develops individual fronds, they grow in a rotational fractural geometry. The fronds are cut back or fall away as the palm grows, creating the palm 'bole'. The resulting geometry comprises raised polygonal shapes arranged on a rotational, diagonal grid (diagrid). The designers adapted this diagrid structure for the façade, based on a 10° inclined surface that springs 900 mm (35.4 in.) above the ground level and wraps continuously around the stadium. The diagrid itself is formed of rectangular hollow sections measuring 250 mm × 150 mm (820 ft × 492 ft), mounted on concrete bases that are designed as benches for the public realm (Figure 17.9).

The palm frond panels sit within the voids of the diagrid. As the panels are entirely external, the construction is simple and lightweight. The panels consist of a polytetrafluoroethylene (PTFE) membrane tensioned over a powder-coated steel frame; a convex boom across the frame provides the characteristic palm frond curvature and required fabric tension. Each panel is mounted along its horizontal axis and held statically in position by rods from the vertices of the diagrid to the top point of the panel. The panels are rotated individually to allow greater porosity where required for increased ventilation or to improve views from the rooms within. The bottom-most panels are rotated to be effectively horizontal to provide shade to the public realm below. Public realm wayfinding is incorporated in the architecture through large-scale signage panels placed within the voids of the diagrid at the base of the façade. The rotational patterning also creates large-scale compositions around entry points to reinforce wayfinding in a subliminal manner.

Architectural lighting is integrated seamlessly into the façade design through the inclusion of elegant and compact stainless-steel light-emitting diode (LED) fixtures mounted at each diagrid vertex to illuminate the panel below. The addressable RGB fittings allow full-colour static and dynamic lighting

Figure 17.10 The stadium and public realm at night. (© Dennis Gilbert, courtesy of Pattern Design Ltd) (This figure is reproduced in the colour-plate section as Plate XXV)

displays for different uses (Figure 17.10). For example, AAFC colours are shown on match days, while the colours of the UAE flag are shown during National Day celebrations.

As the façade transitions from the highest section on the west to the lowest on the north and south, the diagrid expands and compresses as required, creating hundreds of unique connection points between members. The fluctuating size of the diagrid generates panels that vary in width between 2.5 m (8.2 ft) and 6.5 m (21.3 ft). The designers handled the complex design of these elements using parametric 3D modelling. The parametric scripts established the geometric relationships of parts, while creating a dynamic computational system that could be flexed, tested and reviewed iteratively. In this way, the designers tested dozens of options, involving variable panel size, rotation and massing.

The parametric scripts were essential in optimising the dimensions of 695 individual panels to only 82 unique types, achieving significant reductions in cost and construction time. Other relevant data were extracted from the scripts to input into the designers' building information modelling (BIM) database for construction information and exact coordination with specialist subcontractors. Using this information, the various elements of the assembly could be prefabricated and assembled on site. This included the precise fabrication of all the diagrid connection points, where in each case the rectangular hollow section (RHS) members meet at unique angles. In this way, parametric modelling was at the heart of the delivery of the stadium design.

Summary

HBZ Stadium is a unique building of its place and climate. It is the product of a design process that was responsive to context, and developed through advanced parametric geometry. Starting from initial climate analysis, bowl and accommodation massing was developed alongside informed market research and adapted to provide passive shade control. The roof was designed to achieve set shade criteria, balancing conditions for spectator comfort with the requirements of pitch growth. The façade design brought a unity to the elevation, while addressing the many technical requirements of shading, ventilation, lighting and construction through an integrated solution. Design development through parametric scripting and 3D modelling allowed the manipulation and detailing of complex geometry, enabling construction within an 18-month programme.

The result is a building that is beautiful, dramatic and speaks to spectators at all levels. There is a powerful symbolism and visual integrity to the building that fits within its place. It is a technically sophisticated solution to common problems of the climate: heat, shade and airflow. The parametric rigour has produced a legible building that makes sense to the user, and has become a new landmark of Al Ain.

Design and delivery teams

During the design phase the client was the Crown Prince Court of Abu Dhabi. Pattern Design Ltd was the lead designer.

In the preconstruction phase, Broadway Malyan was the lead consultant and master planner. Thornton Thomasetti was the structural and façade engineer, and Hoare Lea was the building services engineer. The architect of record was Mark Habre and Associates, the food and beverage consultant was Tricon, and STRI was the specialist pitch consultant.

Post-contract, BAM International led a design-and-build team under a contract with the new client entity Aafaq. E.C. Harris was the client representative. The design team was led by Pattern Design Ltd. Hoare Lea remained responsible for buildings services. Structural engineering of the roof and façade was undertaken by Schlaich Bergermann und Partner. Structural engineering for the superstructure was completed by WSP Group.

REFERENCE

FIFA (Fédération Internationale de Football Association) (2011) *Football Stadiums: Technical Recommendations and Requirements*, 5th edn. FIFA, Zurich, Switzerland.

Stadium and Arena Design
ISBN 978-0-7277-5790-6

http://dx.doi.org/10.1680/saad.57906.211

Chapter 18
Centre Court, Wimbledon

John Westmuckett Director Westmuckett Hawkes Ltd (former Director Capita Symonds)

Dale Jennins Principal, Populous

Introduction

In January 2004 the All England Lawn Tennis and Croquet Club (AELTC) first revealed plans for a retractable roof over Centre Court (Figure 18.1). The Chairman of the club described the development as being essential to keep the championships at Wimbledon at the forefront of the sports world.

Since 1922, when the championships moved to Church Road from its original home in Worple Road, Wimbledon has almost always set the pace in world tennis while maintaining the standards for which it is renowned. The roof development, as well as providing a solution to the vagaries of the English weather and poor light disrupting play, also increased the capacity of the stadium, while providing wider and more comfortable seats.

The two key issues, however, that determined the whole design and construction process were the protection of the grass and the maintenance of the unique atmosphere of the Centre Court arena. This case study describes how these issues, which were fundamental to the design process, created many technical challenges for the professional team of architects, engineers and contractors, not least of which was the need to plan and phase the construction around the championships each year, which had to take place without loss of quality to the event.

Design brief

The design brief can be summarised as follows:

- To create a new roof with a retractable central section in order that play can continue during the championships whatever weather or light conditions, including the ability to air-condition the court for the court surface, players and spectators.
- To redevelop the east wing to expand the catering and hospitality facilities, including alfresco spaces for dining and drinking to enhance and accentuate the outdoor summer garden-party feel to the tournament within Centre Court (Figure 18.2).
- To improve vertical circulation within both the hospitality and viewing areas of the court, with additional stairs and lifts.
- To expand the terraces surrounding three sides of the stadium to the west, north and east (retaining the existing iconic ivy-covered south façade) to increase seating capacity within Centre Court from 13 800 to 15 000.
- To improve the comfort of spectators by increasing the seat width throughout Centre Court and to replace all seats with 15 000 padded seats.
- To ensure that growing conditions of the grass playing surface are not compromised throughout the year and that sufficient light in terms of photosynthetically active radiation (PAR) is maintained. (PAR is the name given to the band of solar light that is useful to plants in the intake of energy and production of oxygen.)
- To upgrade and increase the commentary positions at the rear of the bowl.
- To increase and improve viewing for spectators with wheelchairs.
- To provide a construction programme which not only ensures minimal disruption to all ongoing Centre Court activities but also ensures that Centre Court and its surroundings are made available for the championships each year without loss of amenity or quality of the event.

Concept design

The concept design was split into three main areas of study, which were run in parallel but closely interlinked:

- east terrace refurbishment and existing structure and seating extension
- fixed roof structural support, profile and geometrical studies
- retractable roof methodologies.

Figure 18.1 Centre Court celebration, May 2009 (Courtesy of AELTC) (This figure is reproduced in the colour-plate section as Plate XXVI)

Figure 18.2 Wimbledon Centre Court east section

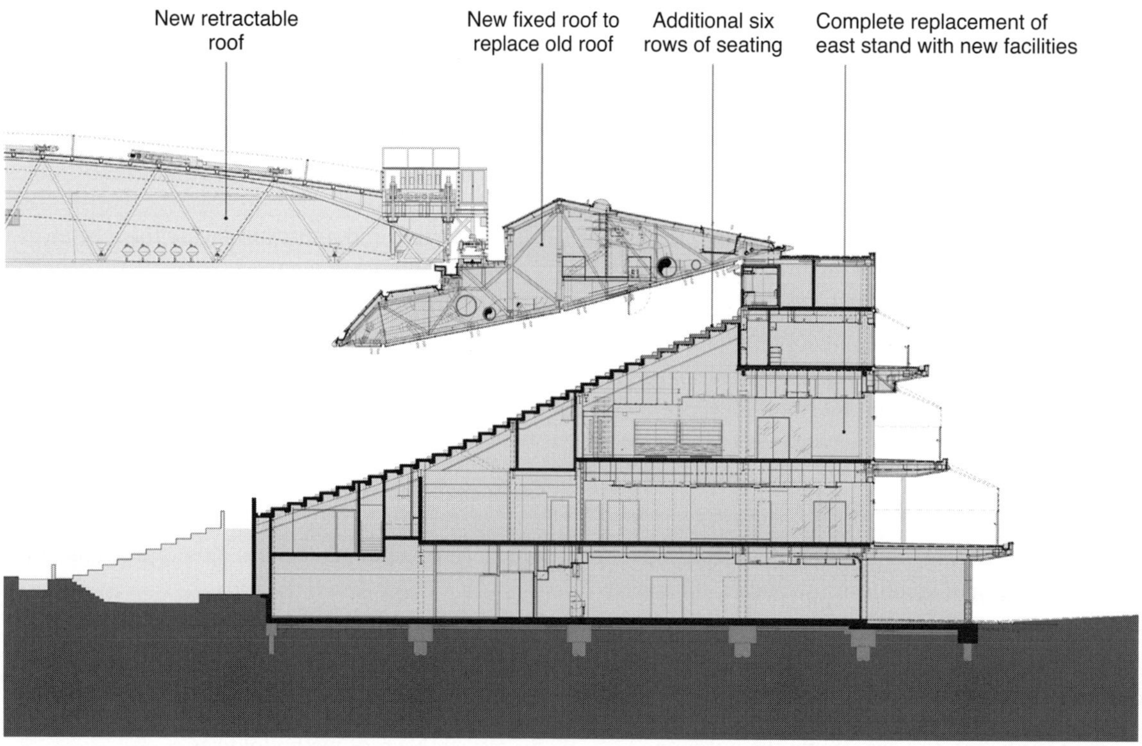

Figure 18.3 3D image of existing structure. (This figure is reproduced in the colour-plate section as Plate XXVII)

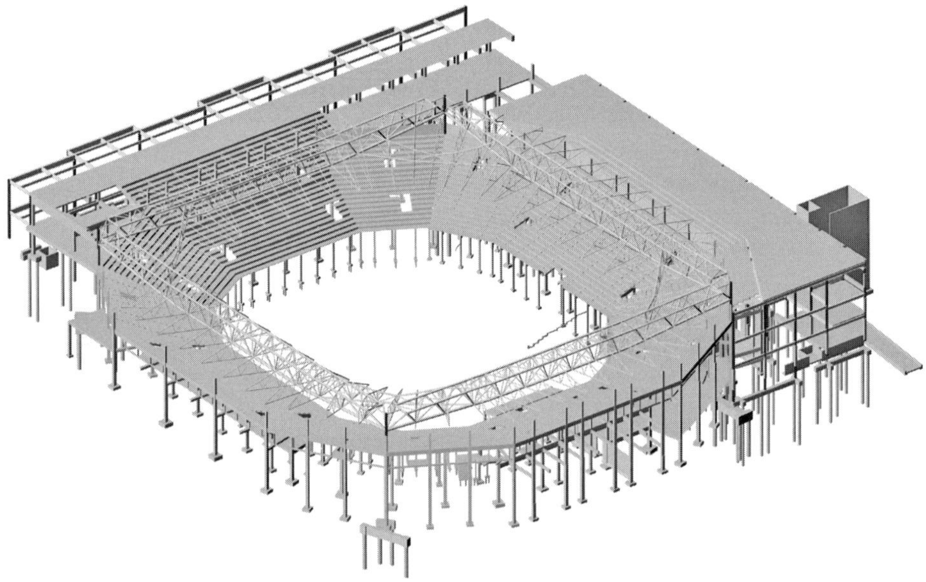

Existing structure

The original Centre Court, designed and built in 1922 by Captain Sir Stanley Peach, is primarily a concrete structure with in situ concrete raking beams and precast concrete terraces.

The stadium has been extended and refurbished many times during its history, with additional terraces added on the west and northern sides together with two new roof arrangements – the latest being in 1990, with a completely new roof supported on four main columns.

A 3D model of the existing structure, based on existing records and survey information, was developed to establish the primary structural elements and load paths as a basis for the new proposals (Figure 18.3).

Roof forms and supporting structures

One of the main drivers for the roof design was the issue of light, not only for the Centre Court grass but also for the adjacent courts surrounding the main stadium. This meant that the options for moving-roof structures based on flat plates were limited by the need to park the plates away from Centre Court without overshadowing the adjacent courts.

It soon became evident that the solution had to be contained within the footprint of the building, and that a stacking or folding method would need to be adopted.

Geometrical studies using PAR studies based on 3D computer models indicated that the retractable roof would have to be four times smaller in its stored position over the fixed roof, than its area when fully deployed. This research led to the resultant folding mechanism using fabric fields, which allow for a tighter compression of the elements of the roof to achieve the set parameters. The use of fabric also enabled the resultant design to feel more 'airy' and maintained the feel of an external environment, important to what is essentially an outdoor 'event'.

To support this 'concertina'-type roof, tracks, built into the top of the two main trusses spanning north to south, would be supported on 'supercolumns' at their ends, located within the existing structure.

While these studies were taking place, the main structural form of both the retractable roof and the fixed roof was developed in conjunction with the mechanical services design needed to air-condition the space. This was critical not only for spectator comfort but also, more importantly, for the playing conditions of the court surface, as additional moisture would be detrimental to play and potentially increase the chances of players slipping.

The extensive air-handling plant and ducts within the fixed-roof form had a major impact on the configuration of the main supporting trusses in terms of coordinating with the prime structural elements, resulting in a structure of high complexity.

East wing redevelopment

General

The existing structures of the east wing were the result of two or more extensions – one in concrete in 1976, followed by a further extension in 1985 in steelwork. Although the club was initially loathe to consider the removal of parts of the original 1922 building, the complexity of these additions was such that a decision was taken to demolish a large part of the east wing and rebuild in structural steelwork with concrete lift and service zones.

The redevelopment of the east wing resulted in decanting the existing facilities, which were primarily the administrative offices and club museum, to a new purpose-built facility located over the turnstile area, which were constructed in 2003 prior to the Centre Court redevelopment. This allowed for extensive improvements to spectator facilities in and around Centre Court, with restaurants/hospitality facilities terraced over four levels, giving spectacular views over the adjacent golf course and the London skyline beyond.

Structure

The design of the east wing structure consists of main raker steelwork supported on columns and linked into a new reinforced concrete lift and stair tower. The terrace units are precast concrete spanning up to 6.0 m (17.7 ft). The main floor plates behind the terraces are of composite construction.

The design loading for the main terrace structure allows for specific point loads from the temporary decking, which was required to support the temporary works for the new roof construction. Openings were left in the terrace units to allow these temporary connections to be made at a later date.

Fixed roof

The new fixed roof consists of four main box trusses located approximately in the centre of each of the wings spanning up to 80 m (262.5 ft) on to new supercolumns. The depth of these trusses is limited by the high sightlines to the underside and the light requirements to the court on the topside, allowing a maximum depth of 4.5 m (14.75 ft) at midspan. Box forms were used for the main trusses to aid stability during erection and to divide the large chord members into two or more manageable sections (Figure 18.4).

Smaller cantilever trusses span at right angles to these trusses and are anchored at the rear, with smaller columns located between the commentary boxes at the topmost level of seating.

Inside the main trusses the fixed roof steps down firstly to accommodate the tracks for the moving roof and then to a lower level to accommodate the moving roof itself and, in particular, the depth necessary for the fold in the fabric when parked. To allow the moving roof to be parked discretely, the zone at this lower level is extended to the north as far as the geometry permits.

Figure 18.4 3D image of truss detail

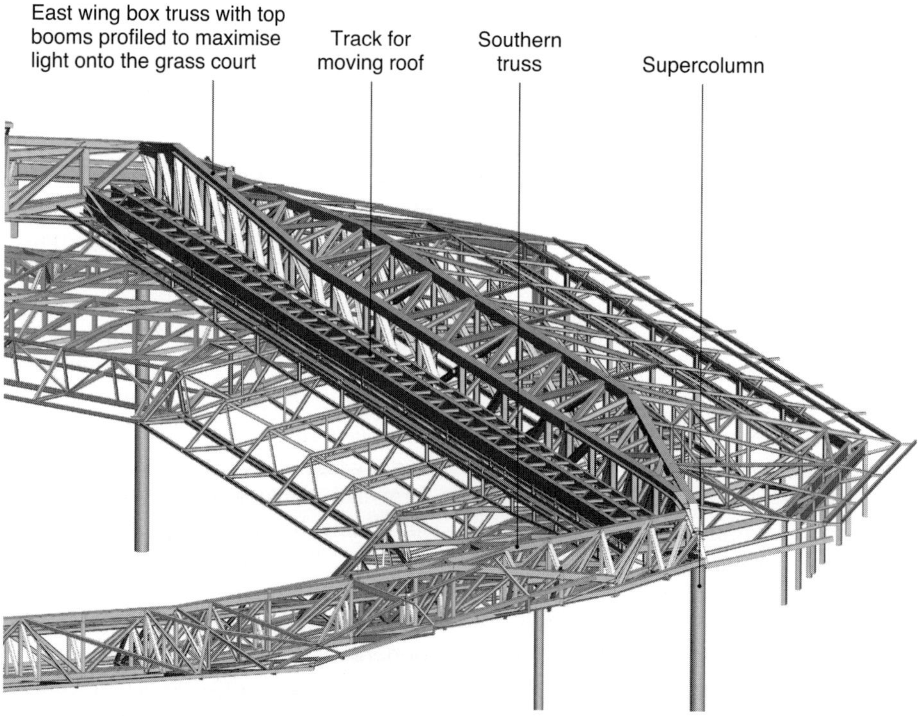

The envelope at the south truss restricts the depth of the truss, and additional support is taken from the two existing roof columns in the south wing to limit deflections. The load transfer in to these columns is limited by providing a spring bearing to ensure that the existing columns and foundations are not overstressed.

There is a tertiary framing of purlins and diagonal plan bracing to support the roof cladding, which comprises metal sheeting and insulation to the top surface and an acoustic surface to its soffit to reduce noise transmission and increase sound absorption in the bowl.

The fixed roof covers a large proportion of the seating area but the central opening has been kept large enough to allow sufficient light to reach the grass throughout the year, as determined by the light studies. The inside edge follows the dodecahedron shape similar to the original 1920s stadium.

Retractable/moving roof

General description and operation

The retractable roof comprises a series of 10 prismatic trusses supported by bogies at each end, with folding fields of fabric spanning between adjacent booms but kept taut by a valley cable anchored at each end by a hinged mechanism.

The fabric is in constant tension even when in the parked mode, and the hinges operate by four electromechanical activators moving in sequence with the electric powered bogies.

The roof is divided into two sections, with five bays in the northern group and four in the southern group. For most of the year, both sections are normally parked together in the north to allow maximum light onto the grass court. Just prior to the championships, the southern group is moved to the south and the southernmost truss is locked to the fixed roof. This movement takes about 20 minutes.

The roof is designed to operate in wind speeds up to 25 m/s (55 mph), and deploying the roof takes about 8 minutes in total.

When rain occurs during play, both sections of the roof deploy and meet in the centre to protect the grass. There are also several iterations, recently introduced, to allow partial deployment of individual sections to allow shading on particularly hot days.

Along the edge of the roof, there is a fabric-clad mechanism (the 'batwing') that opens and closes on the deployment of the roof to prevent wind-driven rain from entering the stadium and helps maintain the controlled atmosphere.

The roof can also be deployed if matches overrun into the evening (currently, play is allowed up to 11.00 p.m.), as there are lights that illuminate the court and allow play during the hours of darkness.

Structural description

The steel trusses span approximately 77 m (252.5 ft) across the court and rise to a maximum height of 6 m (19.7 ft) above the fixed roof eaves line. Between the trusses the structural fabric is prestressed by a valley cable to control its movement under fluctuating wind loading. To ensure free drainage and to provide the necessary curvature in the fabric and valley cable system, the top chords of the trusses have been designed in an arch shape while the bottom chord of the truss is horizontal.

The hinge mechanism

The hinged mechanism links the trusses at their ends and provides the anchorage to the valley cables transferring the loads back to the main trusses (Figure 18.5). At the anchorage point in the hinge, the loads from the cable resolve into vertical and horizontal components, with the horizontal loads transmitted by planar shear in the hinge plate and the vertical loads transmitted through the end arms at each end of the hinge. This separation of load components is effected using a combination of universal and thrust bearing in the hinge connections.

The hinge actuators control and transmit these varying reactive forces as the hinge moves. These hinges are driven and controlled in coordination with the movement of the driven bogies by a computer-based control system that monitors bogie/actuator movements to fractions of a millimetre. The location of the bogies in relation to each side of the stadium is monitored by sensors sending a positioning signal to the drive motors, which can realign the truss. Skew control is thus maintained automatically, but if this exceeds 25 mm the brakes are applied automatically.

Central restraining arms

The trusses are also linked together on top of the fabric fields by four restraining arms spaced across the span. These are also operated by electromechanical actuators operating within the controlling circuitry, which ensures a smooth movement throughout the deployment.

Each folding fabric field operates in turn and is then locked before the next field is deployed, ensuring full stability throughout the system.

The design of the retractable roof trusses was carried out using both static and dynamic design programmes to establish the maximum and fluctuating loads within the system during deployment. The design allows for possible extreme weather conditions if deployed during maintenance periods outside the championships, and also in the unlikely event of the roof being deployed and caught out in a snow storm!

Figure 18.5 Aerial view of the hinge mechanism

The fabric

The fabric used is a comparatively new material called Tenara, introduced by W.L. Gore in 2003, which consists of a high-strength polytetrafluoroethylene (PTFE) yarn with a fluoropolymer coating system and has a 40% light transmission. The fabric form was designed to have a minimum tension sufficient to take up its shape in the folded position. The design was analysed for the additional stresses in the fully deployed position with the addition of environmental loads, including the possibility of snow.

Air-conditioning

The fixed roof is designed to act as an acoustically lined plenum, with supply-air distribution ductwork from air-handling units (AHUs) connected to swirl diffusers in the soffit of the roof (Figure 18.6). Factory-based smoke tests were carried out to ensure that the air-conditioning within the bowl with the roof deployed would spread evenly over the spectators.

The 14 AHUs within the roof are oversized by 10% to ensure redundancy in the system. They are supplied by chilled air from a remote external chiller compound. The AHUs are supported on anti-vibration mountings within the roof to achieve the acoustic design requirements.

A medium-pressure supply air duct forms a complete loop with the fixed roof, combined with low-velocity branch ducts to serve high-level induction jet nozzles aimed directly at the retractable roof steelwork to prevent condensation when the roof is deployed.

The air-conditioning sequence is in two stages:

1. During the deployment of the roof the AHUs operate in recirculation mode, drawing air from the stadium by way of perforated plates on the underside of the roof and directing the air to the roof and to the court to reduce the humidity and excess water vapour.
2. The external louvres are opened and fresh air is drawn in to supply air over the seating bowl and also to the underside of the roof, while at the same time extracting stale air by way of roof-level mechanical extracts.

Trial erection and testing

The AELTC site at Church Road has limited access and working space within the grounds, particularly around Centre Court, so the positioning of cranes and the location of plant and fabrication compounds was crucial in the tight roof-erection programme. The need for a trial erection site combined with the need to test out the mechanism and control systems associated with the retractable roof prior to erection was essential. A suitable site was found outside Sheffield close to the Motive Control suppliers, and a replica of the southern part of the new fixed roof was erected, with the constraints of the Wimbledon site marked out.

This allowed trial erections of three trusses, including the southern truss with the connecting hinges and fabric. These trusses were powered up and tested at various stages to ensure that the design forces and movements, and the monitoring and control devices were fully functioning prior to the erection for real at Wimbledon.

Figure 18.6 Cross-section of Centre Court illustrating airflow from the air-conditioning systems within the fixed roof

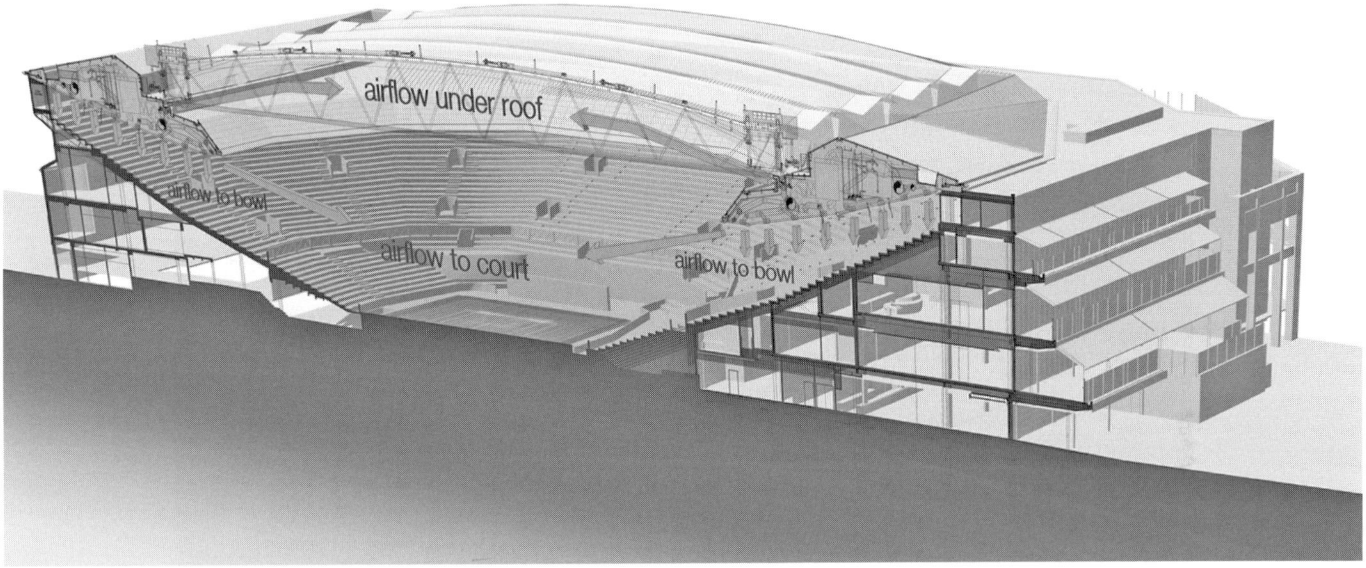

Fabrication and construction

Programme of construction

The key stages in this development were as follows:

- 2005–2006. Enabling work, involving strengthening of the existing structures, underpinning of existing foundations, and provision of new piled foundations for the new supercolumns. Construction of the foundations and walls for the new lift and stair cores to the east wing.
- 2006–2007. Demolition of the existing roof to Centre Court, demolition of the part of the east wing, erection of the new steelwork and floors and raking beams, fabrication and erection of the new terrace units, trial erection and testing at Sheffield, and the first championships held at Wimbledon with no roof on Centre Court.
- 2007–2008. Erection of the new fixed roof and two moving roof trusses in time for the 2008 championships.
- 2008–2009. Completion of the retractable roof, which was fully operational for the 2009 championships.

Fabrication/erection of the fixed roof

The main challenges with the fixed-roof construction were associated with the restricted access to the site and lack of suitable craneage, combined with the complexity of the main box trusses integrated with the AHUs. A working platform with carefully positioned jacking points was provided over the terraces, and the main east and west trusses were delivered to the southern end of the site in sections, with the steelwork fabricated round the AHUs. These were small enough to be lifted with mobile cranes and then skidded into position, in order along the platform, and bolted and welded into place (Figure 18.7).

Fabrication/erection of the retractable/moving roof

The trusses for the retractable roof trusses, complete with the central restraining arms and locking mechanisms, were fabricated at ground level at the southern end of the stadium and then tandem lifted into place onto the bogies positioned on the tracks. The first truss was temporarily propped until the second truss was erected next to it at a predetermined distance, and then temporary frames were inserted to hold them together prior to the installation of the hinges.

The two trusses could then be moved along the tracks together as on a stable unit. The hinges could then be erected separately, but with the temporary bracing in place, until the actuators were commissioned at a later date. The next truss could then be erected adjacent to this pair and the system repeated.

Once the moving roof trusses were commissioned with the hinges moving and the actuators taking load, the fabric field could be installed and the anchorages tensioned to the required prestress.

Figure 18.7 Roof construction

Figure 18.8 Aerial view of Wimbledon during play. (Courtesy of AELTC) (This figure is reproduced in the colour-plate section as Plate XXVIII)

The remainder of the retractable roof trusses could then be erected, including the batwings, locks and control gear.

Conclusions

Although the design and construction programme for this unique project was spread necessarily over a number of years in order to meet the needs of the AELTC championships, the degree of collaboration between the main professional consultants and contractors, and also the suppliers and subcontractors, over this time was remarkable.

To ensure that all elements of the design were manufactured and procured from large castings (for the bearings of the hinge mechanisms) and that the necessary specifically designed elements of computer software were designed in a timely fashion required a special relationship between all parties, and this was evident throughout the whole contract.

The redevelopment of Centre Court, and the retractable roof in particular, was a complex project that has been regarded as a huge success for the championships and also as a cutting-edge piece of kinetic engineering/architecture. It allows for a large percentage of tennis spectators, both at the AELTC grounds as well as the wider TV audience, to experience quality tennis at all times, regardless of the vagaries of the British weather, and keeps Wimbledon at the forefront of the tennis and sporting world (Figure 18.8).

Stadium and Arena Design
ISBN 978-0-7277-5790-6

http://dx.doi.org/10.1680/saad.57906.221

Chapter 19
National Stadium (Bird's Nest), Beijing

Martin Simpson Associate Director, Arup, London, UK

Introduction
The Chinese National Stadium, known as the Bird's Nest, is located in the southern part of the Olympic Green to the north of Beijing and close to the city's central axis (Figure 19.1). Its location was predefined by the master plan, but all other urban and architectural decisions were initiated by the National Stadium project team led by Herzog & de Meuron in consortium with Arup and the China Architectural Design & Research Group (CADG). At the time of the competition, Herzog and de Meuron and Arup were already working together on the Allianz Arena in Munich. With this experience and with CADG providing vital local expertise, the competition team was able to submit a highly developed, fully realisable architectural concept.

At the heart of the scheme is a highly efficient bowl design enclosed within a holistic architectural concept that celebrates the fundamental purpose of this building. From a distance, the stadium looks like a vessel whose undulating form echoes the rising and falling ramps for spectators inside the stadium. The components look like a chaotic thicket of supports, beams and stairs, almost like a bird's nest. Significantly, it was the Chinese themselves who provided this nickname while the initial competition entries were on display in Beijing, essentially assimilating it as their own before it had even left the drawing board.

The bowl
The Bird's Nest was designed inside out, beginning with the bowl. This is because the form of the bowl and the distribution of seating types largely determine all other aspects of a stadium, including the levels and locations of the concourses and premium facilities. The team worked closely with the International Olympic Committee (IOC) and local organising committees to streamline and rationalise the on-field facilities. The result is a more compact bowl with less distance between the spectators and the track than many of its peers.

Bowl design involves a skilful balancing of several key criteria. Most importantly, spectators want to be as close as possible to the action and to have a good view of the field, while the stadium developer needs to accommodate a certain number of seats within a defined budget.

This complex process has been transformed in recent years by parametric relationship modelling. Using custom software, Arup was able to quickly generate and optimise the bowl that would work well for football in legacy mode, but which was optimised for Olympic athletics.

The bowl is split into six segments separated by movement joints. Each segment is between 120 m (393 ft) and 150 m (492 ft) long, and is structurally independent, with its own stability system. Movement joints are continuous through every floor and the terracing. The lower ground level is continuous, acting as a diaphragm tying together the foundations.

The upper floors are reinforced concrete slabs spanning between primary beams. For the middle and upper tiers, the terracing is formed from precast units spanning between the primary frames and supported on raker beams (Figure 19.2). Columns and primary beams comprise frames located on every radial grid line. Under the lower tier, the columns are all vertical. For the middle and upper tiers, the front column is inclined towards the pitch, reducing the cantilever length of the raker beam. At the rear, the columns are inclined both radially and circumferentially, as these are visible from the external concourse and contribute to the overall aesthetic of the stadium (Figure 19.3).

The roof
The overall shape and form of the National Stadium directly responded to two requirements of the initial project brief: it had to have a moving roof and be designed to withstand seismic events twice the magnitude of the 1976 Great Tangshan earthquake that killed more than a quarter of a million people in Beijing.

In addition, the team wanted to reduce the stadium's visual mass, so opted to wrap the roof structure closely to the geometric constraints of the seating bowl and concourses.

Figure 19.1 Typical elevation and the external plaza. (© Gary Wong, Arup)

Having adopted a philosophy for the building's form, the next task was to create a structural solution that conformed to the requirements of brief, location and aesthetics. The answer lay in separating the roof structure from the bowl structure.

Figure 19.2 Inclined tribune beams support the precast units that form the terracing. (© Arup)

- The roof would be a complete entity with no movement joints, providing a stable platform for the moving roof, and thereby greatly simplifying the mechanisation.
- The bowl would be simplified, as the interface with the roof would be minimised.

The challenge was to derive a structure that was sympathetic to the architectural intent but robust enough to deal with vertical loads resulting from the large spans and horizontal loads from seismic events.

The solution was superposition of structural layers. Roof elements are typically assigned to:

- *Primary*. The main structural system. A 3D portalised space truss, enabling the roof to closely follow the bowl structure, remain independent and span the required distances.
- *Secondary*. Required to subdivide the panel size created by the main structural system into manageable cladding panels.
- *Stairs*. Access to the top bowl top tier was integrated into the walls supporting the roof. The continuity of these stair lines over the roof provide further subdivision of cladding panels and additional support to the perimeter stairs on the opposite side (Figure 19.4).

Figure 19.3 Columns are inclined both radially and circumferentially at the back of the bowl. (© Arup)

Figure 19.4 The open structure incorporates stairways, vomitories and passageways all designed to comply with Chinese codes. (© Arup)

The most challenging geometrical conditions, from the fabrication viewpoint, was using a continuous box profile over the whole façade (Figure 19.5). The capacity of the sections to carry different loads was achieved by varying the thickness of the plates that formed the box section, not the overall dimensions. This concept is critical to the visual identity of the Bird's Nest.

This box section was defined using the structure envelope as a control surface. The outer flange of the box is always parallel to the control surface, resulting in a box section that both twists and curves. However, because the geometry was derived from developable surfaces, even the most twisted element could be flattened out, cut from steel plate and then bent to form part of a fabricated box section. This investigation was crucial to proving that, although complex, the structure was buildable.

The use of parametric component-based modelling software (CATIA) was critical to the success of the National Stadium. The ability to deal with a vast number of components allowed the whole stadium to be assembled in a single environment. The advantage of using parametric software is significant when dealing with design that is required to be adjustable and continually changing like stadiums.

The original geometry changed late in the design process due to the omission of the moving roof because the client needed to reduce the resources and overall cost of the Games. It should be noted that the actual cost of the stadium itself was comparing well to its original estimate, but the overall budget for the Games had to be cut. However, due to the advanced software techniques deployed by the team in terms of geometry, analysis,

Figure 19.5 The varying plate thicknesses of the box section members are entirely concealed. (© Arup)

design and optimisation, the project was completed on time, with only a small delay in the construction programme.

Seismic design

The unique structural form, the architectural constraints and the desire to reduce the steel tonnage all posed great challenges to the seismic design of the roof.

The long span meant the seismic design was significantly different from that of a tall building. Seismic design measures that prevent collapse of tall buildings under a level 3 earthquake were insufficient for the Bird's Nest roof structure.

The team adopted a performance-based seismic design and analysis approach for the roof structure. This is the most technically rigorous approach and leads to the most cost-efficient design. To achieve the collapse-prevention performance objective for a level 3 earthquake, Arup established the following performance targets for the structural members:

- primary truss members shall remain elastic or nearly elastic
- secondary truss members are permitted to sustain severe damage.

Arup used Oasys LS-DYNA non-linear finite-element analysis software to demonstrate how the collapse-prevention performance objective was achieved. The non-linear response history analysis captures the time histories of forces and deformations in every primary and secondary truss member in the inelastic range when subjected to earthquake acceleration time histories, representing the ground shaking from a level 3 earthquake.

The innovative concept of nearly elastic design subjected to the level 3 earthquake not only convincingly demonstrated achievement of the collapse-prevention performance objective but also resulted in very significant reduction in the quantity of steel used, creating exceptional value for the client. The thickness of most roof members is substantially lower than that required by the ductile detailing rules specified in many international seismic design codes.

The importance of the National Stadium project meant that the Beijing Municipal government set up an expert panel committee to review the structural design. The expert panel review and approval often requires explicit verification of performance under seismic events, and analysis of the non-linear response history is required to demonstrate that the collapse-prevention performance objective has been achieved. The design passed the review and was endorsed for approval.

Compared with the roof, the seismic design of the bowl structure was straightforward and within the limits and scope of the seismic code GB50011-2001. A dual-level seismic design approach was adopted, with moment frames and core walls sized and proportioned so that member strength capacities and inter-storey drift ratios were satisfied.

Spectator comfort

For spectator comfort the thermal condition inside was critical, especially with up to 91 000 spectators ('Olympic mode'). 'Thermal comfort' is a subjective measure of people's physiological response and cultural adaptation to microclimate.

Givoni's thermal sensation index for thermal comfort assessment was adopted (Givoni, 1998). Temperatures were assessed, especially at the upper tiers where the most uncomfortable conditions were predicted, in order to determine comfort performance. The team evaluated parameters that affect comfort level, including air temperature, humidity, wind speed, solar radiation and surface temperature. A dynamic thermal model (for solar radiation and surface temperature evaluation) and computational fluid dynamics (CFD) model (for air temperature, relative humidity and airflow speed) were used to determine the values of those parameters under design conditions.

With the temperature and relative-humidity distribution and air-velocity vectors evaluated using CFD, the thermal comfort conditions in the spectator area of the stadium were assessed. During the design process, the following optimisations were performed to improve the thermal comfort level:

- the distance between the highest seat and the false ceiling was increased from 2 m (6.5 ft) to 8 m (26 ft) so that the occupants of these seats are below the stratified hot-air layer under the roof
- the area of the outer transparent ethylene tetrafluoroethylene (ETFE) membrane layer at the side was reduced so as to enlarge the opening for natural ventilation.

The thermal sensation index in most areas varied from ~4.0 to ~5.0, which is considered comfortable for a stadium environment, and proved the effectiveness of the proposed design changes.

Building service design

Key issues were established at the outset and the design was carried out in accordance with the Beijing Olympic 2008 Organising Committee's philosophy of 'green Olympics; high-tech Olympics; People's Olympics'. Resilience, reliability, sustainability, advanced technology and user-orientation were the concepts repeatedly emphasised and integrated into the design. Any chance of system failure was inadmissible, and the team undertook risk analyses of the power supply, water supply, heating, ventilation and air-conditioning (HVAC) plant, and drainage systems to ensure that no part of any one system would affect the performance of the whole.

Heating, ventilation and air-conditioning

The HVAC systems included cooling and heating source, ventilation, space heating, ground-source cooling systems, pitch heating, fire protection, pressurisation and smoke-extraction systems, and automatic control systems for air-conditioning.

The design of the HVAC systems had not only to meet the operational requirements of the Games but also those of the post-Olympic commercial operation of the stadium. To fully embody the 'green Olympics' concept, appropriate new techniques and equipment were to be adopted for energy utilisation, the thermal properties of the building envelope, the indoor environment, energy efficiency and environmental protection.

The primary source for space heating and sanitary hot water is the high-temperature supply from Beijing's municipal heating networks. The total heating demand on the municipal networks is 21 576 kW (28 934 hp), that for space heating 19 776 kW (26 520 hp) and that for sanitary hot water 1800 kW (2414 hp).

The total cooling load of the air-conditioning systems during the Games was 14 892.8 kW/4235 RT (971.6 hp/50 820 000 BTU/hr) (refrigeration tonnage) and is 20 993 kW/5970 RT (28 152 hp/71 640 000 BTU/hr) for commercial operation post-Olympics.

Natural ventilation was adopted in the stadium bowl, based on fluid dynamics and thermodynamic analysis. Air-intake vents were located at the lower parts of the stadium and exhaust vents above the highest seating. With the intake and exhaust vents open in summer, air flows through the stadium bowl.

Fan coil or multi-split air-conditioning units were based on the particular room function and purpose. For large spaces, the design was for 100% fresh air free cooling by all air systems in mild seasons.

Plumbing and drainage design

From the outset the design team formulated a water-conservation strategy as there would be significant water consumption due to irrigation, cleaning the car park and running tracks, cooling tower make-up and toilet flushing. Areas of stormwater recapture include the field of play, the roof and the surrounding landscaped area. The maximum quantity of water collectable on the site in 24 hours for a designed 1-year return period is about 12 750 m^3 (450 262 ft^3), which is sufficient for 40 days' average consumption of non-potable water for the whole project.

The design of the stormwater drainage system for the roof combines gravity and siphonic drainage tailored to fit the roof's unique shape. Rainfall runs by gravity to large catch-basins suspended under the roof structure. Siphonic rainwater outlets in these catch-basins then discharge to main stormwater drains.

Electrical services and extra-low-voltage (ELV) systems

As one of the most important facilities in China for welcoming visitors, athletes and political leaders, the National Stadium is classified as Chinese superclass 1 for electricity power supply. The most critical loads for which detailed design reliability assessments were carried out were those from the pitch, royal box, VIP rooms, VIP reception room, pitch lighting, square lighting, time and scoreboard recording systems, computer room, communication equipment room, voice reinforcement service room, TV and broadcasting transfer system, media, emergency lighting, fire-fighting, event information management system, safe and security system, and data network system. Other areas of comparatively lesser importance were designed to different levels of resilience.

The total calculated peak electrical loads were 14 601 kW (19 580 hp) for the Olympics and 15 902 kW (21 325 hp) for post-Olympic operation. Four individual 10 kV power feeders lead into the site from two separate 110 kV substations. The capacity of each incoming power supply feeder was the recommended 10 000 kVA maximum, not exceeding 12 000 kVA. Checks subsequent to the Olympics showed that the maximum load for the whole project during the Games was slightly below 10 000 kW (13 410 hp), well within the capabilities of the electrical system design.

Even with such a high demand on the building services systems during so many important events within just 2 weeks, their design met or even exceeded the expectations of all the athletes, other users and audience, both in the stadium itself and through TV worldwide.

Fire engineering

At the time of design there was no local design code that dealt with evacuation of a stadium this size, so The Green Guide (SGSA, 1997) was adopted. The bowl has been designed to be cleared in not be more than 8 minutes.

The stadium tiers are served by six concourses below the seating areas of tiers 2 and 3. The ground-floor level has direct and open access onto tier 1. The gangways in the seated areas and vomitories are a minimum of 1.2 m (4 ft) wide, and barriers are installed on the exits (see Figure 2.8) to avoid multi-evacuation flows crushing at their entry points.

All viewing accommodation spaces are separated from adjacent areas or voids, and all the stairways, vomitories and passageways were designed to comply with the Chinese codes. The concessions and high-risk areas are protected locally by using the 'cabin' concept, which makes use of sprinklers, smoke barriers and a dynamic smoke-control system (Law, 1990). For structural fire protection, the team adopted a performance-based solution. Most of the structural members of the roof did not require fire protection as it was concluded that additional fire protection was only needed for structural steel roof members within 6 m of the spectators.

Conclusion

The project was highly ambitious, not only in terms of delivering a world-class sports facility and a successful venue for the Games, but also in its conception as an icon of Beijing. It is both a monument for celebrating the great performance of athletes and a great civic building for the local citizens to enjoy for many years after the Olympics.

The project's success was achieved through the dedicated involvement of and seamless collaboration between the various teams of Arup, Herzog & de Meuron, and CADG. In addition, it should be noted that the client, the local authorities, the contractors and the design team's local collaborators all played a big part in the successful delivery. Taken together, the combination of scale, complexity and technology adopted in the National Stadium is unprecedented for a project of this type, and the courage and the commitment of the Beijing government to deliver such a project is truly admirable.

The Bird's Nest was designed and completed in less than 5 years to be ready for the 2008 Games. It was the centre of focus in the Olympic Green, and continues to attract thousands of tourists every day. It provided the perfect venue for athletes to stretch their performance and break new records, and for the designers it was the perfect building to stretch innovative thinking and break new ground in the application of its technologies.

REFERENCES

Arup (2009) The Beijing National Stadium Special Issue. *The Arup Journal* **1/2009**. See publications.arup.com/~/media/.../Files/.../Arup_Journal_1-2009.ashx (accessed 12/02/2015).

Beijing Organising Committee for the Games of the XXIX Olympiad (2011) *Preparation for the Games: New Beijing Great Olympics. Official Report of the Beijing 2008 Olympic Games*, Vol. III, LA84 Foundation, page 170. See http://olympic-museum.de/o-reports/report2008.htm (accessed 11/02/2015).

Givoni B (1998) *Climate Considerations in Building and Urban Design*. Wiley, New York, NY, USA.

Law M (1990) Fire and smoke models: their use in the design of some large buildings. *ASHRAE Transactions* **96(1)**: 963–971.

National Stadium – Bird's Nest (2011) Overview of the National Stadium. See http://www.n-s.cn/en/test/n214619929.shtml (accessed 20/05/2013).

SGSA (Sports Grounds Safety Authority) (1997) *Guide to Safety at Sports Grounds*, 4th edn. See http://www.safetyatsportsgrounds.org.uk/publications/green-guide (accessed 12/02/2015).

FURTHER READING

Arup (2009) The case study is an abridged version of the Beijing National Stadium Special Issue of *The Arup Journal* **44(1)**.

Part V

Appendices

Stadium and Arena Design
ISBN 978-0-7277-5790-6

http://dx.doi.org/10.1680/saad.57906.229

Appendix A
Pitch dimensions, boundaries and layouts

The *raison d'être* of any stadium is its playing area – this is where the audience sees what it has paid to see. In the early days of any sport there may be little but the playing area. In the 1860s and 1870s, baseball club owners raised wooden fences on whatever land they could obtain, erected bench bleachers and maybe built a dugout. Clubhouses were a rarity. Better facilities did not come about until attendances were boosted in the 1890s with the introduction of electric trolley cars, making it easier for people to travel to games. Baseball venues only started to be called 'stadiums' after the cathedral-scale Yankee Stadium was built in 1923.

For soccer grounds in the UK, cantilever stands were designed to give spectators unobstructed views of the playing area. The invention of high-strength steels made it possible for grandstand roofs to be constructed to extend the full width, and even full length, of the pitch. These types of stand left uncovered areas at the pitch corners, which were perfect for locating the floodlighting towers that made stadiums evening as well as daytime venues. In the last 40 years a more integrated approach has been adopted to stadium development and redevelopment. This has seen, for example, 'wraparound' stands creating covered accommodation for spectators located all around the perimeter of the playing area. Now the leading edges of the stand roofs were used to fix the lighting, bringing it closer to the action and eliminating the need for the space-consuming floodlighting towers at the corners of the pitch.

When we think about it, much of this book is about getting people into close proximity to the playing surface, providing them with comfort and safety, and getting them out efficiently at the close of play. So, because the playing surface is of such fundamental importance to the stadium, here are the playing area dimensions for some of the world's most popular sports.

Some of these are simplified versions of more complex official requirements. Dimensions must be checked with relevant governing bodies before final design.

Football

Pitch dimensions for association football (sometimes more commonly called 'football' or 'soccer') are shown in Figure A.1. Safety margins are 6 m behind the goal lines and 3 m along the side touchlines. The overall dimensions may vary according to the level of soccer being played.

Rugby union

Dimensions for the pitch are shown in Figure A.2. There must be no obstructions, such as fencing, within 5 m of the touch line.

Rugby league

Dimensions are shown in Figure A.3 for the pitch size, and there must be a margin of at least 6 m at each end and at least 2 m, preferably 6 m, along each side.

With the above three sports, natural grass is the only acceptable playing surface, but it is anticipated that artificial surfaces may become more widely used.

Hockey

Pitch dimensions are shown in Figure A.4. The surface may be grass or synthetic. For grass, the side margins should be a minimum of 3 m and the end margins 4.57 m, preferably more, so that the pitch can be moved in order to minimise wear. For synthetic pitches there should be a minimum of 5 m between the end lines and any obstruction. Synthetic grass should extend at least 3 m all round the pitch.

Lawn tennis

Dimensions for the court are shown in Figure A.5. Various characteristics are laid down for the requirements of alternative playing surfaces. All markings are in white between 25 mm and 50 mm in width, except the centre service line which must be 50 mm and the baseline which may be up to 100 mm wide.

American football

Pitch dimensions shown in Figure A.6, and there must be a safety margin of a least 1.8 m, but preferably twice that all round the field. Both natural grass and synthetic turf are acceptable. Markings are in white or yellow, and 100 mm wide. The playing area excludes sidelines but includes end zones.

Figure A.1 Pitch size and layout for Association Football or soccer

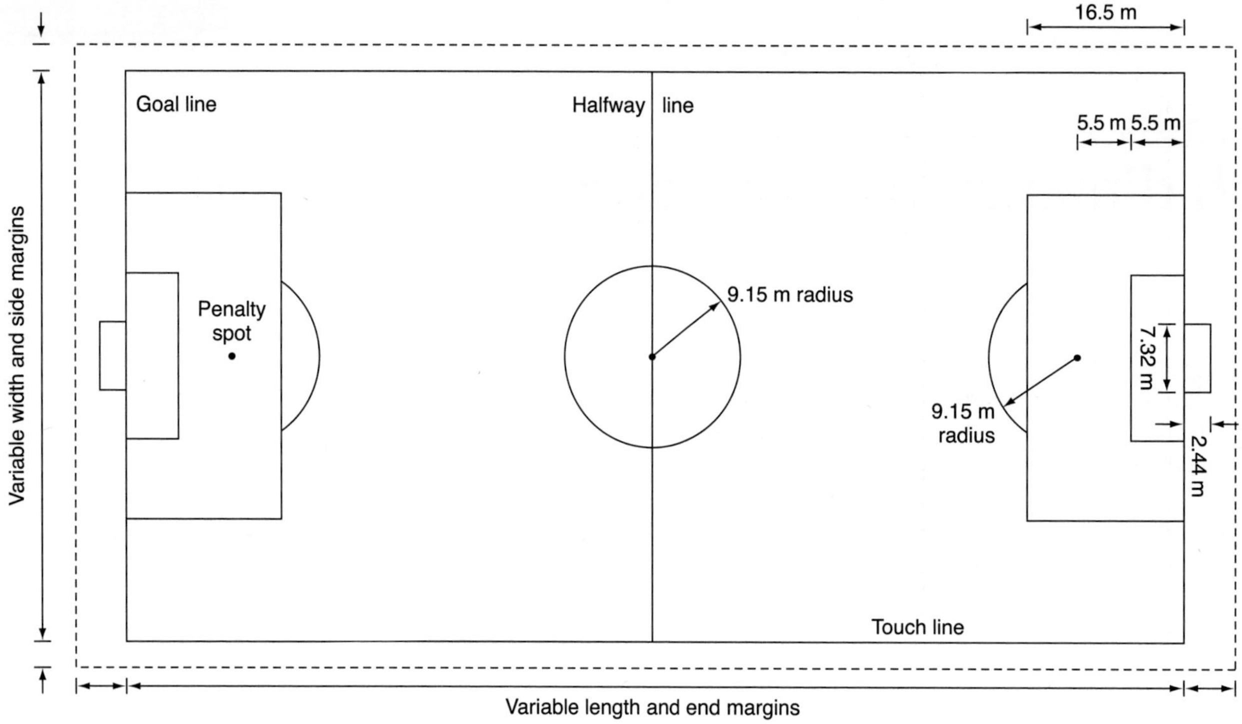

Figure A.2 Pitch size and layout for Rugby Union football

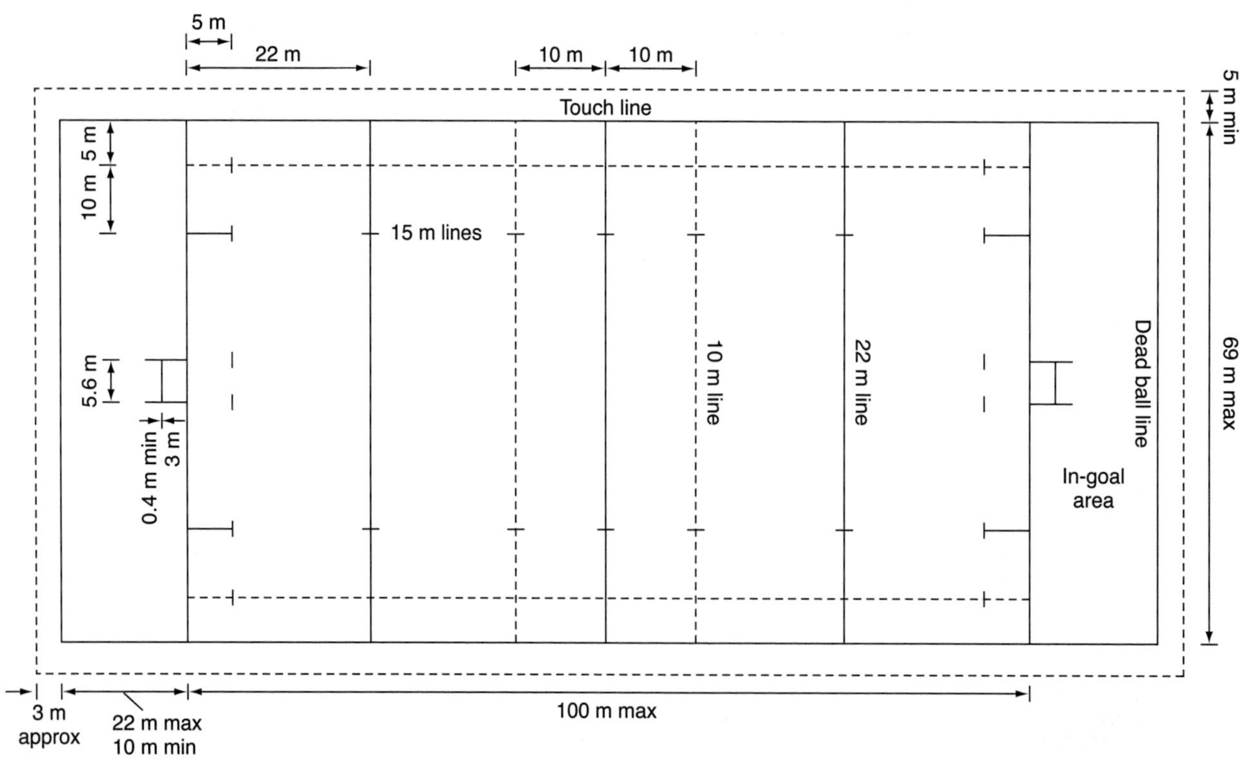

Figure A.3 Pitch size and layout for Rugby League football

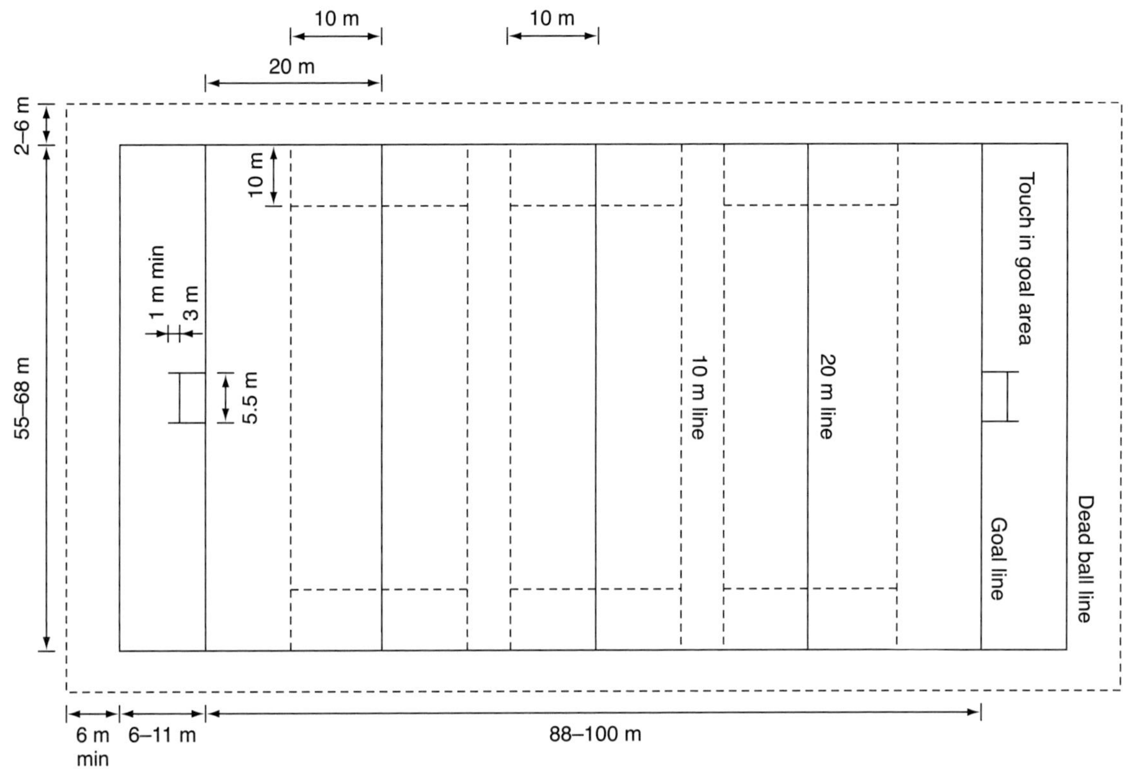

Figure A.4 Pitch size and layout for hockey

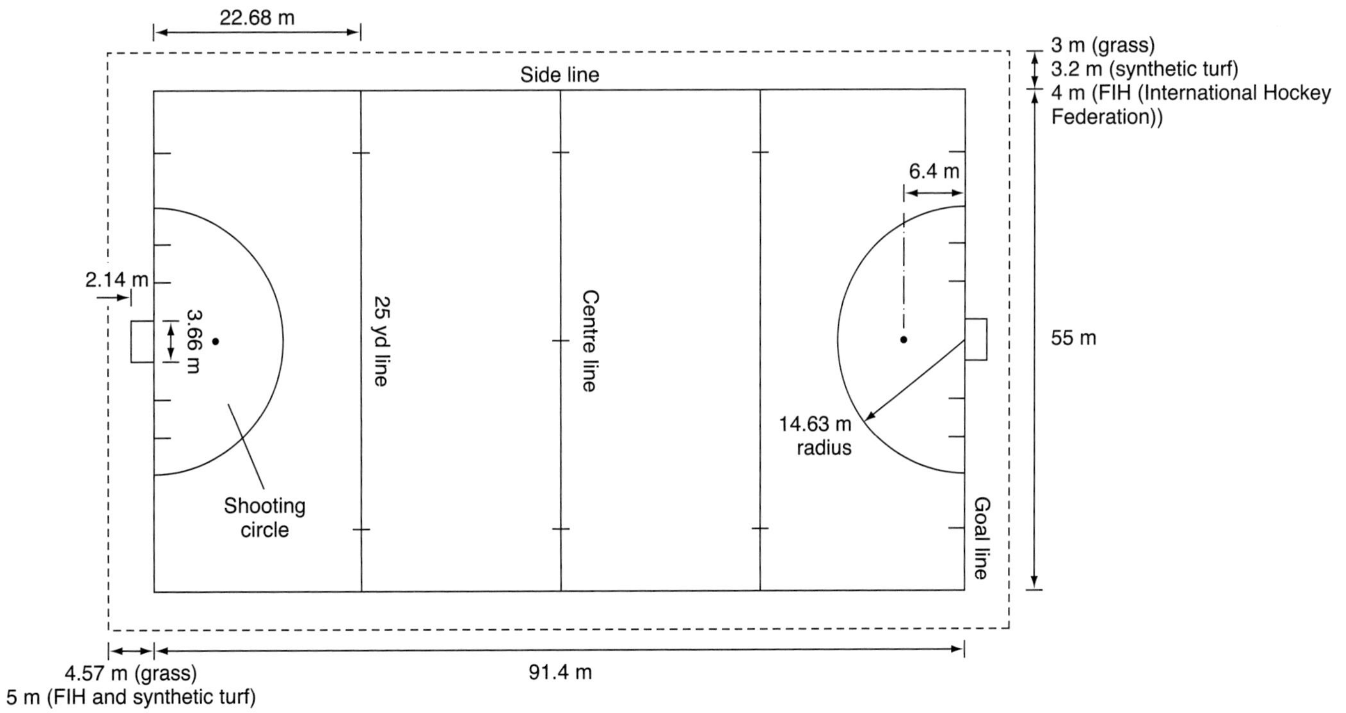

Figure A.5 Pitch size and layout for lawn tennis

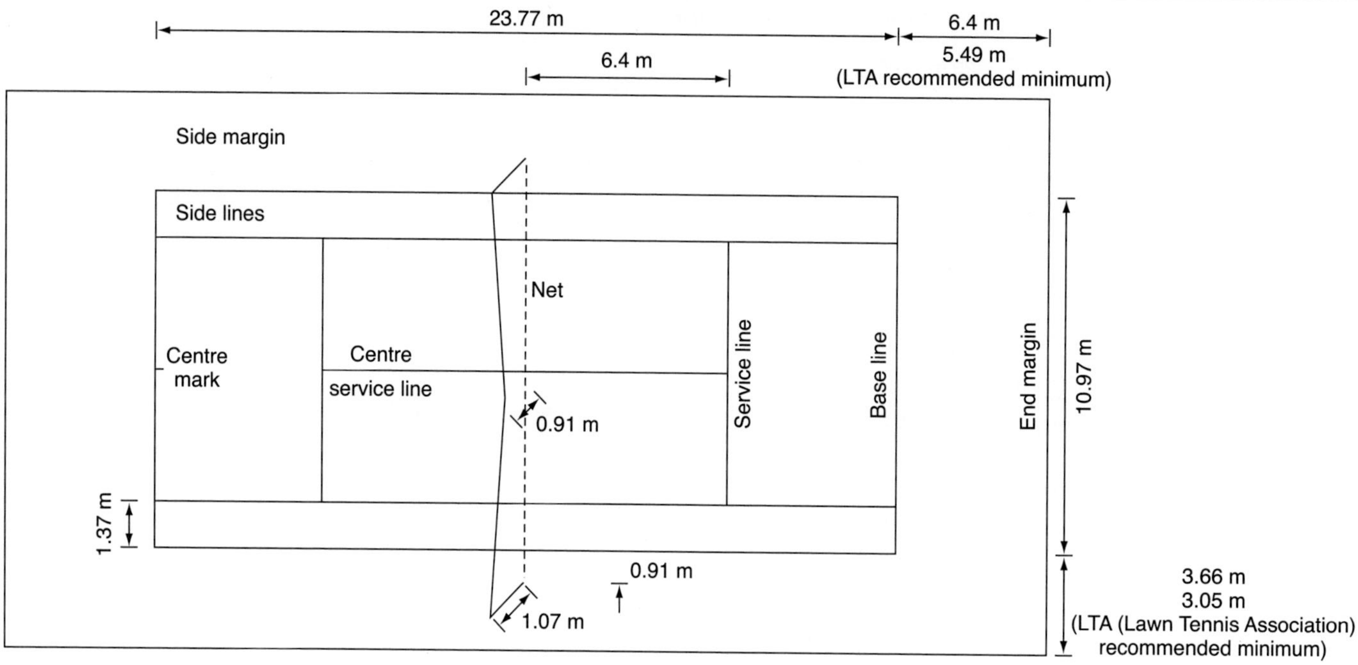

Figure A.6 Pitch size and layout for American football

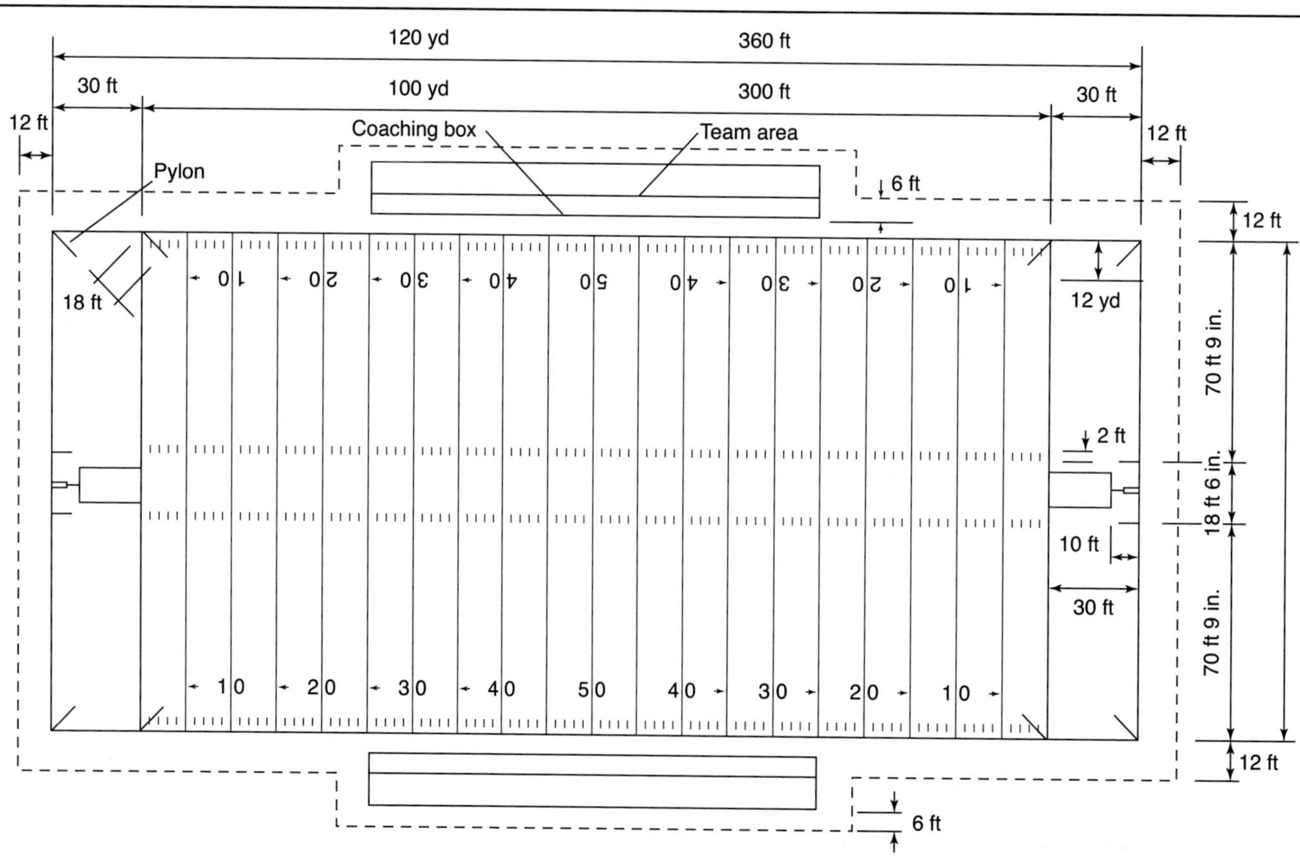

Figure A.7 Pitch size and layout for baseball

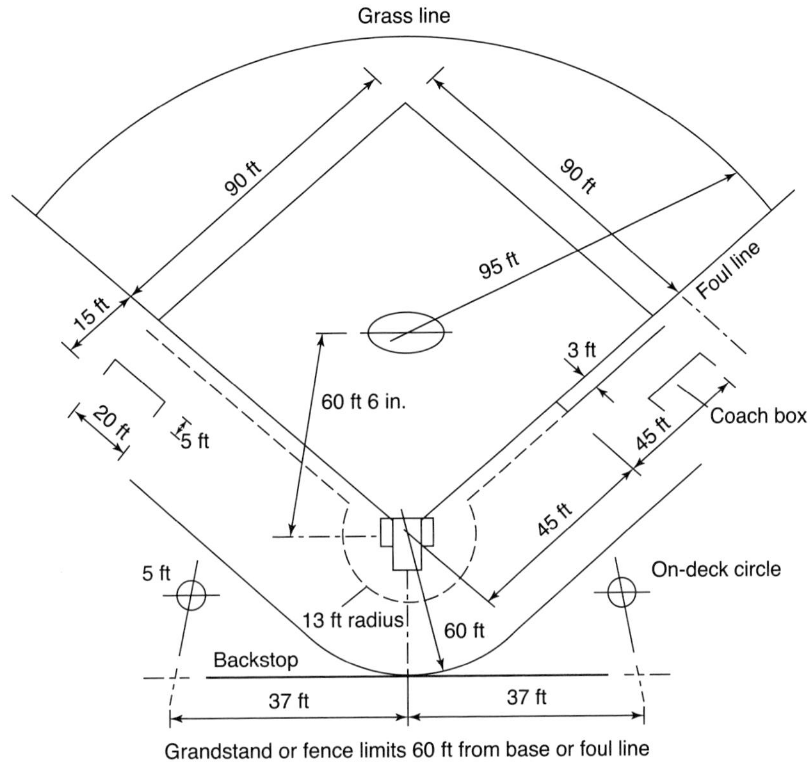

Figure A.8 Pitch size and layout for basketball

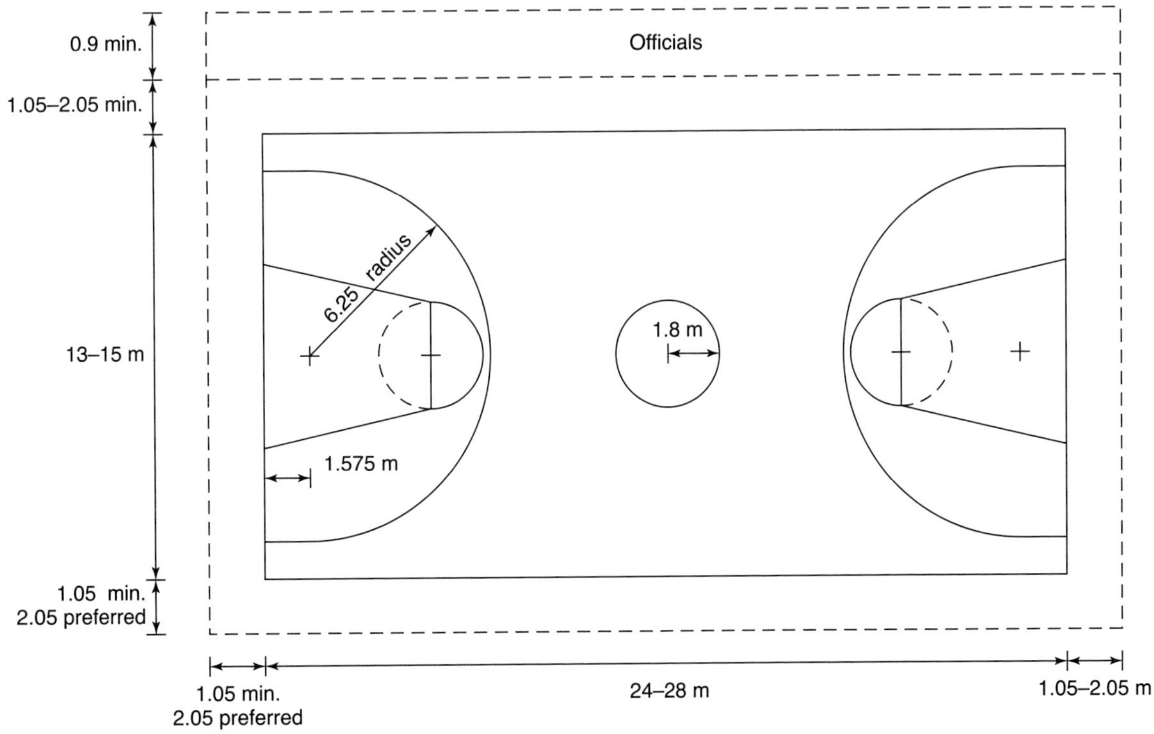

Figure A.9 Pitch size and layout for ice hockey

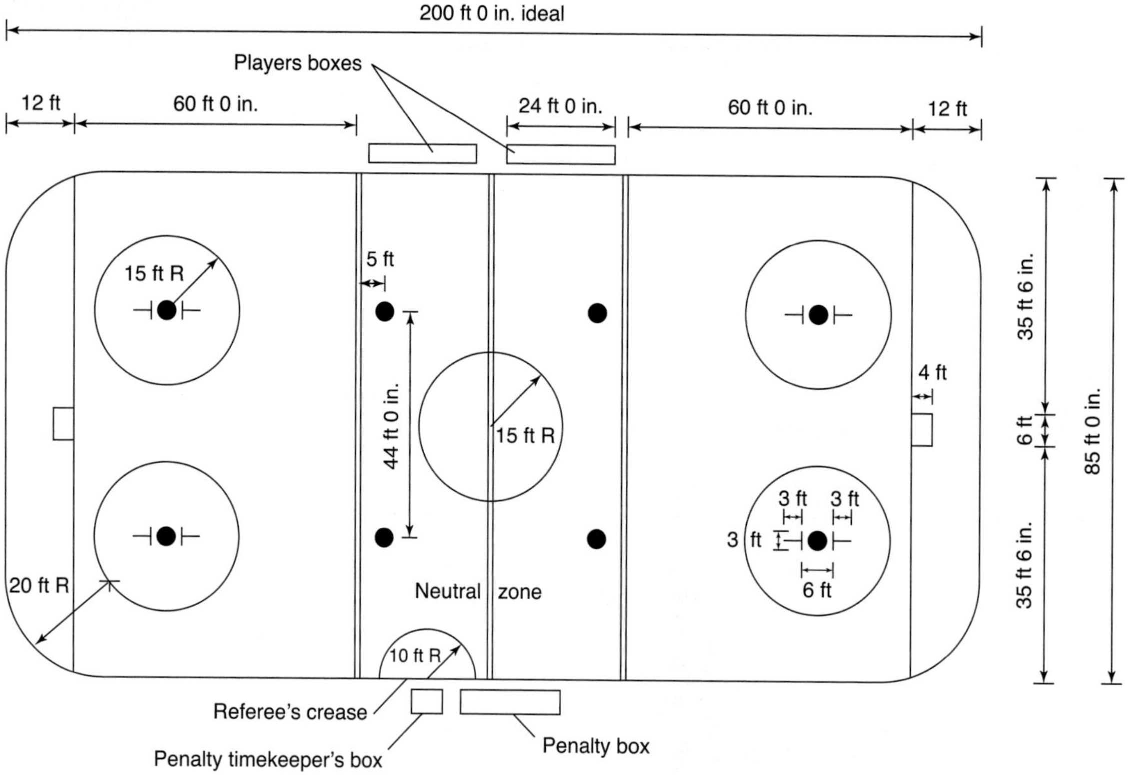

Baseball

Pitch dimensions shown in Figure A.7. The surface can be grass or synthetic. A batting cage is strongly recommended. Markings are in white, 127 mm wide and are included in the playing area.

Basketball

Dimensions shown in Figure A.8. The playing surface consists of a rectangular floor with tiles at each end. In professional organised basketball, especially indoors, the surface is usually made of highly polished maple. Outdoor surfaces can be concrete or asphalt.

Ice hockey

Dimensions shown in Figure A.9. There are two standard sizes for hockey rinks, one used primarily in North America and another used in the rest of the world. 'Rink' is a Scottish word meaning 'course'.

Stadium and Arena Design
ISBN 978-0-7277-5790-6

ICE Publishing: All rights reserved
http://dx.doi.org/10.1680/saad.57906.235

Institution of Civil Engineers
publishing

Appendix B
Medals

Hunting and collecting goes beyond the human condition to something primordial and endemic across the animal kingdom as a whole. Now that we have refrigerators, and do not have to collect food so often, we may collect non-essential things – perhaps newspaper cuttings or postage stamps or coins or medals. I (J.P.) collect stadium medals. Stadiums feature on many sporting achievement medals, and may feature in their own right on facility inauguration or commemoration medals. Examples of all types are shown here. They are arranged left to right, top to bottom, lines 1–5. I have not incorporated a scale, but for a good idea of comparative size the three Beijing medals (bottom left) are each 50 mm in diameter (see Figure B.1).

Figure B.1 Medal collection, John Pascoe. (© Simon Atkinson, 2014) (This figure is reproduced in the colour-plate section as Plate XXIX)

Line 1

Jockey Club Rosario, Santa Fe, Argentina, 1902

Hipodromo de Rosario opened on 18 September 1901. This racecourse was the initiator, in Argentina, of night-time racing with purpose-built lighting. The medal is my oldest. It shows the racecourse with the grandstand beyond. The reverse carries a reference to 'Julio A Roca, Presidente de la Republica', with the date '27 Octobre 1902'.

Stadium Krause, San Juan, Argentina, 1918

Villa Krause is a city, population 107 000, in the province of San Juan. The medal's wording, around the image of the stadium façade, reads 'Inauguracion Stadium Krause 14 Abril 1918'.

Stadio Olimpico, Rome, Italy, 1990

This is a souvenir of the 1990 FIFA World Cup – Italia 1990. Stadio Olimpico was rebuilt for this event and hosted the World Cup Final. The reverse shows, possibly uniquely, an angel carrying a bag of footballs.

Maracana Stadium, Rio de Janeiro, Brazil, 1950

The medal was issued in the year that Brazil hosted the 1950 FIFA World Cup. Its obverse reads 'Estadio Municipal Rio de Janeiro' and the reverse reads 'Campeonato Mundial de Football Brazil 1950'.

Yoyogi National Stadium, Tokyo, Japan, 1964

This souvenir medal depicts the famous suspension roof designed by Kenzo Tange and built between 1961 and 1964 to host events at the Tokyo 1964 Summer Olympics.

Peace and Friendship (SEF) Stadium, Piraeus, Greece, 1985

This is a silver medal from the 16th European Athletics Indoor Championships held at Piraeus, the coastal zone of Athens, in March 1985. The SEF stadium was inaugurated on 16 February 1985 for the first Panhellenic Athletics Indoor Championship. Later it would host indoor volleyball during the Athens 2004 Summer Olympics.

Line 2

Estadio Olimpico Universitario, Mexico City, 1955

This is a Second Place (Secundo Lugar) silver medal from the 2nd Pan American Games held in Mexico City in March 1955. The stadium was built in 1952, and went on to host the Mexico 1968 Summer Olympics. Its Tartan track was the first all-weather running track to be used in the Olympics.

Memorial Stadium (The Rock), Bloomington, Indiana, Oregon, USA, 1960

The wording on the reverse of this inauguration medal reads 'Oregon State. Marquette. Michigan State. Northwestern Dedication Year 1960 Indiana University vs Michigan State Dedication Game October 22, 1960'.

Estadio da Luz, Lisbon, Portugal

I don't know when this medal was struck or what the occasion was. The wording on the other side is 'Sport Lisboa e Benfica Fundado em 1904'. This stadium opened on 1 December 1954 and was demolished in 2002. Its capacity of 120 000 made it Europe's biggest stadium and the world's third biggest.

Dodger Stadium, Los Angeles, California, USA, 1962

The inauguration medal came with a full stadium description beginning 'Official Dodger Stadium Medallion Commemorating the Opening of the Dodger Stadium in Los Angeles on April 10, 1962 ... the most modern ball park in the world today'.

Line 3

Caracas, Venezuala, 1959

This is an 8th Central American and Caribbean Games participation medal dated 1958 (but the events actually took place 6–15 January 1959). I guess the stadium depicted is the Estadio Olimpico de la Universidad Central de Venezuala (UCV), but I don't know for sure which stadium in Caracas hosted the Games.

Estadio das Antas, Porto, Portugal, 1966

This commemoration medal celebrates 60 years of 'Futebol Clube do Porto 1906–1966'. The club was actually founded in 1893 but became inactive before the game of football was revived in the city in 1906. Estadio das Antas served the club from 1952 until 2004, when it was demolished and replaced by Estadio do Dragao.

Olympic Stadium, Athens, Greece, 1982

This is a participation medal from the 13th European Athletics Championships held 6–12 September 1982. The stadium was designed in 1979 and built 1980–1982, in time to host the event. It subsequently hosted the 1997 World Championships in Athletics before being redeveloped for the Greece 2004 Summer Olympics.

Seoul Olympic Stadium (Jamsil Olympic Stadium), Seoul, South Korea, 1988

This is a participation medal from the Seoul 1988 Summer Olympics. The stadium opened on 29 September 1984 and hosted the 10th Asian Games in 1986. The reverse shows a mounted archer together with the Olympic rings and the wording 'XXIVth Olympiad Seoul 1988'.

Line 4

Stadion Poljud, Split, Croatia, 1990

This is a participation medal from the 15th European Athletics Championships, hosted by Stadion Poljud. The image on the obverse is a stylised representation of the stadium, which has a seashell-like structural steel roof spanning 215 m at its widest. The stadium was built for the 8th Mediterranean Games, 1979, and has more recently been the venue for the 2010 IAAF Continental Cup.

Stade Louis II, Monte Carlo, Monaco, 1992

On 6 August 1992, Kevin Young set the current world record of 46.78 s in the final of the 400 m hurdles at the Barcelona 1992 Summer Olympics. Five days later he recorded 47.60 s in winning the 400 m hurdles at the Mobil Grand Prix IAAF Gatorade Herculis Meeting in Monaco. This is the medal he won in Monaco. It shows in the foreground the 18 500 all-seater Stadium Louis II, built in 1985 on land reclaimed from the sea. Because the population of Monaco is approximately 37 000, the stadium capacity/national population ratio is little above 1 : 2.

Gottlieb-Daimler-Stadion, Stuttgart, Germany, 1993

This is a participation medal from the 4th World Championships in Athletics, held 13–22 August 1993. The stadium (now known as the Mercedes-Benz Arena) was named after Gottlieb Wilhelm Daimler, inventor of the high-speed petrol engine, who tested motorcycles and automobiles at this location in the 1880s. Before 1993 the stadium had been called the Neckerstadion, Kampfbahn and Century Stadium. When built originally, in 1933, it was named the Adolf-Hitler-Kampfbahn.

Estadio do Dragao, Porto, Portugal, 2003

'Inaugurado 16-11-2003', the all-seater Estadio do Dragao replaced FC Porto's Estadio das Antas. The inauguration event was a friendly football match between FC Porto and FC Barcelona, with an attendance of 52 000. The name of the stadium comes from the dragon incorporated in the club's crest.

Line 5

National Stadium, National Aquatics Centre & National Indoor Stadium, Beijing, China, 2008

The first three medals are examples of stadium medals issued by event sponsors. The medals came in fitted clear plastic containers on black velvet, in a presentation box with both Chinese and English wording: 'Beijing 2008, PICC, Official Insurance Partner of the Beijing 2008 Olympic Games'. (PICC stands for People's Insurance Company of China.)

Maracana Stadium, Rio de Janeiro, Brazil, 2014

Monnaie de Paris Medaille No. 15, Coupe du Monde de Football 2014 – this big and heavy bronze medal is 81 mm diameter × 8 mm thick.

Stadium and Arena Design
ISBN 978-0-7277-5790-6

http://dx.doi.org/10.1680/saad.57906.239

Appendix C
Stadium projects: selected URLs

Wikipedia Lists of Stadiums: http://en.wikipedia.org/wiki/List_of_stadiums

Sports Venues of the World: http://sketchup.google.com/3dwarehouse/cldetails?clid = b585811ff54dfe5fc0bb9c57d44ec2

World Stadiums: http://www.worldstadiums.com

The Stadium Guide: http://www.stadiumguide.com

Ice Hockey Arenas by Capacity: http://en.wikipedia.org/wiki/List_of_ice_hockey_arenas_by_capacity

National Basketball Association Arenas: http://en.wikipedia.org/wiki/List_of_National_Basketball_Association_arenas

Basketball Arenas by Capacity: http://en.wikipedia.org/wiki/List_of_basketball_arenas_by_capacity

Boxing Venues by Country: http://en.wikipedia.org/wiki/Category:Boxing_venues_by_country

Stadium and Attendances: http://www.stadium-attendances.com

Database of Football Stadiums: http://stadiumdb.com

Major League Baseball Stadiums: http://www.baseballstadiums.net

National Football League Stadiums: http://football.ballparks.com

International Olympic Committee: http://www.olympic.org

Asian Games: http://www.ocasia.org/game/GamesL1.aspx?9QoyD9QEWPeJ2ChZBk5tvA = =

International University Sports Federation (FISU), World Student Games, Universiade: http://www.fisu.net/en/FISU-homepage-3137.html

World Cup 2022: Qatar's Stadiums in Pictures: http://www.guardian.co.uk/football/gallery/2010/dec/02/world-cup-2022-qatar-stadiums-pictures#

2022 FIFA World Cup: http://en.wikipedia.org/wiki/2022_FIFA_World_Cup

Venues of the 2020 Summer Olympics and Paralympics: http://en.wikipedia.org/wiki/Venues_of_the_2020_Summer_Olympics_and_Paralympics

Sporting Venues – Gold Coast City 2018 Commonwealth Games: http://www.commonwealthgames.org.au/page/39/2018-gold-coast-games

2018 Commonwealth Games: http://en.wikipedia.org/wiki/2018_Commonwealth_Games

World Cup 2018: Russia's Stadiums in Pictures: http://www.guardian.co.uk/football/gallery/2010/dec/03/world-cup-2018-russia-stadiums#/?picture = 369353477&index = 5

2018 FIFA World Cup: http://en.wikipedia.org/wiki/2018_FIFA_World_Cup

Venues Map, Rio 2016 Olympics and Paralympics: http://rio2016.com/en/the-games/venues-map

Venues of the 2016 Summer Olympics and Paralympics: http://en.wikipedia.org/wiki/Venues_of_the_2016_Summer_Olympics_and_Paralympics

Rio 2016 Summer Olympics: http://www.olympic.org/rio-2016-summer-olympics

Sports Facilities – 2015 Gwangju Summer Universiade: http://www.gwangju2015.com/sporting-venues

2015 European Games: http://en.wikipedia.org/wiki/2015_European_Games

2015 European Games – Heydar Aliyev Sports and Exhibition Complex: http://en.wikipedia.org/wiki/Heydar_Aliyev_Sports_and_Exhibition_Complex

2014 Asian Games Stadium, Incheon, South Korea: http://worldlandscapearchitect.com/2014-asian-games-stadium-incheon-south-korea-populous

Guide to all the Stadiums to be used at the 2014 FIFA World Cup: http://www.fifa.com/worldcup/destination/stadiums/index.html

XX Commonwealth Games, Glasgow 2014, venues: http://www.glasgow2014.com/your-games/venues

Venues 2012 Olympics London 2012: http://www.london2012.com/spectators/venues

Euro 2012 Stadium Guide – *The Sun*: http://www.thesun.co.uk/sol/homepage/sport/football/euro2012/4266695/Euro-2012-stadium-guide-Pictures-details-of-all-venues-in-Poland-Ukraine.html

Euro 2012 Stadium Guide – *Daily Mail*: http://www.dailymail.co.uk/sport/football/article-2068025/Euro-2012-stadium-guide-Poland-Ukraine.html

XIX Commonwealth Games 2010 Delhi: http://d2010.thecgf.com

Universiade 2011 Shenzhen: http://www.sz2011.org/Universiade

2010 FIFA World Cup Stadiums South Africa: http://www.sa-venues.com/2010/2010-stadium.htm

Euro 2008 Stadiums: http://www.besteuro2008.info/stadiums.html

Melbourne 2006 XVIII Commonwealth Games: http://www.melbourne2006.com.au/Channels

Olympic Venues in Athens: http://www.athensinfoguide.com/olympicvenues.htm

Euro 2004 Venues Guide: http://news.bbc.co.uk/sport1/hi/football/euro_2004/venues_guide/default.stm

Manchester Commonwealth Games 2002: http://www.manchester2002-uk.com/commonwealth.html

2002 FIFA World Cup Venues: http://www.worldcup-2002.co.uk/stadiums.html

Sydney 2000 Summer Olympics: http://www.olympic.org/sydney-2000-summer-olympics

Euro 2000 Stadiums: http://news.bbc.co.uk/hi/english/static/sport/euro2000/stadiums

The Beijing National Stadium, Special Issue of *The Arup Journal*: http://publications.arup.com/Publications/T/The_Arup_Journal/2009/The_Arup_Journal_2009_Issue_1.aspx

Stadium and Arena Design
ISBN 978-0-7277-5790-6

http://dx.doi.org/10.1680/saad.57906.241

Index